Corynebacterium glutamicum

From Systems Biology to Biotechnological Applications

Edited by

Andreas Burkovski

Department of Biology
University of Erlangen-Nuremberg
Erlangen
Germany

Caister Academic Press

Caister Academic Press
Norfolk, UK

www.caister.com

British Library Cataloguing-in-Publication Data
A catalogue record for this book is available from the British Library

ISBN: 978-1-910190-05-0 (hardback)
ISBN: 978-1-910190-06-7 (ebook)

Cover design adapted from Figure 4.1

Contents

Contributors

Michael Bott
Institute of Bio and Geosciences
IBG-1: Biotechnology
Forschungszentrum Jülich GmbH
Jülich
Germany

m.bott@fz-juelich.de

Julia Bürger
Department of Microbiology
Friedrich-Alexander-University Erlangen-Nuremberg
Erlangen
Germany

julia.buerger@fau.de

Andreas Burkovski
Department of Microbiology
Friedrich-Alexander-University Erlangen-Nuremberg
Erlangen
Germany

andreas.burkovski@fau.de

Bernhard J. Eikmanns
Institute of Microbiology and Biotechnology
University of Ulm
Ulm
Germany

bernhard.eikmanns@uni-ulm.de

Roland Freudl
Institute of Bio and Geosciences
IBG-1: Biotechnology
Forschungszentrum Jülich GmbH
Jülich
Germany

r.freudl@fz-juelich.de

Julia Frunzke
Institute of Bio and Geosciences
IBG-1: Biotechnology
Forschungszentrum Jülich GmbH
Jülich
Germany

j.frunzke@fz-juelich.de

Andreas Harst
Department of Plant Biochemistry
Ruhr University Bochum
Bochum
Germany

andreas.harst@ruhr-uni-bochum.de

Masayuki Inui
Research Institute of Innovative Technology for the Earth
Kizugawa
Japan

inui@rite.or.jp

Toru Jojima
Research Institute of Innovative Technology for the Earth
Kizugawa
Japan

tojojima@rite.or.jp

Lennart Leßmeier
CeBiTec
Faculty of Biology
Bielefeld University
Bielefeld
Germany

lennart@cebitec.uni-bielefeld.de

Tang Li
State Key Laboratory of Microbial Resources
Institute of Microbiology
Chinese Academy of Sciences
Beijing
China

pkulitang@yahoo.com.cn

Steffen N. Lindner
CeBiTec
Faculty of Biology
Bielefeld University
Bielefeld
Germany

s_lindner@gmx.de

Shuang-Jiang Liu
State Key Laboratory of Microbial Resources
Institute of Microbiology
Chinese Academy of Sciences
Wuhan; and
Environmental Microbiology Research Center
Institute of Microbiology
Chinese Academy of Sciences
Beijing
China

liusj@im.ac.cn

Ken'ichiro Matsumoto
Division of Biotechnology and Macromolecular Chemistry
Graduate School of Engineering
Hokkaido University
Sapporo
Japan

mken@eng.hokudui.ac.jp

Nurije Mustafi
Institute of Bio and Geosciences
IBG-1: Biotechnology
Forschungszentrum Jülich GmbH
Jülich
Germany

nurije.mustafi@yahoo.de

John Masani Nduko
Division of Biotechnology and Macromolecular Chemistry
Graduate School of Engineering
Hokkaido University
Sapporo
Japan

ikobausy@gmail.com

Katharina Nöh
Institute of Bio and Geosciences
IBG-1: Biotechnology
Forschungszentrum Jülich GmbH
Jülich
Germany

k.noeh@fz-juelich.de

Ansgar Poetsch
Department of Plant Biochemistry
Ruhr University Bochum
Bochum
Germany

ansgar.poetsch@ruhr-uni-bochum.de

Nadine Rehm
Department of Microbiology
Friedrich-Alexander-University Erlangen-Nuremberg
Erlangen
Germany

nrehm@biologie.uni-erlangen.de

Xi-Hui Shen
State Key Laboratory of Crop Stress Biology for Arid Areas
College of Life Sciences
Northwest A&F University
Yangling
China

xihuishen@nwsuaf.edu.cn

Hiroshi Shimizu
Department of Bioinformatic Engineering
Graduate School of Information Science and Technology
Osaka University
Suita
Japan

shimizu@ist.osaka-u.ac.jp

Tomokazu Shirai
Biomass Engineering Program
RIKEN
Yokohama
Japan

tomokazu.shirai@riken.jp

Yuyang Song
Division of Biotechnology and Macromolecular Chemistry
Graduate School of Engineering
Hokkaido University
Sapporo
Japan

yuyangsong@nwsuaf.edu.cn

Seiichi Taguchi
Division of Biotechnology and Macromolecular Chemistry
Graduate School of Engineering
Hokkaido University
Sapporo
Japan

staguchi@eng.hokudai.ac.jp

Volker F. Wendisch
CeBiTec
Faculty of Biology
Bielefeld University
Bielefeld
Germany

volker.wendisch@uni-bielefeld.de

Wolfgang Wiechert
Institute of Bio and Geosciences
IBG-1: Biotechnology
Forschungszentrum Jülich GmbH
Jülich
Germany

w.wiechert@fz-juelich.de

Ying Xu
Key Laboratory of Agricultural and Environmental Microbiology
Wuhan Institute of Virology
Chinese Academy of Sciences
Wuhan
China

ying.xu@sjtu.edu.cn

Hideaki Yukawa
Research Institute of Innovative Technology for the Earth
Kizugawa
Japan

mmg-lab@rite.or.jp

Ahmed Zahoor
CeBiTec
Faculty of Biology
Bielefeld University
Bielefeld
Germany

azahoor@cebitec.uni-bielefeld.de

Elisabeth Zelle
Institute of Bio and Geosciences
IBG-1: Biotechnology
Forschungszentrum Jülich GmbH
Jülich
Germany

e.zelle@fz-juelich.de

Ning-Yi Zhou
Key Laboratory of Agricultural and Environmental Microbiology
Wuhan Institute of Virology
Chinese Academy of Sciences
Wuhan
China

n.zhou@whiov.ac.cn

Current Books of Interest

Preface

Corynebacterium glutamicum is most widely known for its role in the industrial production of L-glutamate and L-lysine and as a platform organism for the production of a variety of fine chemicals, biofuels and polymers. The organism's accessibility to genetic manipulation has resulted in a wealth of data on its metabolism and regulatory networks; this in turn makes *C. glutamicum* the model organism of choice in white biotechnology.

The book provides a comprehensive overview of current knowledge and research on *C. glutamicum* systems biology and biotechnological applications. It summarizes the recent advances of analysis approaches as well as the progress made in respect of new products and applications as well as the utilization of a broader spectrum of nutrient sources by *C. glutamicum*. Topics covered include proteomics, flux analysis of metabolism, metabolic engineering for alternative carbon source utilization, manipulation of nitrogen metabolism, transport, degradation and assimilation of aromatic compounds, engineering for production of organic acids and alcohols, microbial factory for the production of polyesters, biotechnological application oxygen deprivation, the secretory production of heterologous proteins and the development of genetically encoded biosensors.

Trends in *Corynebacterium glutamicum* Research and Application

1

Andreas Burkovski

Abstract

Industrial production of amino acid, a key sector of white biotechnology, is the classical field of application of *Corynebacterium glutamicum*. With a current annual production of more than 2.9 million tons of L-glutamate and more than 1.9 million tons of L-lysine (Ajinomoto, 2013a,b) *C. glutamicum* already represents a key organism for industrial biotechnology. Taking current megatrends in society, such as increasing population and demands for meat in account, it is likely that amino acid production will increase continuously as already estimated previously (Takors *et al.*, 2007). Furthermore, within the last decade, *C. glutamicum* was applied in various production processes for fine chemicals, fuels and polymers. Based on these studies, it becomes increasingly obvious that *C. glutamicum* has an enormous potential for the biotechnological synthesis of many compounds and materials, making *C. glutamicum* a versatile platform organism and biotechnology workhorse.

From glutamate producer to a biotechnology workhorse

In 1957, almost 60 years ago, *Corynebacterium glutamicum* was isolated in a screening programme set up to identify glutamate-producing bacteria (Kinoshita *et al.*, 1957; Udaka, 1960, 2008). The intention was to relieve the malnutrition of Japanese population by better-tasting food (Kinoshita, 2005) and the aim was to apply L-glutamate as flavour enhancer. Despite the obscure rationale behind the screening programme, this was the foundation of Japanese biotechnology industry and the basis of the tremendous success of *C. glutamicum* in industrial biotechnological production. Within a few years, *C. glutamicum* became the leading amino acid producer, not only for L-glutamate, but also for L-lysine. Production of the latter amino acid started in 1958 (Pfefferle *et al.*, 2003; Kelle *et al.*, 2005). L-Lysine is primarily applied in swine and poultry production (Kelle *et al.*, 2005).

In the first years of *C. glutamicum* application, improvements of L-glutamate and L-lysine production strains were achieved by random mutagenesis and screening (Leuchtenberger, 1996; Demain, 2000; Ikeda, 2003). Later main progress was made by rational design, i.e. metabolic engineering of strains, which concentrated on the carbon core metabolism, NADPH supply and regeneration, precursor supply and elimination of competing pathways (de Graaf *et al.*, 2001; Becker and Wittmann, 2012a). Despite the progress made, for almost 50 years the classical strains out-performed the rationally engineered strains, e.g. in case of glutamate classical strains reach titres higher than 100 g/l, while specifically generated mutants produce only about 30 g/l (Ault, 2004; Becker and Wittmann, 2012).

In case of L-lysine, rational strain development was more successful. Ikeda and co-workers demonstrated the power of a global approach successfully in two ground-breaking studies. In 2002, a genome breeding approach was reported as a first global strategy. Beneficial point mutations were identified in a classical strain based on comparative genome sequencing and applied to generate an L-lysine production strain (Onishi *et al.*, 2002). Later Ikeda and co-workers created an efficient, minimally mutated L-lysine producer by site-directed mutagenesis of *C. glutamicum* wild type (Ikeda *et al.*, 2006). In 2011, the first lysine producer, which even outcompetes classical producer strains, was generated based on 12 defined genome-based changes introduced in genes encoding central metabolic enzymes. These target genes were predicted by *in silico* modelling and changes introduced redirected major carbon fluxes towards optimal product formation (Becker *et al.*, 2011).

In the recent years, further progress was made in the development of tailor-made amino acid production strains (Becker and Wittmann, 2012a). For *C. glutamicum*, these efforts led to L-alanine (Jojima *et al.*, 2010), L-arginine and L-citrulline (Ikeda *et al.*, 2009) as well as L-serine (Peters-Wendisch *et al.*, 2005) and L-valine (Radmacher *et al.*, 2002; Halatko *et al.*, 2009; Bartek *et al.*, 2010) producers. With the tools of systems engineering at hand, further products besides amino acids were developed, including D-amino acids (Stäbler *et al.*, 2011), other chemicals and fuels (see below).

Systems biology

The progress made in systems metabolic engineering was mainly based on the development and integration of -omics, systems and synthetic biology approaches (Wendisch *et al.*, 2006; Becker and Wittmann, 2012a). The publication of the genome sequence of *C. glutamicum* by two independent groups (Ikeda and Nakagawa, 2003; Kalinowski *et al.*, 2003) can be considered as the starting point of this new era of *C. glutamicum*-based biotechnology.

Despite the fact that both sequences were determined for the *C. glutamicum* type strain ATCC13032 they show differences in respect to the presence of insertion elements – an observation already made earlier (Quast *et al.*, 1999) – and a prophage (Wendisch *et al.*, 2006; Frunzke *et al.*, 2008; Baumgart *et al.*, 2013). With the development of high-throughput sequencing techniques, genome analyses of production strains and mutants can be easily carried out today, which led to genome breeding approaches. In parallel, genomics provided the basis for transcriptomics and proteomics approaches (Wendisch *et al.*, 2006) and a bioinformatic *in silico* description of *C. glutamicum* metabolism.

Quickly after availability of genome sequence information transcriptome analyses using DNA microarrays were established for *C. glutamicum* (Wendisch, 2003, 2008). In the last 10 years, more than 100 studies reporting the application of global transcriptome analyses were published. This shows that transcriptome techniques became an important approach for *C. glutamicum* research. The technique was for example applied for the analysis of global regulatory networks (e.g. Beckers *et al.*, 2005; Silberbach *et al.*, 2005a,b; Brockmann-Gretza and Kalinowski, 2006; Wennerhold and Bott, 2006; Buchinger *et al.*, 2009; Auchter *et al.*, 2011), the analysis of amino acid producer strains (e.g. Krömer *et al.*, 2004; Stansen *et al.*, 2005; Hayashi *et al.*, 2006a,b; Ikeda *et al.*, 2006; Mitsuhashi *et al.*, 2006), producers of other products (e.g. Kind *et al.*, 2010b; Buschke *et al.*, 2013a; Zahoor *et al.*, 2014) and for general fermentation process analysis (Blombach *et al.*, 2013; Käß *et al.*, 2014).

With the application of high-throughput sequencing on transcriptome analysis and the development of RNA-seq transcriptome an alternative to DNA microarrays was established (Pfeifer-Sancar *et al.*, 2013), which allows deeper insight in the occurrence and function of small RNAs (Mentz *et al.*, 2013), transcriptional organization and regulation mechanisms (Neshat *et al.*, 2014). The techniques was also applied in the analysis of production strains as a basis of genome breeding (Kim *et al.*, 2013) and in the analysis of adaptive evolution of *C. glutamicum* (Lee *et al.*, 2013).

Two-dimensional gel electrophoresis (2-D PAGE) has been the first method developed for high-resolution proteome separation and also the first method applied to *C. glutamicum*. In the pre-genome era, protein identification relied on Edman sequencing and only very limited number of protein spots were identified by this method. Publication of the *C. glutamicum* genome sequence (Ikeda and Nakagawa, 2003; Kalinowski *et al.*, 2003) allowed the application of matrix-assisted laser desorption/ionization mass spectrometry (MALDI-TOF MS), which dramatically improved protein identification. A 2-D PAGE reference map with a comprehensive coverage of the cytoplasmic and membrane-associated proteins of *C. glutamicum* was presented already in 2001 (Schaffer *et al.*, 2001). Later, *C. glutamicum* was also used as model organism for the development of new proteomic technologies such as phosphoproteomics (Bendt *et al.*, 2003) and the analysis of membrane proteins. The membrane proteome of *C. glutamicum* became accessible by the development of new techniques for sample pre-fractionation, protein and peptide separation as well as protein digestion (Schluesener *et al.*, 2005; Fischer *et al.*, 2006; Fränzel *et al.*, 2009; Rietschel *et al.*, 2009a,b; Arrey *et al.*, 2010). A landmark in this respect was the identification of more than 50% of all predicted *C. glutamicum* membrane proteins (Fischer *et al.*, 2006). The proteomics methods established were applied in a large number of studies focusing on different aspects of *C. glutamicum* metabolism (for review, see Poetsch *et al.*, 2011, and Chapter 2).

Modelling of metabolic networks provides a powerful tool to understand microbial production hosts in order to optimize production processes and to exploit their industrial potential. Driven by this industrial interest, in 2009 two genome-scale models of *C. glutamicum* metabolism were published (Kjeldsen and Nielsen, 2009; Shinfuku *et al.*, 2009). Both models are based on the genome sequences published and differ only in a few aspects such as inclusion of protons and water within reactions. The metabolic network developed by Kjeldsen and Nielsen contains 411 metabolites and 446 reactions, while the model of Shinfuku and co-workers contains 423 metabolites and 502 reactions. The current state of stoichiometric modelling for *C. glutamicum* is presented in Chapter 4. Based on the metabolic network reconstruction by Kjeldsen and Nielsen, a new, curated network containing 475 reaction steps and 408 metabolites is described.

Genome scale metabolic network reconstruction is relying on information provided not only by -omics approaches, but also by metabolic flux analyses. This techniques was applied for more than 20 years in *C. glutamicum* (e.g. Wiechert and de Graaf, 1996; Sahm *et al.*, 2000). By application of this technique, it is possible to assess the intracellular metabolisms of a bacterium before and after gene modification. As outlined in Chapter 3, not only was metabolic flux analysis applied to describe and analyse amino acid metabolism and other production pathways in *C. glutamicum*, but *C. glutamicum* was used as a model organism to improve precision of metabolic flux analysis. By using ^{13}C trace experiments, ^{13}C labelling information and extracellular measurements together with bioinformatics approaches, highly precise estimation of fluxes in intracellular branching and reversible reactions can be obtained. These lead to better understanding of *C. glutamicum* physiology, and at the same time enable the identification of modification targets to improve the production of valuable compounds (Chapter 3).

Corynebacterium glutamicum as a synthetic biology platform

While systems biology focuses on the 'complete' description of a system by global analysis techniques (-omics) and integration of this information, synthetic biology deals with the design and construction of novel biological parts, devices, modules and systems as well as the redesign of existing biological systems (Roberts *et al.*, 2013). For an optimization in respect to time and effort reduction, a standardization of biological parts, modules and circuits is proposed (Endy, 2005; J. W. Lee *et al.*, 2011). Such a collection of biological components, tailor-made for one platform organism, allows expansion of the capacity of a production host and the assembly of new production pathways. This concept was recently also employed for *C. glutamicum*. As a standardized toolbox Kang and co-workers developed CoryneBrick vectors (Kang *et al.*, 2014) based on BglBrick standard using four unique restriction enzymes (T. S. Lee *et al.*, 2011). These plasmids can be used in combination with traditional *C. glutamicum* expression vectors. As a proof of principle, CoryneBrick vectors were used for expression of red fluorescent protein and to express the xylolytic pathway of *Escherichia coli* in *C. glutamicum* (T. S. Lee *et al.*, 2011).

Corynebacterium glutamicum-based green technology

Today's bioeconomy concepts include the replacement of oil-derived chemicals and products by equivalent or new products relying – at least partially – on biomass (Philp *et al.*, 2013a). With its transition from an amino acid producer to a versatile biotechnology platform organism *C. glutamicum* plays an important role in this development. Within the last decade enormous progress has been made in respect to an extended feedstock as well as production of novel products and materials by *C. glutamicum* (Becker and Wittmann, 2012b; Buschke *et al.*, 2013b).

Besides new nutrient sources and products, which will be discussed below, also new production processes might contribute to a *C. glutamicum*-based green technology. High-temperature production, which reduces the need to cool and might be especially beneficial in tropical regions with high temperatures but low costs for carbon sources, can minimize the energy input for production. Corresponding strains were already described (Onishi *et al.*, 2003; Ikeda *et al.*, 2006). Additionally, anaerobic production was already tested to decrease energy costs and even more important to direct the metabolic flux to the desired product. Production of ethanol and organic acids such as D-lactic acid and succinic acid under oxygen deprivation was established previously (Inui *et al.*, 2004; Okino *et al.*, 2008a,b; Litsanov *et al.*, 2012b). In future, also application in bioremediation of polluted soils or contaminated water might be an interesting application taking the high potential of *C. glutamicum* in respect to accumulation of metalloids (Mateos *et al.*, 2006) and metabolism of aromatic compounds (see Chapter 7) into account. Moreover, a number of studies was presented applying *C. glutamicum* for protein production and protein surface display (Date *et al.*, 2003, 2004, 2006; Kikuchi *et al.*, 2008; An *et al.*, 2013; Scheele *et al.*, 2013; Song *et al.*, 2013; Yim *et al.*, 2014; see also Chapter 11).

A number of studies were carried out to extend the spectrum of carbon sources used by *C. glutamicum* (Becker and Wittmann, 2012; Zahoor *et al.*, 2013; see also Chapter 5). The majority of work concentrated on the use of plant or industry waste materials as alternative carbon sources and sources of renewable feedstock. For example, metabolism of xylose, the main component of hemicellulose, was achieved by heterologous expression of xylose utilization genes from *E. coli* (Kawaguchi *et al.*, 2006), while arabinose metabolism was achieved by expression of the *E. coli* arabinose degradation pathway (Kawaguchi *et al.*, 2008; Schneider *et al.*, 2011). More recently, the possibility of also using crude hemicellulose hydrolysates for biotechnological production with recombinant *C. glutamicum* was reported (Buschke *et al.*, 2011; Gopinath *et al.*, 2011; Meiswinkel *et al.*, 2013a; Wang *et al.*, 2014). Besides strains able to utilize processed agricultural waste products such as corn cob, rice straw or wheat bran hydrolysates, strains with metabolic pathways for utilization of crude glycerol, a side and waste product of biodiesel production, were engineered (Rittmann *et al.*, 2008; Litsanov *et al.*, 2013; Meiswinkel *et al.*, 2013b). Recently, even the indirect use of CO_2 by the production of succinate from CO_2-grown microalgal biomass was published (Lee *et al.*, 2014). Adaptation of the *C. glutamicum* nutrient spectrum in respect to new nitrogen sources were also tested. These studies will be discussed elsewhere in this book (Chapter 6).

While first attempt to expand the *C. glutamicum* product spectrum were already made after its discovery (see L-lysine production described above), developments such as rational strain design, genome breeding and systems metabolic engineering accelerated the process of product diversification dramatically. Two major groups of products can be distinguished, bio-based chemicals and materials and bio-based fuels (Becker and Wittmann, 2012). The most important chemicals are diamines (Kind and Wittmann, 2011; Kind *et al.*, 2010a,b, 2011) and organic acids such as pyruvate (Schreiner *et al.*, 2005; Blombach *et al.*, 2007), lactic acid (Song *et al.*, 2012) or succinate (Litsanov *et al.*, 2012a,b, 2014; Wang *et al.*, 2014; Zhu *et al.*, 2013, 2014; for recent review, see Chapter 8). Bio-based fuels produced by *C. glutamicum* are ethanol (Inui *et al.*, 2004; Sakai *et al.*, 2007) and isobutanol (Smith *et al.*, 2010; Blombach *et al.*, 2011; Yamamoto *et al.*, 2013; for recent review, see Chapter 8). An especially interesting sector in this respect is the production of bio-based plastics. Plastics have a tremendous socio-economic impact. With their enormous flexibility and usefulness, but also rising environmental problems, e.g. ocean garbage patches, plastics are a prime target of green technology approaches (Philp *et al.*, 2013b). Several attempts were made to employ *C. glutamicum* for bioplastics production. These can be based on polyhydroxyalkanoates (Jo *et al.*, 2006, 2007, 2009; Liu *et al.*, 2007; Matsumoto *et al.*, 2011), poly-(lactic acid) (Auras *et al.*, 2004; Tokiwa and Calabia, 2006; Song *et al.*, 2013) or the polymerization of diamines to bio-nylon (Shimizu, 2013; Kind *et al.*, 2014; for review, see also Chapter 9).

Acknowledgements

Work in the author's lab was funded by various programmes of the Bundesministerium für Bildung und Forschung (Proteomanalyse, GenoMik, GenoMik+, Genomik Transfer, e:bio).

References

Ajinomoto Co., Inc. (2013a) Feed-use amino acids business. Available at: http://www.ajinomoto.com/en/ir/pdf/Feed-useAA-Oct2013.pdf.

Ajinomoto Co., Inc. (2013b) First quarter-FY2013 market and other information. Available at: http://www.ajinomoto.com/en/ir/pdf/FY13_Data_E.pdf.

An, S. J., Yim, S. S., and Jeong, K. J. (2013) Development of a secretion system for the production of heterologous proteins in *Corynebacterium glutamicum* using the porin B signal peptide. *Protein Expr. Purif.* **89**, 251–257.

Arrey, T. N., Rietschel, B., Papasotiriou, D. G., Bornemann, S., Baeumlisberger, D., Karas, M., and Meyer, B. (2010) Approaching the complexity of elastase-digested membrane proteomes using off-gel IEF/nLC-MALDI-MS/MS. *Anal. Chem.* **82**, 2145–2149.

Auchter, M., Cramer, A., Hüser, A., Rückert, C., Emer, D., Schwarz, P., Arndt, A., Lange, C., Kalinowski, J., Wendisch, V. F., and Eikmanns, B. J. (2011) RamA and RamB are global transcriptional regulators in *Corynebacterium glutamicum* and control genes for enzymes of the central metabolism. *J. Biotechnol.* **154**, 126–139.

Ault, A. (2004) The monosodium glutamate story: the commercial production of MSG and other amino acids. *J. Chem. Ed.* **81**, 347–355.

Auras, R., Harte, B., and Selke, S. (2004) An overview of polylactides as packaging materials. *Macromol. Biosci.* **4**, 835–864.

Bartek, T., Blombach, B., Zönnchen, E., Macus, P., Lang, S., Eikmanns, B. J., and Oldiges, M. (2010) Importance of NADPH supply for improved L-valine formation in *Corynebacterium glutamicum. Biotechnol.* Prog. **26**, 361–371.

Baumgart, M., Unthan, S., Rückert, C., Sivalingam, J., Grünberger, A., Kalinowski, J., Bott, M., Noack, S., and Frunzke, J. (2013) Construction of a prophage-free variant of *Corynebacterium glutamicum* ATCC 13032 for use as a platform strain for basic research and industrial biotechnology. *Appl. Environ. Microbiol.* **79**, 6006–6015.

Becker, J., and Wittmann, C. (2012a) Systems and synthetic metabolic engineering for amino acid production – the heartbeat of industrial strain development. *Curr. Opin. Biotechnol.* **23**, 718–726.

Becker, J., and Wittmann, C. (2012b) Bio-based production of chemicals, materials and fuels – *Corynebacterium glutamicum* as versatile cell factory. *Curr. Opin. Biotechnol.* **23**, 631–640.

Becker, J., Zelder, O., Häfner, S., Schröder, H., and Wittmann, C. (2011) From zero to hero – design-based systems metabolic engineering of *Corynebacterium glutamicum* for L-lysine production. *Metab. Eng.* **13**, 159–168.

Beckers, G., Strösser, J., Hildebrandt, U., Kalinowski, J., Farwick, M., Krämer, R., and Burkovski, A. (2005) Regulation of AmtR-controlled gene expression in *Corynebacterium glutamicum*: mechanism and characterization of the AmtR regulon. *Mol. Microbiol.* **58**, 580–595.

Bendt, A. K., Krämer, R., Burkovski, A., Schaffer, S., Bott, M., Busker, E., Hermann, T., Pfefferle, W., and Farwick, M. (2003) Towards a phospho-proteome map of *Corynebacterium glutamicum. Proteomics* **3**, 1637–1646.

Blombach, B., Schreiner, M. E., Holatko, J., Bartek, T., Oldiges, M., and Eikmanns, B. J. (2007) L-Valine production with pyruvate dehydrogenase complex-deficient *Corynebacterium glutamicum. Appl. Environm. Microbiol.* **73**, 2079–2084.

Blombach, B., Riester, T., Wieschalka, S., Ziert, C., Youn, J. W., Wendisch, V. F., and Eikmanns, B. J. (2011) *Corynebacterium glutamicum* tailored for efficient isobutanol production. *Appl. Environ. Microbiol.* **77**, 3300–3310.

Blombach, B., Buchholz, J., Busche, T., Kalinowski, J., and Takors, R. (2013) Impact of different CO_2/HCO_3^- levels on metabolism and regulation in *Corynebacterium glutamicum. J. Biotechnol.* **168**, 331–340.

Brockmann-Gretza, O., and Kalinowski, J. (2006) Global gene expression during stringent response in *Corynebacterium glutamicum* in presence and absence of the *rel* gene encoding (p)ppGpp synthase. *BMC Genomics* **8**, 230.

Buchinger, S., Strösser, J., Rehm, N., Hänßler, E., Hans, S., Bathe, B., Schomburg, D., Krämer, R., and Burkovski, A. (2009) A combination of transcriptome and metabolome analyses reveals new targets of the *Corynebacterium glutamicum* nitrogen regulator AmtR. *J. Biotechnol.* **140**, 68–74.

Buschke, N., Schröder, H., and Wittmann, C. (2011) Metabolic engineering of *Corynebacterium glutamicum* for production of 1,5-diaminopentane from hemicellulose. *Biotechnol. J.* **6**, 306–317.

Buschke, N., Becker, J., Schäfer, R., Kiefer, P., Biedendieck, R., and Wittmann, C. (2013a) Systems metabolic engineering of xylose-utilizing *Corynebacterium glutamicum* for production of 1,5-diaminopentane. *Biotechnol. J.* **8**, 557–570.

Buschke, N., Schäfer, R., Becker, J., and Wittmann, C. (2013b) Metabolic engineering of industrial platform microorganisms for biorefinery applications – Optimization of substrate spectrum and process robustness by rational and evolutive strategies. *Biores. Technol.* **135**, 544–554.

Date, M., Yokoyama, K.-i., Umezawa, Y., Matsui, H., and Kikuchi, Y. (2003) Production of native-type *Streptoverticillium mobaraense* transglutaminase in *Corynebacterium glutamicum*. *Appl. Environ. Microbiol.* **69**, 3011–3014.

Date, M., Yokoyama, K.-i., Umezawa, Y., Matsui, H., and Kikuchi, Y. (2004) High level expression of *Streptomyces mobaraensis* transglutaminase in *Corynebacterium glutamicum* using a chimeric pro-region from *Streptomyces cinnamoneus* transglutaminase. *J. Biotechnol.* **110**, 219–226.

Date, M., Itaya, H., Matsui, H., and Kikuchi, Y. (2006) Secretion of human epidermal growth factor by *Corynebacterium glutamicum*. *Lett. Appl. Microbiol.* **42**, 66–70.

De Graaf, A. A., Eggeling, L., and Sahm, H. (2001) Metabolic engineering for L-lysine production by *Corynebacterium glutamicum*. *Adv. Biochem. Eng. Biotechnol.* **73**, 9 29.

Demain, A. L. (2000) Microbial biotechnology. *Trends Biotechnol.* **18**, 26–31.

Endy, D. (2005) Foundations for engineering biology. *Nature* **438**, 449–453.

Fischer, F., Wolters, D., Rögner, M., and Poetsch, A. (2006) Toward the complete membrane proteome: high coverage of integral membrane proteins through transmembrane peptide detection. *Mol. Cell. Proteomics* **5**, 444–453.

Fränzel, B., Fischer, F., Trötschel, C., Poetsch, A., and Wolters, D. (2009) The two-phase partitioning system a powerful technique to purify integral membrane proteins of *Corynebacterium glutamicum* for quantitative shotgun analysis. *Proteomics* **9**, 2263–2272.

Frunzke, J., Bramkamp, M., Schweitzer, J.-E., and Bott, M. (2008) Population heterogeneity in *Corynebacterium glutamicum* ATCC 13032 caused by prophage CGP3. *J. Bacteriol.* **190**, 5111–5119.

Gopinath, V., Meiswinkel, T. M., Wendisch, V. F., and Nampoothiri, K. M. (2011) Amino acid production from rice straw and wheat bran hydrolysates by recombinant pentose-utilizing *Corynebacterium glutamicum*. *Appl. Microbiol. Biotechnol.* **92**, 985–996.

Halatko, J., Elisakova, V., Prouza, M., Sobotka, M., Nesvera, J., and Patek, M. (2009) Metabolic engineering of L-valine biosynthesis pathway in *Corynebacterium glutamicum* using promoter activity modulation. *J. Biotechnol.* **139**, 203–210.

Hayashi, M., Ohnishi, J., Mitsuhashi, S., Yonetani, Y., Hashimoto, S., and Ikeda, M. (2006a) Transcriptome analysis reveals global expression changes in an industrial L-lysine producer of *Corynebacterium glutamicum*. *Biosci. Biotechnol. Biochem.* **70**, 546–550.

Hayashi, M., Mizuguchi, H., Ohnishi, J., Mitsuhashi, S., Yonetani, Y., Hashimoto, S., and Ikeda, M. (2006b) A *leuC* mutation leading to increased L-lysine production and *rel*-independent global expression changes in *Corynebacterium glutamicum*. *Appl. Microbiol. Biotechnol.* **72**, 783–789.

Ikeda, M. (2003) Amino acid production processes. *Adv. Biochem. Eng. Biotechnol.* **79**, 1–35.

Ikeda, M., and Nakagawa, S. (2003) The *Corynebacterium glutamicum* genome: features and impacts on biotechnological processes. *Appl. Microbiol. Biotechnol.* **62**, 99–109.

Ikeda, M., Onishi, J., Hayashi, M., and Mitsuhashi, S. (2006) A genome-based approach to create a minimally mutated *Corynebacterium glutamicum* strain for efficient L-lysine production. *J. Ind. Microbiol. Biotechnol.* **33**, 610–615.

Ikeda, M., Mitsuhashi, S., Tanaka, K., and Hayashi, M. (2009) A genome-based approach to create a minimally mutated *Corynebacterium glutamicum* strain for efficient L-lysine production. *J. Ind. Microbiol. Biotechnol.* **33**, 610–615.

Inui, M., Kawaguchi, H., Murakami, S., Vertes, A. A., and Yukawa, H. (2004) Metabolic engineering of *Corynebacterium glutamicum* for fuel ethanol production under oxygen-deproviation conditions. *J. Mol. Microbiol. Biotechnol.* **8**, 243–254.

Jo, S. J., Maeda, M., Ooi, T., and Taguchi, S. (2006) Production system for biodegradable polyester polyhydroxybutyrate by *Corynebacterium glutamicum*. *J. Biosci. Bioeng.* **102**, 233–236.

Jo, S. J., Matsumoto, K., Leong, C. R., Ooi, T., and Taguchi, S. (2007) Improvement of poly(3-hydroxybutyrate) [P(3HB)] production in *Corynebacterium glutamicum* by codon optimization, point mutation and gene dosage of P(3HB) biosynthetic genes. *J. Biosci. Bioeng.* **104**, 457–463.

Jo, S. J., Leong, C. R., Matsumoto, K., and Taguchi, S. (2009) Dual production of poly (3-hydroxybutyrate) and glutamate using variable biotin concentrations in *Corynebacterium glutamicum*. *J. Biosci. Bioeng.* **107**, 409–411.

Jojima, T., Fujii, M., Mori, E., Inui, M., and Yukawa, H. (2010) Engineering of sugar metabolism of *Corynebacterium glutamicum* for production of amino acid L-alanine under oxygen deprivation. *Appl. Microbiol. Biotechnol.* **87**, 159–165.

Käß, F., Hariskos, I., Michel, A., Brandt, H. J., Spann, R., Junne, S., Wiechert, W., Neubauer, P., and Oldiges, M. (2014) Assessment of robustness against dissolved oxygen/substrate oscillations for *C. glutamicum* DSM1933 in two-compartment bioreactor. *Bioprocess Biosyst. Eng.* **37**, 11-51–1162.

Kalinowski, J., Bathe, B., Bischoff, N., Bott, M., Burkovski, A., Dusch, N., Eggeling, L., Eikmanns, B. J., Gaigalat, L., Goesmann, A., *et al.* (2003) The complete *Corynebacterium glutamicum* ATCC 13032 genome sequence and its impact on the production of L-aspartate-derived amino acids and vitamins. *J. Biotechnol.* **104**, 5–25.

Kang, M.-K., Lee, J., Um, Y., Lee, T. S., Bott, M., Park, S. J., and Woo, H. M. (2014) Synthetic biology platform of CoryneBrick vectors for gene expression in *Corynebacterium glutamicum* and its application to xylose utilization. *Appl. Microbiol. Biotechnol.* **98**, 5991–6002.

Kawaguchi, H., Vertes, A. A., Okino, S., Inui, M., and Yukawa, H. (2006) Engineering of a xylose metabolic pathway in *Corynebacterium glutamicum*. *Appl. Environ. Microbiol.* **72**, 3418–3428.

Kawaguchi, H., Sasaki, M., Vertes, A. A., Inui, M., and Yukawa, H. (2008) Engineering of an L-arabinose metabolic pathway in *Corynebacterium glutamicum*. *Appl. Environ. Microbiol.* **77**, 1053–1062.

Kelle, R., Hermann, T., and Bathe, B. (2005) L-Lysine production. In: Bott, M., and Eggeling, L. (eds.) Handbook of *Corynebacterium glutamicum*. CRC Press LLC, Boca Raton, FL, USA, pp. 465–488.

Kikuchi, Y., Itaya, H., Date, M., Matsui, K., and Wu, L.-F. (2008) Production of *Chryseobacterium proteolyticum* protein-glutaminase using the twin-arginine translocation pathway in *Corynebacterium glutamicum*. *Appl. Microbiol. Biotechnol.* **78**, 67–74.

Kim, H. I., Nam, J. Y., Cho, J. Y., Lee, C. S., and Park, Y. J. (2013) Next-generation sequencing-based transcriptome analysis of L-lysine-producing *Corynebacterium glutamicum* ATCC 21300 strain. *J. Microbiol.* **51**, 877–880.

Kind, S., and Wittmann, C. (2011) Bio-based production of the platform chemical 1,5-diaminopentane. *Appl. Environ. Microbiol.* **91**, 1287–1296.

Kind, S., Jeong, W. K., Schröder, H., and Wittmann, C. (2010a) Systems-wide metabolic pathway engineering for in *Corynebacterium glutamicum* for bio-based production of diaminopentane. *Metab. Eng.* **12**, 341–351.

Kind, S., Jeong, W. K., Schröder, H., Zelder, O., and Wittmann, C. (2010b) Identification and elimination of the competing N-acetyldiaminopentane pathway for improved production of diaminopentane by *Corynebacterium glutamicum*. *Appl. Environ. Microbiol.* **76**, 5175–5180.

Kind, S., Kreye, S., and Wittmann, C. (2011) Metaboilic engineering of cellular transport for overproduction of the platform chemical 1,5-diaminopentane in *Corynebacterium glutamicum*. *Metab. Eng.* **13**, 617–627.

Kinoshita, S. (2005) A short history of the birth of the amino acid industry in Japan. In: Bott, M., and Eggeling, L. (eds.) Handbook of *Corynebacterium glutamicum*. CRC Press LLC, Boca Raton, FL, USA, pp. 3–5.

Kinoshita, S., Udaka, S., and Shimono, M. (1957) Studies on the amino acid fermentation part.1. Production of L-glutamic acid by various microorganisms. J. *Gen. Appl. Microbiol.* **3**, 193–205.

Kjeldsen, K. R., and Nielsen, J. (2009) *In silico* genome-scale reconstruction and validation of the *Corynebacterium glutamicum* metabolic network. *Biotechnol. Bioeng.* **102**, 583–597.

Krömer, J. O., Sorgenfrei, O., Klopprogge, K., Heinzle, E., and Wittmann, C. (2004) In-depth profiling of lysine-producing *Corynebacterium glutamicum* by combined analysis oft he transcriptome, metabolome, and fluxome. *J. Bacteriol.* **186**, 1769–1784.

Lee, J. W., Kim, T. Y., Jang, Y.-S., Choi, S., and Lee, S. Y. (2011) Systems metabolic engineering for chemicals and materials. *Trends Biotechnol.* **29**, 370–378.

Lee, J. Y., Seo, J., Kim, E. S., Lee, H. S., and Kim, P. (2013) Adaptive evolution of *Corynebacterium glutamicum* resistant to oxidative stress and its global gene expression profiling. *Biotechnol. Lett.* **35**, 709–717.

Lee, J., Sim, S. J., Bott, M., Um, Y., Oh, M.-K., anf Woo, H. M. (2014) Succinate production from CO_2-grown microalgal biomass as carbon source using engineered *Corynebacterium glutamicum* through consolidated bioprocessing. *Sci. Rep.* **4**, 5819.

Lee, T. S., Krupka, R. A., Zhang, F., Hajimorad, M., Holtz, W. J., Prasad, N., Lee, S. K., Keasling, J. D. (2011) BglBrick vectors and datasheets: A synthetic biology platform for gene expression. *J. Biol. Eng.* **5**, 12.

Leuchtenberger, W. (1996) Amino acids – technical production and use. In: Roehr, M. (ed.) Biotechnology, 2nd edition, Vol. 6, Products of Primary Metabolism. VCH, Weinheim, Germany, pp. 465–502.

Litsanov, B., Brocker, M., and Bott, M. (2012a) Toward homosuccinate fermentation: metabolic engineering of *Corynebacterium glutamicum* for anaerobic production of succinate from glucose and formate. *Appl. Environm. Microbiol.* **78**, 3325–3337.

Litsanov, B., Kabus, A., Brocker, M., and Bott, M. (2012b) Efficient aerobic succinate production from glucose in minimal medium with *Corynebacterium glutamicum*. *Microb. Biotechnol.* **5**, 116–128.

Litsanov, B., Brocker, M., and Bott, M. (2013) Glycerol as a substrate for aerobic succinate production in minimal medium with *Corynebacterium glutamicum*. *Microb. Biotechnol.* **6**, 189–195.

Litsanov, B., Brocker, M., Oldiges, M., and Bott, M. (2014) Succinic acid. In: Kondo A., and Bisaria V. (eds.) Bioprocessing of Renewable Resources to Commodity Bioproducts. John Wiley & Sons, Hoboken, NJ, USA, pp. 437–474.

Liu, Q., Ouyang, S. P., Kim, J., and Chen, G. Q. (2007) The impact of PHB accumulation on L-glutamate production by recombinant *Corynebacterium glutamicum*. *J. Biotechnol.* **132**, 273–279.

Mateos, L. M., Ordonez, E., Letek, M., and Gil, J. A. (2006) *Corynebacterium glutamicum* as a model bacterium for bioremediation of arsenic. *Int. Microbiol.* **9**, 207–215.

Matsumoto, K., Kitagawa, K., Jo, S. J., Song, Y., and Tageuchi, S. (2011) Production of poly(3-hydroxybutyrate-co-3-hydroxyvalerate) in recombinant *Corynebacterium glutamicum* using propionate as a precursor. *J. Biotechnol.* **152**, 144–146.

Meiswinkel, T. M., Gopinath, V., Lindner, S. N., Nampoothiri, K. M., and Wendisch, V. F. (2013a) Accelerated pentose utilization by *Corynebacterium glutamicum* for accelerated production of lysine, glutamate, ornithine and putrescine. *Microb. Biotechnol.* **6**, 131–140.

Meiswinkel, T. M., Rittmann, D., Lindner, S. N., and Wendisch, V. F. (2013b) Crude glycerol-based production of amino acids and putrescine by *Corynebacterium glutamicum*. *Bioresour Technol.* **145**, 254–258.

Mentz, A., Neshat, A., Pfeifer-Sancar, K., Pühler, A., Rückert, C., and Kalinowski, J. (2013) Comprehensive discovery and characterization of small RNAs in *Corynebacterium glutamicum* ATCC 13032. *BMC Genomics* **14**, 714.

Mitsuhashi, S., Hayashi, M., Ohnishi, J., and Ikeda, M. (2006) Disruption of malate:quinone oxidoreductase increases L-lysine production by *Corynebacterium glutamicum*. *Biosci. Biotechnol. Biochem.* **70**, 2803–2806.

Neshat, A., Mentz, A., Rückert, C., and Kalinowski, J. (2014) Transcriptome sequencing revealed the transcriptional organization at ribosome-mediated attenuation sites in *Corynebacterium glutamicum* and identified a novel attenuator involved in aromatic amino acid biosynthesis. *J. Biotechnol.* [Epub ahead of print].

Okino, S., Suda, M., Fujikura, K., Inui, M., and Yukawa, H. (2008a) Production of D-lactic acid by *Corynebacterium glutamicum* under oxygen deprivation. *Appl. Microbiol. Biotechnol.* **78**, 449–454.

Okino, S., Noburyu, R., Suda, M., Jojima, T., Inui, M., and Yukawa, H. (2008b) An efficient succinic acid production process in a metabolically engineered *Corynebacterium glutamicum* strain. *Appl. Microbiol. Biotechnol.* **81**, 459–464.

Onishi, J., Mitsuhashi, S., Hayashi, M., Ando, S., Yokoi, H., Ochiai, K., and Ikeda, M. (2002) A novel methodology employing *Corynebacterium glutamicum* genome information to generate a new L-lysine producing mutant. *Appl. Microbiol. Biotechnol.* **58**, 217–223.

Onishi, J., Hayashi, M., Mitsuhashi, S., and Ikeda, M. (2003) Efficient 40°C fermentation of L-lysine by a new *Corynebacterium glutamicum* mutant developed by genome breeding. *Appl. Microbiol. Biotechnol.* **62**, 69–75.

Peters-Wendisch, P., Stolz, M., Etterich, H., Kennerknecht, N., Sahm, H., and Eggeling, L. (2005) Metabolic engineering of *Corynebacterium glutamicum* for L-serine production. *Appl. Environ. Microbiol.* **71**, 7139–7144.

Pfefferle, W., Möckel, B., Bathe, B., and Marx, A. (2003) Biotechnological manufacture of L- lysine. In: Scheper, T. (ed.) Advances in Biochemical Engineering/ Biotechnology, Springer, Berlin, Germany, p. 59.

Pfeifer-Sancar, K., Mentz, A., Rückert, C., and Kalinowski, J. (2013) Comprehensive analysis of the *Corynebacterium glutamicum* transcriptome using an improved RNAsep technique. *Genomics* **14**, 888.

Philp, J. C., Ritchie, R. J., and Allan, J. E. M. (2013a) Biobased chemicals: the convergence of green chemistry with industrial biotechnology. *Trends Biotechnol.* **31**, 219–221.

Philp, J. C., Ritchie, R. J., and Guy, K. (2013b) Biobased plastics in a bioeconomy. *Trends Biotechnol.* **31**, 65–67.

Poetsch, A., Haußmann, U., and Burkovski, A. (2011) Proteomics of corynebacteria: from biotechnology workhorses to pathogens. *Proteomics* **11**, 3244–3255.

Quast, K., Bathe, B., Pühler, A., and Kalinowski, J. (1999) The *Corynebacterium glutamicum* insertion sequence ISCg2 prefers conserved target sequences located adjacent to genes involved in aspartate and glutamate metabolism. *Mol. Gen. Genet.* **262**, 568–578.

Radmacher, E., Vaitsikova, A., Burger, U., Krumbach, K., Sahm, H., and Eggeling, L. (2002) Linking central metabolism with increased pathway flux: L-valine accumulation by *Corynebacterium glutamicum*. *Appl. Environ. Microbiol.* **68**, 2246–2250.

Rietschel, B., Arrey, T. N., Meyer, B., Bornemann, S., Schuerken, M., Karas, M., and Poetsch, A. (2009a) Elastase digests: new ammunition for shotgun membrane proteomics. *Mol. Cell. Proteomics* **8**, 1029–1043.

Rietschel, B., Bornemann, S., Arrey, T. N., Baeumlisberger, D., Karas, M., and Meyer, B. (2009b) Membrane protein analysis using an improved peptic in-solution digestion protocol. *Proteomics* **9**, 5553–5557.

Rittmann, D., Lindner, S. N., and Wendisch, V. F. (2008) Engineering of a glycerol utilization pathway for amino acid production by *Corynebacterium glutamicum*. *Appl. Environ. Microbiol.* **74**, 6216–6222.

Roberts, M. A., Cranenburgh, R. M., Stevens, M. P., and Oyston, P. C. F. (2013) Synthetic biology: biology by design. *Microbiology* **159**, 1219–1220.

Sahm, H., Eggeling, L., and de Graaf, A. A. (2000) Pathway analysis and metabolic engineering in *Corynebacterium glutamicum*. *Biol. Chem.* **381**, 899–910.

Sakai, S., Tsuchida, Y., Nakamoto, H., Okino, S., Ichihashi, O., Kawaguchi, H., Watanabe, T., Inui, M., and Yukawa, H. (2007) Effect of lignocellulose-derived inhibitors on growth of and ethanol production by growth-arrested *Corynebacterium glutamicum* R. *Appl. Environm. Microbiol.* **73**, 6328–6328.

Schaffer, S., Weil, B., Nguyen, V. D., Dongmann, G., Günther, K., Nickolaus, M., Hermann, T., and Bott, M. (2001) A high-resolution reference map for cytoplasmic and membrane-associated proteins of *Corynebacterium glutamicum*. *Electrophoresis* **22**, 4404–4422.

Scheele, S., Oertel, D., Bongaerts, J., Evers, S., Hellmuth, H., Maurer, K.-H., Bott, M., and Freudl, R. (2013) Secretory production of an FAD cofactor-containing cytosolic enzyme (sorbitol-xylitol oxidase from *Streptomyces coelicolor*) using the twin-arginine translocation (Tat) pathway of *Corynebacterium glutamicum*. *Microb. Biotechnol.* **6**, 202–206.

Schluesener, D., Fischer, F., Kruip, J., Rögner, M., and Poetsch, A. (2005) Mapping the membrane proteome of *Corynebacterium glutamicum*. *Proteomics* **5**, 1317–1330.

Schneider, J., Niermann, K., and Wendisch, V. F. (2011) Production of the amino acids L-glutamate, L-lysine, L-ornithinne and L-arginine from arabinose by recombinant *Corynebacterium glutamicum*. *J. Biotechnol.* **154**, 191–198.

Schreiner, M. E., Fiur, D., Holatko, J., Patek, M., and Eikmanns, B. (2005) E1 enzyme of the pyruvate dehydrogenase complex in *Corynebacterium glutamicum*: molecular analysis of the gene and phylogenetic aspects. *J. Bacteriol.* **187**, 6005–6018.

Shimizu, H. (2013) Systems metabolic engineering for the production of bio-nylon precursor. *Biotechnol. J.* **8**, 513–514.

Shinfuku, Y., Sorpitiporn, N., Sono, M., Furusawa, C., Hirasawa, T., and Shimizu, H. (2009) Development and experimental verification of a genome-scale metabolic model for *Corynebacterium glutamicum*. *Microb. Cell Fact.* **8**, 43.

Silberbach, M., Hüser, A., Kalinowski, J., Pühler, A., Walter, B., Krämer, R., and Burkovski, A. (2005a) DNA microarray analysis of the nitrogen starvation response of *Corynebacterium glutamicum*. *J. Biotechnol.* **119**, 357–367.

Silberbach, M., Schäfer, M., Hüser, A., Kalinowski, J., Pühler, A., Krämer, R., and Burkovski, A. (2005b) Adaptation of *Corynebacterium glutamicum* to ammonium-limitation: a global analysis using transcriptome and proteome techniques. *Appl. Environ. Microbiol.* **71**, 2391–2402.

Smith, K. M., Cho, K. M., and Liao, J. C. (2010) Engineering *Corynebacterium glutamicum* for isobutanol production. *Appl. Microbiol. Biotechnol.* **87**, 1044–1055.

Song, Y. Y., Matsumoto, K., Yamada, M., Gohda, A., Brigham, C. J., Sinskey, A. J., and Taguchi, S. (2012) Engineered *Corynebacterium glutamicum* as an endotoxin-free platform strain for lactate-based polyester production. *Appl. Microbiol. Biotechnol.* **93**, 1917–1925.

Song, Y., Matsumoto, K., Tanaka, T., Kondo, A., and Taguchi, S. (2013) Single-step production of polyhydroxybutyrate from starch by using α-amylase cell-surface displaying system of *Corynebacterium glutamicum*. *J. Biosci. Bioeng.* **115**, 12–14.

Stäbler, N., Oikawa, T., Bott, M., and Eggeling, L. (2011) *Corynebacterium glutamicum* as a host for synthesis and export of D-amino acids. *J. Bacteriol.* **193**, 1702–1709.

Stansen, C., Uy, D., Delauny, S., Eggeling, L., Goergen, J. L., and Wendisch, V. F. (2005) Characterization of a *Corynebacterium glutamicum* lactate utilization operon induced during temperature-triggered glutamate production. *Appl. Environ. Microbiol.* **71**, 5920–5928.

Takors, R., Bathe, B., Rieping, M., Hans, S., Kelle, R., and Huthmacher, K. (2007) Systems biology for industrial strains and fermentation processes – Example: Amino acids. *J. Biotechnol.* **129**, 181–190.

Tokiwa, Y., and Calabia, B. (2006) Biodegradability and biodegradation of poly(lactide). *Appl. Microbiol. Biotechnol.* **72**, 244–251.

Udaka, S. (1960) Screening method for microorganisms accumulating metabolites and its use in isolation of *Micrococcus glutamicus*. *J. Bacteriol.* **79**, 754–755.

Udaka, S. (2008) The discovery of *Corynebacterium glutamicum* and birth of amino acid fermentation industry in Japan. In: Burkovski, A. (ed.) Corynebacteria – Genomics and Molecular Biology. Caister Academic Press Norfolk, UK, pp. 1–6.

Wang, C., Zhang, H., Cai, H., Zhou, Z., Chen, Y., Chen, Y., and Ouyang, P. (2014) Succinic acid production from corn cob hydrolysates by genetically engineered *Corynebacterium glutamicum*. *Appl. Biochem. Biotechnol.* **172**, 340–350.

Wendisch, V. F. (2003) Genome-wide expression analysis in *Corynebacterium glutamicum* using DNA microarrays. *J. Biotechnol.* **104**, 273–285.

Wendisch, V. F. (2008) DNA microarray-based transcriptome analysis in *C. glutamicum*. In: Burkovski, A. (ed.) Corynebacteria – Genomics and Molecular Biology. Caister Academic Press Norfolk, UK, pp. 33–54.

Wendisch, V. F., Bott, M., Kalinowski, J., Oldiges, M., and Wiechert, W. (2006) Emerging *Corynebacterium glutamicum* systems biology. *J. Biotechnol.* **124**, 74–92.

Wennerhold, J., and Bott, M. (2006) The DtxR regulon of *Corynebacterium glutamicum*. *J. Bacteriol.* **188**, 2907–2918.

Wiechert, W., and de Graaf, A. A. (1996) *In vivo* stationary flux analysis by ^{13}C labeling experiments. *Adv. Biochem. Eng. Biotechnol.* **54**, 109–154.

Yamamoto, S., Suda, M., Niimi, S., Inui, M., and Yukawa, H. (2013) Strain optimization for efficient isobutanol production using *Corynebacterium glutamicum* under oxygen deprivation. *Biotechnol. Bioeng.* **110**, 2938–2948.

Yim, S. S., An, S. J., Choi, J. W., Ryu, A. J., and Jeong, K. J. (2014) High-level secretory production of recombinant single-chain variable fragment (scFv) in *Corynebacterium glutamicum*. *Appl. Microbiol. Biotechnol.* **98**, 273–284.

Zahoor, A., Lindner, S. N., and Wendisch, V. F. (2012) Metabolic engineering of *Corynebacterium glutamicum* aimed at alternative carbon sources and new products. *Comput. Struct. Biotechnol. J.* **3**, e201210004.

Zahoor, A., Otten, A., and Wendisch, V. F. (2014) Metabolic engineering of *Corynebacterium glutamicum* for glycolate production. *J. Biotechnol.* [Epub ahead of print].

Zhu, N. Q., Xia, H. H., Wang, Z. W., Zhao, X. M., and Chen, T. (2013) Engineering of acetate recycling and citrate synthase to improve aerobic succinate production in *Corynebacterium glutamicum*. *Plos One* **8**, e60659.

Zhu, N., Xia, H., Yang, J., Zhao, X., and Chen, T. (2014) Improved succinate production in *Corynebacterium glutamicum* by engineering glyoxylate pathway and succinate export system. *Biotechnol. Lett.* **36**, 553–560.

Proteomics of *Corynebacterium glutamicum*

2

Andreas Harst and Ansgar Poetsch

Abstract

Proteomics is an important technique to study the biology of *Corynebacterium glutamicum* with focus on processes related to the biotechnological production of amino acids and other chemicals. Within this review, a comprehensive overview is given about the methods established for *C. glutamicum*. The methods developed in *C. glutamicum* have been and can be applied to other organisms. Technologies for quantification, membrane enrichment and MS analysis tested in *C. glutamicum* might lead to higher coverage for membrane proteins or improve the accuracy and reliability of quantification based on isotope labelling. In summary, proteomics data are important for systems biology approaches in *C. glutamicum*.

Introduction

Proteomics is a branch of biomolecule analysis, which strives to detect, quantify and study the proteome of an organism in a certain biological condition. Considering that the proteome is defined as the full complement of proteins contained in a cell, proteomics relies on the accurate and reliable identification of proteins present in a sample. The method of choice to achieve this is biological mass spectrometry.

Owing to its role as amino acid producer *Corynebacterium glutamicum* is of great importance for biotechnology. Therefore, a large number of studies investigating the genome and transcriptional dynamics of this organism have been published. Still genomics and transcriptomics can only lead to partial understanding of an organism's biological processes in a cell. To gauge the presence of proteins and to accurately estimate their synthesis and degradation rates, proteomics is needed. Only this method can determine with high sensitivity the actual cellular amount of enzymes, or calculate the rates of central physiological reactions and investigate the role of post-translational modifications.

This section will give an overview over the different branches of proteomics, which have been applied to *C. glutamicum* up to now and also highlight possible avenues for future research.

Understanding *Corynebacterium glutamicum* physiology with proteomics: application examples

Its biotechnological importance has stimulated a considerable amount of works towards a better understanding of *C. glutamicum* physiology with proteomics. In Table 2.1, these studies are summarized and categorized according to their research question. The response of *C. glutamicum* to many stressors was studied, as well as impact of amino acid production on physiology and utilization of different nitrogen and carbon sources. Furthermore, defects in energy metabolism, caused by ATP synthase inactivity, were investigated.

Methods of *Corynebacterium glutamicum* proteomics

2D gels

The first method of proteomics applied to *C. glutamicum* was two-dimensional polyacrylamide gel

Table 2.1 Summary of proteome studies for *Corynebacterium glutamicum* physiology

Research question	Reference
Protein localization	Hermann *et al.* (2001), Hansmeier *et al.* (2006), Franzel and Wolters (2011)
Protein modification	Bendt *et al.* (2003), Chi *et al.* (2014), Kuberl *et al.* (2014)
Ammonium utilization	Beckers *et al.* (2004, 2005), Silberbach *et al.* (2005)
Heat shock response	Barreiro *et al.* (2005), Trötschel *et al.* (2012)
L-Glutamate biosynthesis	Li *et al.* (2007a)
L-Lysine biosynthesis	Schluesener *et al.* (2007), Franzel *et al.* (2010a)
ATP synthase defect	Li *et al.* (2007b), Koch-Koerfges *et al.* (2012)
Metal stress	Fanous *et al.* (2008, 2010)
Aromatics catabolism	Qi *et al.* (2007), Haußmann *et al.* (2009), Haußmann and Poetsch (2012)
Central carbon metabolism	Polen *et al.* (2007), Voges and Noack (2012)
Acidic stress	Barriuso-Iglesias *et al.* (2008), Follmann *et al.* (2009), Vasco-Cardenas *et al.* (2013)
Salt stress	Fränzel *et al.* (2010b)
Oxidative stress	Kromer *et al.* (2008)
Antibiotic/herbicide stress	Fanous *et al.* (2007), Franzel *et al.* (2010c)

electrophoresis (2D-PAGE). The combination of the sieving effect of acrylamide and immobilized pH gradients enables a gel-based separation of proteins sequentially by their isoelectrical point (pI) as well as by their molecular weight. Proteins are electrophoretically separated along these pH gradients, in which proteins move to the point where they have no net charge. In the second step, proteins are separated according to their molecular weight using SDS-PAGE; the proteins thus separated are then amenable to analysis using mass spectrometry. Thus, it was possible to look at the proteomes of different biological conditions. Several stresses as oxidative and pH stress have been analysed using this approach. The results of these studies were summarized by Wendisch and Polen (2013).

Protein identification with 2D-PAGE depends on a number of technological aspects. Imaging of 2D gels is fundamental for protein identification and quantification, and a large variety of staining techniques are available ranging from classic methods as Coomassie/colloidal CBB-G250 (Candiano *et al.*, 2004) and silver-based staining to highly sensitive fluorescence stains, which are most suitable for quantitative image analysis. Most powerful is the 2D-DIGE technique (Karp *et al.*, 2004), where samples are covalently labelled with Cy-dyes and separated together on a single gel to avoid gel to gel variation. The protein spots thus visualized are excised and proteins are digested using amino acid sequence specific proteases, commonly trypsin. This method has been applied to *C. glutamicum* for analysis of the effect of industrially relevant C4 and C5 dicarboxylic acids on its proteome (Vasco-Cárdenas *et al.*, 2013). Peptides can then be subjected to matrix-assisted laser desorption ionization mass spectrometry (MALDI-MS) or liquid chromatography mass spectrometry (LC-MS). Usually MALDI is the method of choice, due to its robustness, high sample throughput, and the low sample complexity obtained after 2D separation. An excellent summary concerning the technological aspect of 2D-PAGE has been published in (Burkovski, 2008).

The use of 2D proteomics greatly enhanced our knowledge about the proteome of *C. glutamicum*. Apart from generally lower bioinformatics and mass spectrometry demands, the high protein resolution power of 2D proteomics makes it an excellent tool to address scientific questions favouring the analysis of intact proteins, such as identification of proteolysis events or covalent modifications. However, many questions of proteome research are only partially amenable to 2D-PAGE proteomics. This method cannot identify proteins containing more than one trans-membrane helix, or which are very hydrophobic, because these proteins tend to precipitate during IEF (isoelectric focusing) based electrophoresis (Klein *et al.*, 2005). Furthermore, quantification of proteins is only based on densitometric comparison of gel pictures, which is hampered by protein co-migration, whereas for LC-MS a number of very powerful methods for relative quantification and absolute quantification have been developed. Additionally, LC-MS workflows are more sensitive and still show considerable increase in throughput with new instrument developments.

Methods of membrane proteomics

When the first genome sequence of *C. glutamicum* was published, the number of putative membrane proteins was estimated to be 660 (Kalinowski *et al.*, 2003). The predicted molecular weight of membrane proteins ranges from 2 to 150 kDa and the pIs from 3 to 12. Overall, 18% are transport or binding proteins and 6% were predicted to constitute cell envelope proteins. A further 3% are involved in protein fate or energy metabolism; the residual 57% are proteins to which no function has been assigned up to now. Software such as TMHMM predicts that most membrane proteins contain fewer than 10 trans-membrane (TM) helices.

The TM helices of integral membrane proteins (IMPs) can have a length ranging from 14 to 36 hydrophobic amino acids (Bowie, 1997), and rarely contain lysine or arginine. In MS-based proteomics, tryptic digests are preferred, but with IMPs these digests result in reduced sequence coverage usually missing TM domains due to the long length of the obtained peptides, which are difficult to fragment with MS. In a 2006 study combinations of trypsin and cyanogen bromide, and trypsin and chymotrypsin were compared with a tryptic digest for *C. glutamicum*, *Halobacterium* sp. and *Saccharomyces cerevisiae* clearly showing the superiority of the combined digests (Fischer and Poetsch, 2006) in terms of coverage and number of identified IMPs, because cleavages inside TM domains can be obtained with these digestion conditions.

To improve the efficiency of membrane protein digests also the elastase protease was tested. This protease cleaves semi-specific after a number of amino acids, many of which are hydrophobic, hence the enzyme is a good candidate for membrane digests (Rietschel *et al.*, 2009). The direct comparison of trypsin and elastase in a digest of *C. glutamicum* membranes showed that both enzymes yield a similar number of protein identifications. Importantly, each protease identifies a different set of proteins with rather weak overlap, demonstrating high complementarity. The number of peptides originating from TM parts of IMPs was higher in elastase digests than for trypsin, thus proving its ability to increase the coverage of membrane proteins (Rietschel *et al.*, 2009).

For the resolution of cellular compartments, in initial *C. glutamicum* proteome studies the cytoplasm was separated from the membrane fraction before 2D-PAGE (Burkovski, 2002). Even though some membrane-associated proteins were resolved, IMPs were not detected, most likely owing to their irreversible precipitation during isoelectric focusing (IEF). The use of liquid agarose gels for IEF could not alleviate this problem. In contrast, it is well known that SDS-PAGE is perfectly compatible with membrane proteins, and SDS-PAGE pre-separation of proteins with subsequent chromatography of their proteolysis products coupled to MS (LC-MS) showed successful identification of very hydrophobic and/or multiple TM domain containing IMPs with *C. glutamicum*. Of note, it has been reported proteins with basic pI can be underrepresented (Klein *et al.*, 2005) with this approach.

An approach which elegantly circumvents the loss of IMP's in IEF is the use of Offgel separation also displayed in Fig. 2.1. Here, instead of proteins, their peptides upon digestion with proteases such as trypsin are separated using a compartmentalized liquid upper phase and an IPG strip as lower phase. Peptides are added to every compartment and migrate in the pH gradient of the solution, the mechanism of this migration is based on a similar concept as separation in the pH dimension in 2D gels. Here fractionation by IEF is provided without precipitation of membrane proteins occurring, because peptides are in solution all the time and generally are less prone to aggregation than proteins (Righetti *et al.*, 2003).

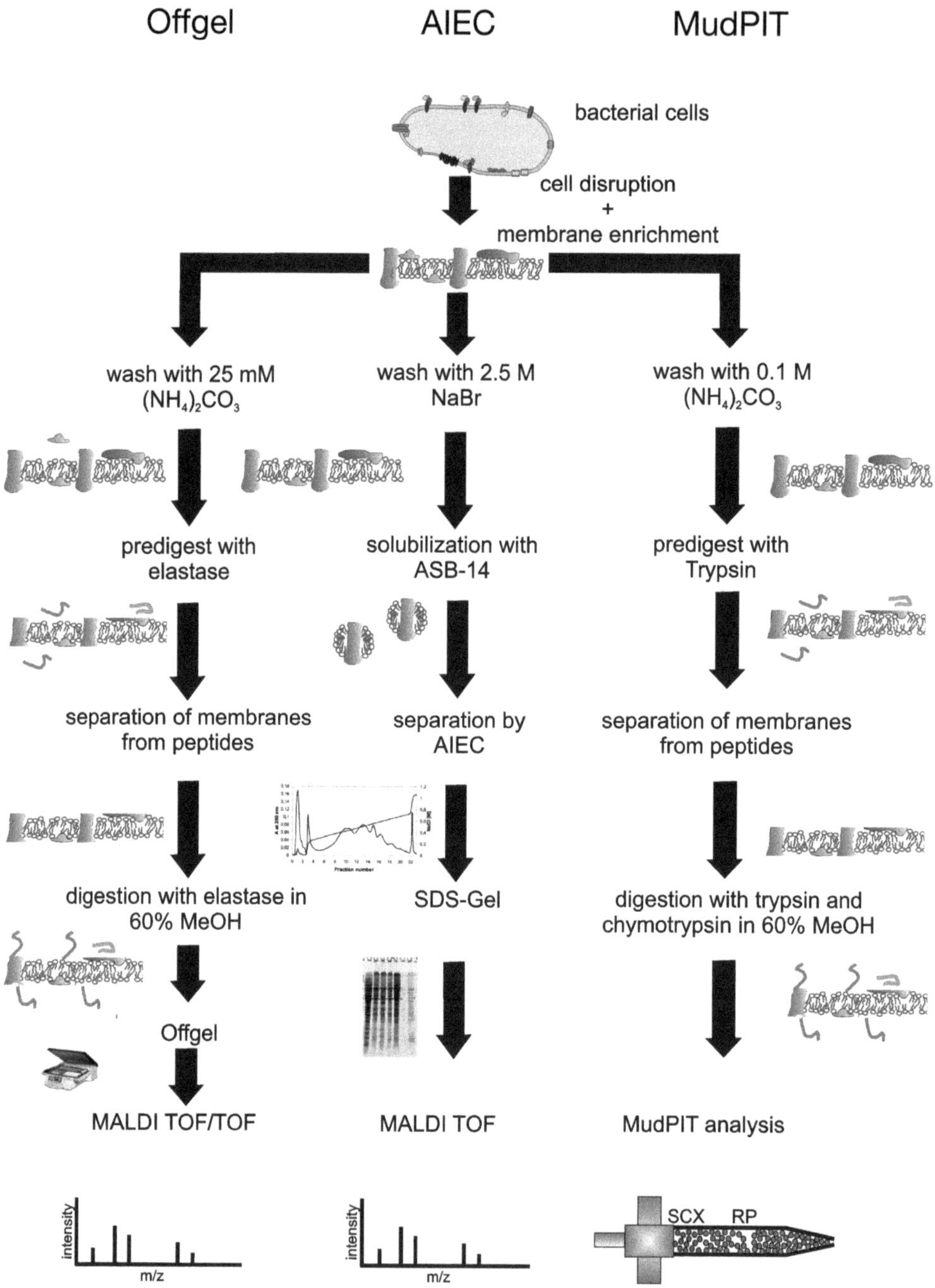

Figure 2.1 Schematic of different workflows for membrane enrichment and mass spectrometry-based analysis including peptide separation using Offgel, membrane protein separation with anion exchange chromatography (AIEC) and multidimensional protein identification technology (MudPIT). A colour version of this figure is available in the plate section at the back of the book.

This method provides a yield of 124 membrane proteins in *C. glutamicum* as well as for purple membranes from *Halobacterium salinarum* after digestion with elastase, a number much higher than usually achieved with 2D-PAGE (Arrey *et al.*, 2010).

Coupling anion exchange chromatography (AIEC) to SDS-PAGE also increases the number of identified membrane proteins. AIEC as IEF also separates proteins by their charge but without precipitation (Schluesener *et al.*, 2005). Moreover, several membrane protein solubilization methods containing different detergents were compared in this study (Fig. 2.1). The results point to ASB 14 as the detergent, which solubilizes the largest portion of membrane proteins without complete denaturation, which is important for subsequent AIEC. This technique has been adapted in several studies up to date (Ludke *et al.*, 2007; Polen *et al.*, 2007). It may still be a worthwhile approach for separation of intact membrane proteins, but has been replaced in our group with other methods (SDS-PAGE/LC-MS/MS, MudPIT) that offer several advantages, such as higher robustness and speed and show a similar membrane proteome coverage.

Comprehensive proteome coverage of membranes for *C. glutamicum* has been achieved by multidimensional protein identification technology (MudPIT). Here 326 membrane proteins were identified, among them many IMPs containing several TM helices (Fischer *et al.*, 2006). MudPIT is based on the separation of peptides in two dimensions. In the here employed 'online' MudPIT technique, the same column (displayed in Fig. 2.1) contains an strong cation exchange (SCX) resin followed by reversed phase material. Protein digests are usually loaded offline with pressure bombs. This is followed by consecutive displacement of peptides from SCX resin with increasing salt concentrations, alternating with reversed phase chromatography until all peptides are eluted (Washburn *et al.*, 2001). Even though this chromatography technique is extremely powerful, for deep coverage of the membrane proteome this setup needs to be complemented by improved protein digests and/or membrane protein enrichment steps. Such membrane protein enrichment can be achieved with the SIMPLE method; here membrane fractions are pre-digested with trypsin, then proteins are again subjected to a chymotrypsin/trypsin digest in the presence of organic solvents, usually 60% methanol, before MudPIT (Fischer *et al.*, 2006). Indeed, enrichment of membranes and/or membrane proteins is crucial, and if carefully developed, can substantially improve membrane protein identification: using the previously introduced MudPIT workflow and combining it with a biphasic protein extraction system 297 IMPs were identified, the hitherto most comprehensive membrane proteome coverage for *C. glutamicum*. Here, a hydrophobic PEG phase and a hydrophilic dextran phase were used to subject membranes to a partitioning equilibrium, which efficiently separates them from cytoplasmic fraction. By this means membrane proteins were additionally enriched before using SIMPLE (Fränzel *et al.*, 2009).

The publications presented in this section show the progress of membrane proteomics for *C. glutamicum*. By now about 50% of the estimated total membrane proteome can be identified using techniques as SIMPLE and MudPIT, which corresponds to the majority of all expressed membrane proteins under a given growth condition (Fischer *et al.*, 2006) – most of the missed proteins are very small, which challenge MS-based identification with their low number of obtainable peptide sequences. For the Gram-positive bacterium *Bacillus subtilis* a similar proportion of the membrane proteome has been identified (Becher *et al.*, 2011). The best methods presented here all provide an increased identification of peptides originating from the trans-membrane part of IMPs and enrichment of the membrane fraction.

Absolute and relative protein quantification

Proteomics is used to quantify the absolute amount of a protein or the relative ratio of proteins. For absolute quantification, the signal of a protein has to be compared to a spiked-in standard. Relative quantification between one or several conditions is based on label free quantification as well as stable isotope labelling, either by chemical coupling or by metabolic incorporation. When peptides from labelled and unlabelled samples are combined their peaks have a defined mass difference and their ratios can be easily extracted. The probably best-established and easy-to-use label free method quantifies proteins based on the number of their detected MS/MS spectra, this method can be improved by using the integral over the signal of a peptide (Cox and Mann, 2011). For *C. glutamicum* proteomics various quantitative studies have been published ranging from stable isotope labelling to a new technological platform for proteome turnover analysis. The combination of high membrane proteome coverage achieved and the large number of established methods for quantitative analysis make this organism very interesting for modern quantitative proteomics.

An indirect stable isotope-based quantification method was used to elucidate the differences in the membrane proteome of *C. glutamicum* cells grown on glucose or benzoate (Haußmann *et al.*, 2009), proteins such as BenE and BenA, which are involved in benzoate degradation and transport increased in abundance. A ^{15}N-labelled membrane fraction was obtained by cultivation with $^{15}NH_4Cl$ as sole nitrogen source and used as a protein standard; both conditions were compared by dividing their regulation factors by each other (Fig. 2.2a). The indirect relative quantitative comparison between two or more samples via an internal standard has several benefits, in particular the possibility to select different growth conditions as sample or respectively as standard and to circumvent any possible effect of the stable isotope on physiology and protein abundance. However, this experimental design requires measuring of light/heavy peptide pairs for all proteins in a sample to allow their comparison, which can never be achieved completely in reality. In addition to this study, a further publication (Fränzel

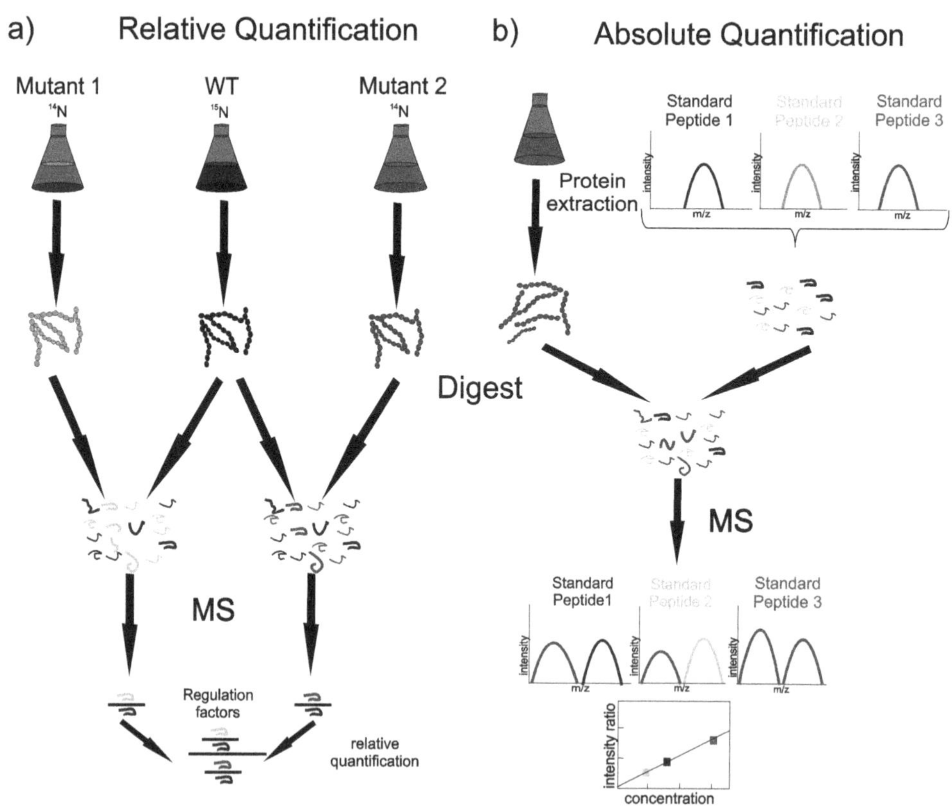

Figure 2.2 Schematic of workflows for quantification methods used for *Corynebacterium glutamicum*. (a) For relative quantification, mutant cultures are grown on ^{14}N medium and a wild-type (WT) culture grown on ^{15}N serves as internal standard. Cell extracts are mixed 1:1 and analysed using high-resolution mass spectrometry (MS). The ratios of unlabelled proteins to labelled proteins are calculated and called regulation factors. These factors can be used to elucidate if a protein is differentially regulated in two conditions. (b) Absolute quantification using standard peptides. Labelled peptides of the same sequence as found in the sample are synthesized and constant amounts are added to a digested sample of interest. Abundance of proteins is estimated using the peptides intensity ratios. A colour version of this figure is available in the plate section at the back of the book.

et al., 2010b) investigates salt stress with SIMPLE and metabolic label-based quantification using a similar set-up. Quantification of membrane proteomes from the wild-type DM21 strain and the production strain DM1730 was used in the following study (Fränzel *et al.*, 2009). Here this method has been combined with the enrichment of membrane proteins using biphasic protein extraction. Quantification showed that for both strains there is an equal partitioning coefficient and data are reliable when compared with prior studies (Schluesener *et al.*, 2007). Concerning the regulation of membrane proteins the study compared quantification based on spectral counting and by metabolic labelling with ^{15}N, then subsequent relative quantification using the software ProRata (Pan *et al.*, 2006).

Additionally, proteins from the cytoplasm are of interest and their changes have been quantified under different conditions. A study has been published where the cytoplasm and membrane proteome of cells grown on the aromatic carbon source protocatechuate was compared with that from cells grown on glucose (Haußmann and Poetsch, 2012). This approach gives a global overview of biological processes involved in transition of the bacterium from one energy source to another. By comparing this dataset with earlier results for benzoate, increasing abundance of respiratory chain components – probably due to eventual degradation of aromatics in the TCA cycle – was in common, whereas some other protein abundance changes were unique for the respective aromatic carbon source.

To perform relative quantification for metabolic labelling proteomics experiments a number of bioinformatics tools were developed. For the first publications based on metabolic labelling with ^{15}N of *C. glutamicum* the algorithm embedded in ProRata was used (Pan *et al.*, 2006). Here the labelled and unlabelled version of the peak profile of a detected peptide are extracted; based on this profile peptide abundance is calculated using principal component analysis (PCA). The abundances of two or more peptides are then used to assign the protein ratio with the highest probability. A subsequent study (Fränzel *et al.*, 2009) found that ProRata attenuates regulation of proteins in comparison to the aforementioned approach of counting MS/MS spectra and therefore should be considered for finely tuned regulation, whereas spectral counting should be preferred for proteins regulated in a high dynamic range. Though ProRata can only deal with fixed isotope incorporation ratios, it can be used for proteome turnover calculations as long as the assumption that all peptides are fully labelled or unlabelled is accurate (Trotschel *et al.*, 2013).

Besides the comparison of given states, metabolic labelling with stable isotopes can be used to determine the synthesis and degradation rates of proteins in pulse-chase experiments (Trötschel *et al.*, 2012). A method for quantification of synthesis and degradation rates has been developed and integrated into the internet based software platform QuPE by (Albaum *et al.*, 2009). Whereas most available quantification programs as ProRata need fixed incorporation rates, the algorithm in QuPE accounts for partial incorporation of stable isotopes into a peptide. Before the stable isotope pulse all peptides have the same composition of stable isotopes, proteins synthesized after an isotope pulse or switch will show a different distribution of stable isotopes. The software searches for the peaks of the new peptides in the chromatogram and calculates the abundance ratio of these to the peptides synthesized before the pulse. In this study (Trötschel *et al.*, 2012) abundance ratios were acquired for *C. glutamicum* at different time points after the pulse and at different temperatures to investigate effects of heat shock, pointing to a decreased protein synthesis, but unaffected protein degradation after heat shock.

An additional method of metabolic labelling which has risen to prominence is stable isotope labelling by amino acids in cell culture (SILAC). Here cell cultures are grown in media containing an amino acid, which is essential for the cultured organism, usually lysine and/or arginine. This essential amino acid can be substituted by an analogue labelled with stable isotopes, thus enabling the comparison of several culture conditions using differently labelled amino acids (Ong *et al.*, 2002). Quantification is based on the determination of labelled to unlabelled peptide ratios. For the specific purpose of quantification in SILAC experiments the software MaxQuant has been developed (Cox and Mann, 2008). Though very useful for proteomics of animal cell lines, in microbial proteomics a disadvantage of this method is its reliance on strains, which are auxotrophic for the used amino acid. Up to date no study using SILAC for *C. glutamicum* has been published, which would require *C. glutamicum* strain auxotrophic for natural amino acids.

Besides untargeted quantification as described in the previous sections, proteins can be quantified in a targeted manner. An example is quantification of benzoate transporters BenE and BenK (Haußmann *et al.*, 2009). A defined amount of synthetic peptides labelled with stable isotopes was added to SIMPLE digests of *C. glutamicum* cultures. The peptide sequences had been derived from preceding SIMPLE digests of BenE and BenK. The masses of these peptides were targeted in MS survey scans, and if found, fragmented for sequence validation. To quantify the proteins the peak intensity sums of labelled and unlabelled peptides were extracted from the MS survey scans. Then the regulation of proteins was determined by first calculating the ratios of unlabelled to labelled peptides, followed by division of these ratios with each other.

Targeted quantification can also be performed using single reaction monitoring (SRM). For an SRM experiment, peptides from a protein of interest are chosen and peptides of this protein are fragmented. For most sensitive quantification the fragments with the highest intensities are targeted during fragmentation. This approach was used to relatively quantify the dynamics of enzymes involved in central carbon metabolism for *C. glutamicum* either growing on acetate or glucose. Cells grown on ^{15}N were mixed one to one with cells labelled with ^{14}N cultured under the same conditions. Targeted fragments resulting from ^{15}N-labelled peptides were used as an internal standard to compare the change in the abundance of 19 enzymes for both conditions (Voges and Noack, 2012). A hallmark of this MS/MS based targeted quantification method is the combination of extreme sensitivity with high dynamic range at the expense of extensive method optimization and limited proteome coverage.

Defined amounts of synthesized peptides can be used for quantification of a protein's copy number, i.e. absolute quantification. Peptides are synthesized having the same sequence as ones found in the sample; a constant amount of the synthesized peptides is then added as an internal standard. For quantification the ratio of internal standard peptides to sample peptides is calculated, this ratio is then used to estimate the absolute concentration of the analyte (Fig. 2.2b) (Villanueva *et al.*, 2014). Label-free approaches of absolute quantification can be performed by summing up precursor peak areas and comparing them to calculated protein abundance estimates found in databanks as MitoCarta. A 2012 paper compares protein abundances calculated by precursor peak areas with RNA expression data and found good correlation (Ning *et al.*, 2012).

In the last decade, quantitative proteomics has seen a rapid development from the first established methods to its full integration into modern systems biology. By now quantitative proteomics stands almost at par with genomics and transcriptomics in regards to sensitivity and quantification power. Especially ongoing developments in absolute quantification of proteins will enable researchers to more accurately investigate metabolic pathways and to unravel mechanisms of their regulation.

Post-translational modifications

Covalent modifications of proteins after translation are known as post-translational modifications (PTM). Modifications as phosphorylation or acetylation are usually catalysed by enzymes and are involved in maintaining protein functions and cell signalling. Other modifications result from reactions of the proteins with reactants in the cellular environment. One example for such modifications are oxidative modifications, which are also good measures of the redox state of a cell. By now a number of biochemical and MS-based methods are available to monitor their incidence in the proteome (Mermelekas *et al.*, 2013). In bacteria analysis of PTMs on a global scale has been facilitated by PTM enrichment and proteomics, especially acetylation and phosphorylation are widespread in microbial species, even cross-talk between both modifications has been observed (Soufi *et al.*, 2012). To date only a few publications have elucidated the PTM landscape of *C. glutamicum* in a global approach, but without a doubt, PTMs play an important role in the biology of this bacterium. Here the state of the art will be presented.

Oxidative modifications

The response of *C. glutamicum* to oxidative stress is of high interest especially because of its use as an industrial strain. Oxidative stress leads to modification of cellular molecules and subsequently to inactivation of proteins and DNA damage, severely restraining the ability of the bacteria to reproduce. Therefore, *C. glutamicum* shows adaptation of the transcriptome to oxidative stress in long-term chemostat cultures under constant addition of H_2O_2 (Lee *et al.*, 2013). Free thiol residues are most commonly oxidized and can be used as a marker to identify the redox state of a cell (Mermelekas *et al.*, 2013). Modern MS is able to quantitatively interrogate these states.

The analysis of cysteine thiol oxidations with proteomics depends on suitable enrichment workflows, due to the low abundance of oxidized thiols in living cells. A method constituting a combination of enrichment and labelling for MS is ICAT (isotope coded affinity tag). ICAT tags consist of three components: a reactive group which binds to free unoxidized thiol groups; an isotope labelled linker and a biotin residue for affinity purification. Proteins are first digested, then peptides are labelled by binding of ICAT to free thiol groups, the tagged peptides were enriched using avidin affinity chromatography. More strongly oxidized samples will show increased incidence of the tag, thus this method enables relative quantification of the oxidation state of a sample (Mermelekas *et al.*, 2013). A variation of this method is called OxICAT where at first proteins are tagged with light labelled ICAT on free thiols. Then thiol groups, which are reversibly modified by oxidations, are reduced and tagged by heavy ICAT. Both technologies show that labelling of peptides and proteins on free thiol residues enables the specific analysis of the redox state of a cell. (Leichert *et al.*, 2008).

To interrogate the redox level of cysteines in *C. glutamicum* the covalent modification of cysteines by mycothiols (MSHs) can be analysed. In actinomycetes MSH is the major thiol, its functions are the detoxification of reactive oxygen species (ROS) and provision of a thiol buffer for maintaining the reductive environment of the cytoplasm. A recent study identified 25 proteins modified with MSH in cells stressed with NaOCl, loss of a mycothiol residue was indicated by a mass shift of 225 Da in MS/MS spectra of peptides. Furthermore, 16 of the identified proteins were also characterized as being especially sensitive to NaOCl stress. The study demonstrates that mycothiolation regulates the activity of the peroxidase Tpx because mycothiolated Tpx lost its peroxidase activity as shown in an *in vitro* experiment reconstructing the TRX electron transfer pathway (Chi *et al.*, 2014).

Isobaric tag for relative and absolute quantification (ITRAQ) is a technology for protein quantification, which also has shown great versatility in interrogating the redox state of samples. For quantification a synthesized tag molecule containing an *N*-hydroxysuccinimide group which covalently binds to free amino residues of lysines. Eight versions of this tag can be synthesized though all have the same weight they differ in their isotope composition and can thus be distinguished after MS fragmentation (Wiese *et al.*, 2007). Samples were first enriched for oxidized thiols, enriched proteins were then digested and labelled using ITRAQ. This combination facilitates the analysis of oxidations in a sample even allowing the use of multiplex analysis (Leichert *et al.*, 2008).

A number of MS-based methods for analysis of oxidative modifications were developed. Their future application to *C. glutamicum* strains would greatly enhance our understanding of the mechanisms of ROS protection in this organism and under fermentation conditions.

Phosphorylation

Phosphorylation is the most common post-translational modification in eukaryotes and prokaryotes. The identification of phosphoproteins was for a long time based solely on the use of phosphorylation specific antibodies and phosphor-protein stains for 2D gel proteomics. Recently the methodological focus has moved to gel free phospho-proteomics (Macek and Mijakovic, 2011). In bacteria early on mainly phosphorylation of histidine and aspartate residues catalysed by two component systems was the target of research (Mitrophanov and Groisman,2008) while attention has only recently shifted to phosphorylation of serine, threonine and tyrosine residues, which are commonly phosphorylated by Hanks kinases (Macek and Mijakovic, 2011).

For *C. glutamicum* the first analysis of the phosphor-proteome was prepared using both radioactive labelling with ^{33}P and immunoblotting based on 2D gels. With radioactive labelling 60 proteins were identified, after immunostaining 90 proteins were identified. Thirty-one proteins were identified using both methods (Bendt *et al.*, 2003). This study shows that 2D gels can successfully be used to identify phosphoproteins, but it also shows that this method only leads to small yields of phosphorylated proteins.

Phosphosite identification by 2D gels is restrained by the low recovery of peptides from gel based methods. For analysis of phosphorylation networks in bacteria, a high yield of identified phospho-peptides is needed. This is only possible if gel free approaches in combination with phospho-peptides enrichment methods are used. The application of enrichment methods such as IMAC (immobilized metal affinity chromatography), where the phosphate groups are enriched by their high affinity to trivalent metal ions, or chromatography based on TiO_2 beads, which also capture and enrich phosphopeptides as well as SCX chromatography, enabled the abundant identification of phosphopeptides in bacteria (Macek and Mijakovic, 2011). These technologies facilitated the global analysis of phosphosites in bacterial proteomes as for mycobacteria where 301 phosphoproteins have been identified containing 516 phosphosites (Prisic *et al.*, 2010). We expect the application and adaptation of these techniques to *C. glutamicum* in the future, because a much larger number than currently known phosphorylated serine and threonine residues can be assumed. Especially if one considers that four serine threonine protein kinases (STPKs) have been identified in the genome and characterized in *C. glutamicum* by now (Fiuza *et al.*, 2008). These proteins, namely PknA, PknB, PknL and PknG, constitute a phosphorylation cascade involved in the regulation of a number of biological processes as glutamine utilization, cell division and the synthesis of the bacterial cell wall (Wendisch and Polen, 2013). These studies emphasize the importance of phosphorylation for the regulation of central physiological processes in *C. glutamicum* and further studies will show the stimuli activating these STPK's and also additional targets.

Conclusion and outlook

Proteomics has become an important means in the endeavour to elucidate the biology of *C. glutamicum* with focus on processes related to the fermentative production of amino acids. For stress conditions major adaptations on the proteome level have been elucidated using quantitative approaches. Targeted proteomics examining the dynamics of enzymes involved in important metabolic pathways as well as the examination of PTMs have led to a better apprehension of the changes and regulations taking place during fermentation. A distinct characteristic of this organism is the strong focus of the field on membranes, owing to the importance of transport processes for amino acid production.

The methods developed in *C. glutamicum* have been and can be applied to other organisms. Technologies for quantification, membrane enrichment and MS analysis tested here might lead to higher coverage for membrane proteins or improve the accuracy and reliability of quantification based on isotope labelling. New avenues of proteomics in *C. glutamicum* will attempt to integrate the large number of data attainable by modern high resolution MS with systems biology. Pre-eminently determination of copy numbers for proteins will give new insights into synthesis rates and metabolic fluxes.

References

Albaum, S. P., Neuweger, H., Fränzel, B., Lange, S., Mertens, D., Trötschel, C., Wolters, D., Kalinowski, J., Nattkemper, T. W., and Goesmann, A. (2009) Qupe—a rich internet application to take a step forward in the analysis of mass spectrometry-based quantitative proteomics experiments. *Bioinformatics* **25**, 3128–3134.

Arrey, T. N., Rietschel, B., Papasotiriou, D. G., Bornemann, S., Baeumlisberger, D., Karas, M., and Meyer, B. (2010) Approaching the complexity of elastase-digested membrane proteomes using off-gel IEF/nLC-MALDI-MS/MS. *Anal. Chem.* **82**, 2145–2149.

Barreiro, C., Gonzalez-Lavado, E., Brand, S., Tauch, A., and Martin, J. F. (2005) Heat shock proteome analysis of wild-type *Corynebacterium glutamicum* ATCC 13032 and a spontaneous mutant lacking GroEL1, a dispensable chaperone. *J. Bacteriol.* **187**, 884–889.

Barriuso-Iglesias, M., Schluesener, D., Barreiro, C., Poetsch, A., and Martin, J. F. (2008) Response of the cytoplasmic and membrane proteome of *Corynebacterium glutamicum* ATCC 13032 to pH changes. *BMC Microbiol.* **8**, 225.

Becher, D., Büttner, K., Moche, M., Heßling, B., and Hecker, M. (2011) From the genome sequence to the protein inventory of *Bacillus subtilis*. *Proteomics* **11**, 2971–2980.

Beckers, G., Bendt, A. K., Kramer, R., and Burkovski, A. (2004) Molecular identification of the urea uptake system and transcriptional analysis of urea transporter- and urease-encoding genes in *Corynebacterium glutamicum*. *J. Bacteriol.* **186**, 7645–7652.

Beckers, G., Strosser, J., Hildebrandt, U., Kalinowski, J., Farwick, M., Kramer, R., and Burkovski, A. (2005) Regulation of AmtR-controlled gene expression in *Corynebacterium glutamicum*: mechanism and characterization of the AmtR regulon. *Mol. Microbiol.* **58**, 580–595.

Bendt, A. K., Burkovski, A., Schaffer, S., Bott, M., Farwick, M., and Hermann, T. (2003) Towards a phosphoproteome map of *Corynebacterium glutamicum*. *Proteomics* **3**, 1637–1646.

Bowie, J. U. (1997) Helix packing in membrane proteins. *J. Mol. Biol.* **272**, 780–789.

Burkovski, A. (2002) Proteomanalyse von *Corynebacterium glutamicum*. *Biospektrum* **8**, 496–497.

Burkovski, A. (2008) Corynebacteria: Genomics and Molecular Biology. Caister Academic Press, Wymondham, Norfolk, UK.

Candiano, G., Bruschi, M., Musante, L., Santucci, L., Ghiggeri, G. M., Carnemolla, B., Orecchia, P., Zardi, L., and Righetti, P. G. (2004) Blue silver: a very sensitive colloidal Coomassie G-250 staining for proteome analysis. *Electrophoresis* **25**, 1327–1333.

Chi, B. K., Busche, T., Van Laer, K., Basell, K., Becher, D., Clermont, L., Seibold, G. M., Persicke, M., Kalinowski, J., Messens, J., and Antelmann, H. (2014) Protein S-mycothiolation functions as redox-switch and thiol protection mechanism in *Corynebacterium glutamicum* under hypochlorite stress. *Antiox. Redox Signal.* **20**, 589–605.

Cox, J., and Mann, M. (2008) MaxQuant enables high peptide identification rates, individualized p.p.b.-range mass accuracies and proteome-wide protein quantification. *Nat. Biotech.* **26**, 1367–1372.

Cox, J., and Mann, M. (2011) Quantitative, high-resolution proteomics for data-driven systems biology. *Annu. Rev. Biochem.* **80**, 273–299.

Fanous, A., Weiland, F., Luck, C., Gorg, A., Friess, A., and Parlar, H. (2007) A proteome analysis of *Corynebacterium glutamicum* after exposure to the herbicide 2,4-dichlorophenoxy acetic acid (2,4-D). *Chemosphere* **69**, 25–31.

Fanous, A., Weiss, W., Gorg, A., Jacob, F., and Parlar, H. (2008) A proteome analysis of the cadmium and mercury response in *Corynebacterium glutamicum*. *Proteomics* **8**, 4976–4986.

Fanous, A., Hecker, M., Gorg, A., Parlar, H., and Jacob, F. (2010) *Corynebacterium glutamicum* as an indicator for environmental cobalt and silver stress – a proteome analysis. *J. Environ. Sci. Health B* **45**, 666–675.

Fischer, F., and Poetsch, A. (2006) Protein cleavage strategies for an improved analysis of the membrane proteome. *Proteome Sci.* **4**, 2.

Fischer, F., Wolters, D., Rögner, M., and Poetsch, A. (2006) Toward the complete membrane proteome: high coverage of integral membrane proteins through transmembrane peptide detection. *Mol. Cell. Proteomics* **5**, 444–453.

Fiuza, M., Canova, M. J., Zanella-Cléon, I., Becchi, M., Cozzone, A. J., Mateos, L. M., Kremer, L., Gil, J. A., and Molle, V. (2008) From the characterization of the four serine/threonine protein kinases (PknA/B/G/L) of *Corynebacterium glutamicum* toward the role of PknA and PknB in cell division. *J. Biol. Chem.* **283**, 18099–18112.

Follmann, M., Ochrombel, I., Kramer, R., Trotschel, C., Poetsch, A., Ruckert, C., Huser, A., Persicke, M., Seiferling, D., Kalinowski, J., *et al.* (2009) Functional genomics of pH homeostasis in *Corynebacterium glutamicum* revealed novel links between pH response, oxidative stress, iron homeostasis and methionine synthesis. *BMC Genomics* **10**, 621.

Fränzel, B., and Wolters, D. A. (2011) Advanced MudPIT as a next step toward high proteome coverage. *Proteomics* **11**, 3651–3656.

Fränzel, B., Fischer, F., Trötschel, C., Poetsch, A., and Wolters, D. (2009) The two-phase partitioning system – a powerful technique to purify integral membrane proteins of *Corynebacterium glutamicum* for quantitative shotgun analysis. *Proteomics* **9**, 2263–2272.

Fränzel, B., Poetsch, A., Trotschel, C., Persicke, M., Kalinowski, J., and Wolters, D. A. (2010a) Quantitative proteomic overview on the *Corynebacterium glutamicum* L-lysine producing strain DM1730. *J Proteomics* **73**, 2336–2353.

Fränzel, B., Trötschel, C., Rückert, C., Kalinowski, J., Poetsch, A., and Wolters, D. A. (2010b) Adaptation of *Corynebacterium glutamicum* to salt-stress conditions. *Proteomics* **10**, 445–457.

Fränzel, B., Frese, C., Penkova, M., Metzler-Nolte, N., Bandow, J. E., and Wolters, D. A. (2010c) *Corynebacterium glutamicum* exhibits a membrane-related response to a small ferrocene-conjugated antimicrobial peptide. *J. Biol. Inorg. Chem.* 15, 1293–1303.

Hansmeier, N., Chao, T. C., Puhler, A., Tauch, A., and Kalinowski, J. (2006) The cytosolic, cell surface and extracellular proteomes of the biotechnologically important soil bacterium Corynebacterium efficiens YS-314 in comparison to those of *Corynebacterium glutamicum* ATCC 13032. *Proteomics* **6**, 233–250.

Haußmann, U., and Poetsch, A. (2012) Global proteome survey of protocatechuate- and glucose-grown *Corynebacterium glutamicum* reveals multiple physiological differences. *J. Proteomics* **75**, 2649–2659.

Haußmann, U., Qi, S.-W., Wolters, D., Rögner, M., Liu, S.-J., and Poetsch, A. (2009) Physiological adaptation of *Corynebacterium glutamicum* to benzoate as alternative carbon source – a membrane proteome-centric view. *Proteomics* **9**, 3635–3651.

Hermann, T., Pfefferle, W., Baumann, C., Busker, E., Schaffer, S., Bott, M., Sahm, H., Dusch, N., Kalinowski, J., Puhler, A., *et al.* (2001) Proteome analysis of *Corynebacterium glutamicum*. *Electrophoresis* **22**, 1712–1723.

Kalinowski, J., Bathe, B., Bartels, D., Bischoff, N., Bott, M., Burkovski, A., Dusch, N., Eggeling, L., Eikmanns, B. J., Gaigalat, L., *et al.* (2003) The complete *Corynebacterium glutamicum* ATCC 13032 genome sequence and its impact on the production of L-aspartate-derived amino acids and vitamins. *J. Biotechnol.* **104**, 5–25.

Karp, N. A., Kreil, D. P., and Lilley, K. S. (2004) Determining a significant change in protein expression with DeCyder™ during a pair-wise comparison using two-dimensional difference gel electrophoresis. *Proteomics* **4**, 1421–1432.

Klein, C., Garcia-Rizo, C., Bisle, B., Scheffer, B., Zischka, H., Pfeiffer, F., Siedler, F., and Oesterhelt, D. (2005) The membrane proteome of *Halobacterium salinarum*. *Proteomics* **5**, 180–197.

Koch-Koerfges, A., Kabus, A., Ochrombel, I., Marin, K., and Bott, M. (2012) Physiology and global gene expression of a *Corynebacterium glutamicum* DeltaF(1) F(O)-ATP synthase mutant devoid of oxidative phosphorylation. *Biochim. Biophys. Acta* **1817**, 370–380.

Kromer, J. O., Bolten, C. J., Heinzle, E., Schroder, H., and Wittmann, C. (2008) Physiological response of *Corynebacterium glutamicum* to oxidative stress induced by deletion of the transcriptional repressor McbR. *Microbiology* **154**, 3917–3930.

Kuberl, A., Franzel, B., Eggeling, L., Polen, T., Wolters, D. A., and Bott, M. (2014) Pupylated proteins in *Corynebacterium glutamicum* revealed by MudPIT analysis. *Proteomics* **14**, 1531–1542.

Lee, J. Y., Seo, J., Kim, E. S., Lee, H. S., and Kim, P. (2013) Adaptive evolution of *Corynebacterium glutamicum* resistant to oxidative stress and its global gene expression profiling. *Biotechnol. Lett.* **35**, 709–717.

Leichert, L. I., Gehrke, F., Gudiseva, H. V., Blackwell, T., Ilbert, M., Walker, A. K., Strahler, J. R., Andrews, P. C., and Jakob, U. (2008) Quantifying changes in the thiol redox proteome upon oxidative stress *in vivo*. *Proc. Natl. Acad. Sci. U.S.A.* **105**, 8197–8202.

Li, L., Wada, M., and Yokota, A. (2007a) Cytoplasmic proteome reference map for a glutamic acid-producing *Corynebacterium glutamicum* ATCC 14067. *Proteomics* **7**, 4317–4322.

Li, L., Wada, M., and Yokota, A. (2007b) A comparative proteomic approach to understand the adaptations of an H^+-ATPase-defective mutant of *Corynebacterium glutamicum* ATCC14067 to energy deficiencies. *Proteomics* **7**, 3348–3357.

Ludke, A., Kramer, R., Burkovski, A., Schluesener, D., and Poetsch, A. (2007) A proteomic study of AAA+ protease FtsH. *BMC Microbiol.* **7**, 6.

Macek, B., and Mijakovic, I. (2011) Site-specific analysis of bacterial phosphoproteomes. *Proteomics* **11**, 3002–3011.

Mermelekas, G., Makridakis, M., Koeck, T., and Vlahou, A. (2013) Redox proteomics: from residue modifications to putative biomarker identification by gel- and LC-MS-based approaches. *Exp. Rev. Proteomics* **10**, 537–549.

Mitrophanov, A. Y., and Groisman, E. A. (2008) Signal integration in bacterial two-component regulatory systems. *Genes Devel.* **22**, 2601–2611.

Ning, K., Fermin, D., and Nesvizhskii, A. I. (2012) Comparative analysis of different label-free mass spectrometry based protein abundance estimates and their correlation with RNA-Seq gene expression data. *J. Proteome Res.* **11**, 2261–2271.

Ong, S.-E., Blagoev, B., Kratchmarova, I., Kristensen, D. B., Steen, H., Pandey, A., and Mann, M. (2002) Stable isotope labeling by amino acids in cell culture, SILAC, as a simple and accurate approach to expression proteomics. *Mol. Cell. Proteomics* **1**, 376–386.

Pan, C., Kora, G., McDonald, W. H., Tabb, D. L., VerBerkmoes, N. C., Hurst, G. B., Pelletier, D. A., Samatova, N. F., and Hettich, R. L. (2006) ProRata: a quantitative proteomics program for accurate protein abundance ratio estimation with confidence interval evaluation. *Anal. Chem.* **78**, 7121–7131.

Polen, T., Schluesener, D., Poetsch, A., Bott, M., and Wendisch, V. F. (2007) Characterization of citrate utilization in *Corynebacterium glutamicum* by transcriptome and proteome analysis. *FEMS Microbiol. Lett.* **273**, 109–119.

Prisic, S., Dankwa, S., Schwartz, D., Chou, M. F., Locasale, J. W., Kang, C.-M., Bemis, G., Church, G. M., Steen, H., and Husson, R. N. (2010) Extensive phosphorylation with overlapping specificity by *Mycobacterium tuberculosis* serine/threonine protein kinases. *Proc. Natl. Acad. Sci. U.S.A.* **107**, 7521–7526.

Qi, S. W., Chaudhry, M. T., Zhang, Y., Meng, B., Huang, Y., Zhao, K. X., Poetsch, A., Jiang, C. Y., Liu, S., and Liu, S. J. (2007) Comparative proteomes of *Corynebacterium glutamicum* grown on aromatic compounds revealed novel proteins involved in aromatic degradation and a clear link between aromatic catabolism and gluconeogenesis via fructose-1,6-bisphosphatase. *Proteomics* **7**, 3775–3787.

Rietschel, B., Arrey, T. N., Meyer, B., Bornemann, S., Schuerken, M., Karas, M., and Poetsch, A. (2009) Elastase digests: new ammunition for shotgun membrane proteomics. *Mol. Cell. Proteomics* **8**, 1029–1043.

Righetti, P. G., Castagna, A., Herbert, B., Reymond, F., and Rossier, J. S. (2003) Prefractionation techniques in proteome analysis. *Proteomics* **3**, 1397–1407.

Schluesener, D., Fischer, F., Kruip, J., Rögner, M., and Poetsch, A. (2005) Mapping the membrane proteome of *Corynebacterium glutamicum*. *Proteomics* **5**, 1317–1330.

Schluesener, D., Rögner, M., and Poetsch, A. (2007) Evaluation of two proteomics technologies used to screen the membrane proteomes of wild-type

Corynebacterium glutamicum and an L-lysine-producing strain. *Anal. Bioanal. Chem.* **389**, 1055–1064.

Silberbach, M., Schafer, M., Huser, A. T., Kalinowski, J., Puhler, A., Kramer, R., and Burkovski, A. (2005) Adaptation of *Corynebacterium glutamicum* to ammonium limitation: a global analysis using transcriptome and proteome techniques. *Appl. Environ. Microbiol.* **71**, 2391–2402.

Soufi, B., Soares, N. C., Ravikumar, V., and Macek, B. (2012) Proteomics reveals evidence of cross-talk between protein modifications in bacteria: focus on acetylation and phosphorylation. *Curr. Opin. Microbiol.* **15**, 357–363.

Trötschel, C., Albaum, S. P., Wolff, D., Schröder, S., Goesmann, A., Nattkemper, T. W., and Poetsch, A. (2012) Protein turnover quantification in a multilabeling approach: from data calculation to evaluation. *Mol. Cell. Proteomics* **11**, 512–526.

Trotschel, C., Albaum, S. P., and Poetsch, A. (2013) Proteome turnover in bacteria: current status for *Corynebacterium glutamicum* and related bacteria. *Microb. Biotechnol.* **6**, 708–719.

Vasco-Cárdenas, M. F., Baños, S., Ramos, A., Martín, J. F., and Barreiro, C. (2013) Proteome response of *Corynebacterium glutamicum* to high concentration of industrially relevant C_4 and C_5 dicarboxylic acids. *J. Proteomics* **85**, 65–88.

Villanueva, J., Carrascal, M., and Abian, J. (2014) Isotope dilution mass spectrometry for absolute quantification in proteomics: concepts and strategies. *J. Proteomics* **96**, 184–199.

Voges, R., and Noack, S. (2012) Quantification of proteome dynamics in Corynebacterium glutamicum by (15)N-labeling and selected reaction monitoring. *J. Proteomics* **75**, 2660–2669.

Washburn, M. P., Wolters, D., and Yates, J. R. (2001) Large-scale analysis of the yeast proteome by multidimensional protein identification technology. *Nat. Biotech.* **19**, 242–247.

Wendisch, V. F., and Polen, T. (2013) *Corynebacterium glutamicum*: biology and biotechnology. In: Yukawa, H., and Inui, M. (eds.), Microbiology Monographs, Springer, Heidelberg, Germany, pp. 173–215.

Wiese, S., Reidegeld, K. A., Meyer, H. E., and Warscheid, B. (2007) Protein labeling by iTRAQ: a new tool for quantitative mass spectrometry in proteome research. *Proteomics* **7**, 340–350.

Developing Interpretation of Intracellular Metabolism of *Corynebacterium glutamicum* by Using Flux Analysis Technology

3

Tomokazu Shirai and Hiroshi Shimizu

Abstract

Corynebacterium glutamicum is often used as a model organism because of its ability to produce various useful substances. For achievement of high productivity, metabolic flux analysis (MFA) is a powerful technique, which is used to determine the state of intracellular metabolism in microorganisms. In order to develop the precise MFA, it is necessary to obtain ^{13}C labelling information of intracellular metabolites, in addition to the data of extracellular measurements such as substrate consumption rates, cell growth rates, and production rates. Metabolic flux can be estimated accurately through a mathematical approach by using computer calculations to analyse ^{13}C labelling information of intracellular metabolites by performing nuclear magnetic resonance and/or gas chromatography–mass spectrometry. The accurate determination of intracellular metabolisms by using MFA permits a deeper understanding of *C. glutamicum* physiology and can produce information for improving the production of useful compounds by using this bacterium.

Introduction

Corynebacterium glutamicum was isolated in 1957. Since the establishment of glutamate production by using fermentation (Kinoshita *et al.*, 2004), many amino acids have been produced by using this bacterium. For example, the screening of auxotrophic mutants and lysine analogue-resistant mutants has enabled the accumulation of lysine at high concentrations in culture medium (Shiio *et al.*, 1970). Although there have been many successful examples of good strains screening by treatment with mutagens, it is difficult to establish the rational methodology to modify metabolic pathways for increase productivity of the target metabolites because this method is based on the random mutation. The development of analyses of changes in intracellular metabolisms are highly desired.

Recently, researches on the production of useful chemicals from biomass by using *C. glutamicum*'s fermentation have got attention as an alternative to having to use fossil fuels: isobutanol (Blombach *et al.*, 2011; Smith *et al.*, 2010), ethanol (Inui *et al.*, 2004a), lactate and succinate (Inui *et al.*, 2004b), polyhydroxybutyrate (Jo *et al.*, 2006), lysine (Kind *et al.*, 2010), and cadaverine (Tateno *et al.*, 2009). In order to improve the production of such chemicals, it has become essential that the intracellular metabolisms of the bacterium be accurately assessed both before and after gene modification. Many intracellular metabolic reactions includes reversible reactions and cycles in metabolic pathways. One example in *C. glutamicum* is the anaplerotic pathway which has a large impact on amino acid and organic acid production: phosphoenol pyruvate carboxylase (PEPc), pyruvate carboxylase (Pc), phosphoenolpyruvate carboxykinase (PEPck), the malic enzyme reaction, and the glyoxylate shunt (Eikmanns *et al.*, 1989; Gubler *et al.*, 1994; Jetten and Sinskey, 1993; Peters-Wendisch *et al.*, 1993, 1998). There is no doubt that the accurate quantification of these intracellular fluxes would aid in the identification of target gene modifications for improving the production of these substances by using *C. glutamicum*. Metabolic flux analysis (MFA) is a powerful tool for understanding the metabolic conditions in cells and for designing cells that can produce useful compounds efficiently. Generally, analytical results are displayed on a map as actual values of each individual flux, or as a ratio of each flux relative to the substrate uptake flux (Fig. 3.1). Since MFA is performed by using the actual measured values of intracellular or extracellular metabolites, *in vivo* conditions of metabolism is accurately visualized. Moreover, the intracellular metabolic reaction system facilitates our understanding of the metabolic network. Comparing the results of MFA between wild-type strains and mutant strains, different species, or culture conditions, enables us to rationally design cells that produce useful compounds becomes possible. Given these criteria, the most important point to consider when performing MFA is the precision of the analysis. There are four main factors that affect the precision of MFA: (1) designing culture experiments with high reproducibility; (2) establishing analytical techniques with high accuracy for metabolites; (3) designing metabolic models that can precisely describe intracellular metabolisms; and (4) an appropriate mathematical calculation method that can be employed by using a computer. Imprecision in any of these factors can often lead to unsatisfactory MFA results. Until now, many approaches have been developed to improve the precision of MFA. This chapter overviews how flux analysis techniques contributes to understanding of intracellular metabolisms in *C. glutamicum* and consequently to improvements in the production of useful compounds.

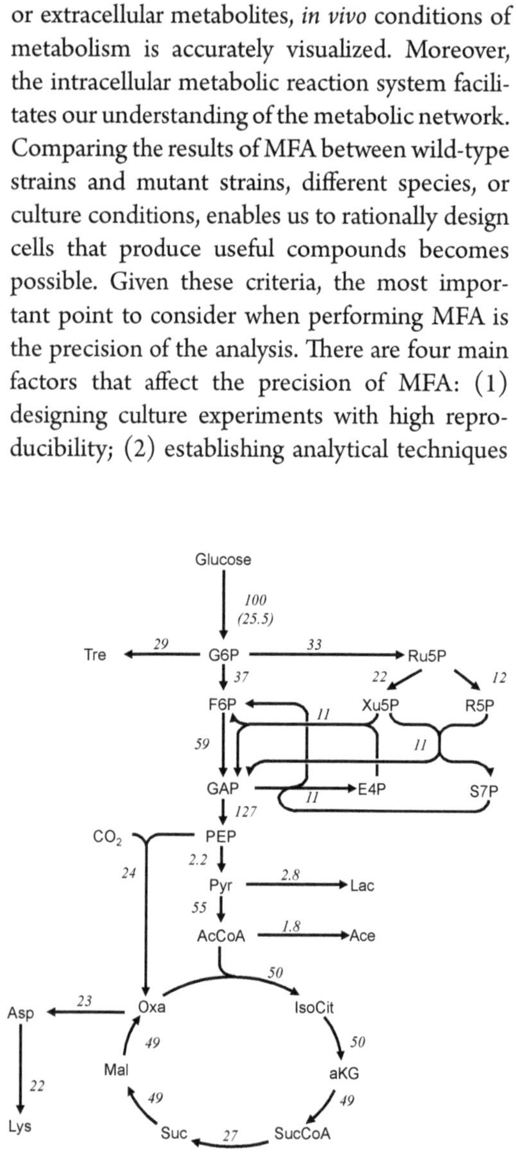

Figure 3.1 Flux map from *Corynebacterium glutamicum* under lysine-producing conditions. Data taken from Vallino and Stephanopoulos (1993).

Understanding the physiology of *Corynebacterium glutamicum* lysine production and glutamate production through conventional metabolic flux analysis

MFA starts with the design of stoichiometry that describes how substrates are converted into metabolites and components for biosynthesis. Stephanopoulos *et al.* (1998) first established a methodology for determining metabolic flux. The material balance between each metabolic product that they used is shown by using the equation below:

$$Ar_c = r_m \tag{3.1}$$

where A is the coefficient matrix of $m \times n$, r_c is the vector for the flux of the nth degree, and r_m is the vector for the rate of accumulation for each metabolic product to the mth degree. In actual, the rates of accumulation for intracellular metabolites are zero due to the pseudo-steady state assumption. The flux (r_c) that needs to be obtained can be calculated by using the weighted least-squares solution (Lawson and Hanson, 1995). Kiss and Stephanopoulos used this method to investigate the relationship between specific growth rate and the fluxes of anaplerotic pathways that are used to produce lysine in *C. glutamicum* (Kiss and Stephanopoulos, 1992). They analysed the relationship between each flux and specific growth rate (0.033–0.3/h) by obtaining 10 types of specific growth rates for continuous cultures. As a result, they found that when the yield for lysine production is relatively low (<0.25 mol Lys/mol glucose), the ratio of the flux of anaplerotic pathways to that of glycolysis, and the ratio of flux of oxaloacetate synthesis to that of lysine production, did not depend on the specific growth rate and remained constant. Further MFA helped analyse the time course of fluxes in lysine production in *C. glutamicum* batch cultures (Vallino and Stephanopoulos, 1993). NADPH is necessary as a reducing agent during lysine production using *C. glutamicum*. The question of where NADPH is produced inside the cell was answered by performing MFA. It was estimated that the fluxes in the pentose phosphate pathway and anaplerotic pathways increase 2-fold as *C. glutamicum* transitions from the growth phase to the lysine production phase. In particular, MFA results also suggested a consistency between changes in flux of anaplerotic pathways and changes to the specific activity of PEPc. In addition, the MFAs obtained from fed-batch culture experiments by Takiguchi *et al.* (1997) showed that a positive linear relationship exists between the flux for the pentose phosphate pathway and that for lysine production.

With respect to lysine production in *C. glutamicum*, the fluxes of anaplerotic pathways have an important role, in addition to the pentose phosphate pathway required for NADPH production. Included in the anaplerotic pathways is PEPc, which catalyses the reaction of phosphoenolpyruvate (PEP) to oxaloacetate, and pyruvate carboxylase Pc, which catalyses the reaction of pyruvate to oxaloacetate. The physiology associated with each of these enzymatic reactions in the cell has been analysed by using MFA (Park *et al.*, 1997b). A reduction in PEPc leads to increase the flux of PEP kinase (Pyk) catalysing the conversion of PEP into pyruvate. The flux catalysed by Pc also increases simultaneously. A reduction in Pyk leads to an increase in flux of PEPc and a decrease in flux of Pc. Moreover, in the mutant strain that lack both PEPc and Pyk, the flux through the pentose phosphate pathway increases 2-fold, relative to the wild-type condition.

The physiology corresponding to the intracellular metabolism of glutamate production in *C. glutamicum* has also been examined through MFA approaches. Glutamate production in *C. glutamicum* can be induced by causing the depletion of biotin (Shiio *et al.*, 1962a,b), through the addition of penicillin (Nunheimer *et al.*, 1970), through the addition of surfactants such as Tween 40 (Takinami *et al.*, 1966), or via shifts to higher culture temperatures (Delaunay *et al.*, 2002). Changes in the flux around 2-oxoglutarate (αKG) were analysed when glutamate production was induced through biotin depletion or the addition of Tween 40 (Shirai *et al.*, 2005). Although glutamate production was induced, there were no large changes in the influx into the tricarboxylic acid (TCA) cycle. However, it was revealed that a reduction in flux from αKG to succinyl-CoA catalysed by 2-oxoglutarate dehydrogenase complex (ODHC) led to an increase in the flux of glutamate production. A similar change was observed in the flux for glutamate production when the culture temperatures were shifted (Uy *et al.*, 2005). Shimizu *et al.* (2003) analysed enzymatic reactions that affect glutamate production by applying metabolic control analysis (MCA) in addtion to MFA. MCA provides a useful framework for determining the relative effect of each reaction in this pathway. The effect is expressed by the magnitude of the flux control coefficient (FCC) of each reaction for the pathway flux, which is defined as the ratio of the relative change in the pathway flux from a steady-state value to the relative change in the activity of the corresponding enzyme (Kacser and Burns, 1973).

FCC C_i^k

is defined as the relative effects of infinitesimal changes in the activities of enzymes upon changes in steady-state fluxes represented by Equation 3.2 as:

$$C_i^k = \frac{\partial J_k / J_k}{\partial e_i / e_i} \quad (3.2)$$

where e_i and J_k are the enzyme activity of reaction i and the flux of reaction k, respectively. At a branch point such as αKG in glutamate production, 9 FCCs are defined because there are three enzyme activities and three fluxes at the branched point: isocitrate dehydrogenase (ICDH), glutamate dehydrogenase (GDH), and αKG dehydrogenase (ODHC). Through MCA, the reduction in ODHC enzymatic activity was determined to be the largest factor that affected glutamate production. This was determined by analysing the MFA results and the specific activity of each enzyme in wild-type, GDH, and ICDH mutants of *C. glutamicum*.

Fractional ^{13}C enrichment-based metabolic flux analysis

The MFA described in the previous section is a useful tool that can contribute to understanding of intracellular metabolism, which has been considered to be a black box until recently. However, there are several big problems associated with the conventional MFA. For example, the material balance models for each metabolite have to be constructed based on values obtained in past experiments, such as those for ATP and other energy substances that are required for cellular synthesis, and a balance between redox substances (Shimizu *et al.*, 2003; Vallino and Stephanopoulos, 1993). Furthermore, it is difficult to estimate the fluxes of reversible reactions, and only net fluxes can be estimated. Finally, mathematically equivalent reaction fluxes, such as those catalysed by PEPc and Pc in anaplerotic pathways, are indistinguishable from each other (Park *et al.*, 1997b). A more detailed flux analysis can be performed by performing trace experiments that use stable isotopes, and by analysing labelled amino acids and intermediate metabolites by using nuclear magnetic resonance (NMR) and mass spectrometry (MS). Owing to the high information content of the labelling data, neither the assumptions about the stoichiometry of the energy metabolism (e.g. Y_{ATP}, P/O ratio, and H^+/e^- ratio), which is a matter of controversial debate (Neijssel and Teixeira de Mattos, 1994; Russell and Cook, 1995), nor those that are based on any *in vitro* enzyme activity, have to be taken into account. Because the stable isotope ^{13}C is incorporated in a substrate used in the culture, this method is called ^{13}C metabolic flux analysis (^{13}C-MFA).

In 1993, the distribution ratios of fluxes for lysine synthesis pathways in *C. glutamicum* were determined by ^{13}C-MFA (Sonntag *et al.*, 1993). In *C. glutamicum*, lysine is synthesized either via a dehydrogenase variant or a succinylase variant of the diaminopimelate pathway (Schrumpf *et al.*, 1991) (Fig. 3.2). Glucose labelled with ^{13}C at the 6th position, [6–^{13}C] glucose, was used for

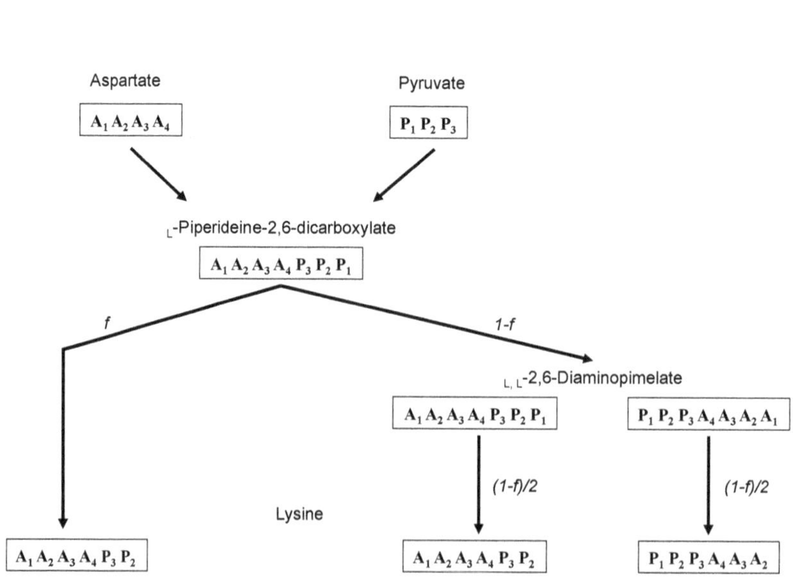

Figure 3.2 Scheme of carbon atom distribution from aspartate and pyruvate through the two split lysine pathways. A_n is the ^{13}C enrichment in aspartate carbons, and P_n that in pyruvate.

C. glutamicum cultivation, and from the analysis of lysine by ^{1}H-NMR and ^{13}C-NMR the degree of fractional ^{13}C enrichments were determined. Based on this analysis, the flux distribution ratio of the 2 pathways that contribute to lysine production could be calculated as follows:

$$L_2 = 0.5(1+f)A_2 + 0.5(1-f)P_2 \qquad (3.3)$$

$$L_6 = 0.5(1+f)P_2 + 0.5(1-f)A_2 \qquad (3.4)$$

where L_2 and L_6 are enrichments in C-2 and C-6 of lysine, A_2 is the enrichment in C-2 in aspartate, and P_2 is the enrichment in C-2 in pyruvate (see Fig. 3.2). The solution for Equations 3.3 and 3.4 yields the following:

$$f = \frac{L_6 - L_2}{2 \times P_2 - L_6 - L_2} \qquad (3.5)$$

In the early phase of lysine production, the flux due to diaminopimelate dehydrogenase is 70%. However, this value gradually decreases to 0% during the late phase of production.

^{13}C enrichment data obtained by performing NMR analyses also contributed to our understanding of intracellular metabolism in *C. glutamicum* during glutamate production. Experiments with [5-^{13}C]glucose and ^{1}H-NMR allowed for the estimation of flux distribution ratios between anaplerotic pathways and the TCA cycle (45% and 55%, respectively), which is responsible for producing oxaloacetate during the growth phase of *C. glutamicum*, based on the ^{13}C enrichments of intracellular glutamate (Wendisch *et al.*, 1997). Experiments with [1-^{13}C]glucose and [6-^{13}C]glucose led to the establishment of flux distribution estimation methods that are related to the flux distribution ratio of glycolysis and the pentose phosphate pathway, the fluxes of the glyoxylate shunt and anaplerotic pathways, which are important for the production of glutamate, and the number of turns (single or multiple) of the TCA cycle (Rollin *et al.*, 1995). This method was also applied to a flux analysis of *C. glutamicum* fructose metabolism. In this case, it was revealed that the flux distribution ratio for the pentose phosphate pathway was half (40% to 20%) relative to that for glucose metabolism, but was unchanged (about 60%) relative to the fluxes of the anaplerotic pathways (Dominguez *et al.*, 1998). Under biotin-depleted conditions, the reaction catalysed by PEPc in the anaplerotic pathways is predicted to affect the yield of glutamate production based on culture data by using mutant *C. glutamicum* strains (Sato *et al.*, 2008). This prediction is supported from the flux level by using the methods that were mentioned above established by Rollins *et al.* (1995). In this manner, trace experiments that use ^{13}C-labelled substrates have enabled the estimation of flux distribution ratios of branching pathways, which is a difficult analytical task for conventional MFA.

However, it is necessary to simplify the metabolic model, especially for ^{13}C-MFA, where only forward calculations without iterative algorithm can be performed by using the ^{13}C enrichments of specific amino acids. Because of the existence of many reversible reactions in intracellular enzymatic reactions, it is possible that reversibility affects the ^{13}C enrichments of metabolites. For this reason, fluxes that are estimated by using a simplified model may differ from actual intracellular fluxes.

In order to develop a detailed metabolic model that includes reversible reactions and cycles, and to estimate fluxes accurately, it is necessary to obtain large amounts of ^{13}C labelling information for amino acids. ^{13}C labelling information can be used for flux analyses of metabolic networks by comparing simulated labelling distributions with experimental data. The set of fluxes that exhibit minimum deviation from the experimental data and still fulfil the stoichiometric constraints of conventional MFAs will give the best estimates of intracellular metabolic fluxes. In this process, a model is required that can be used to simulate steady-state isotopomer distribution as a function of a given set of steady-state fluxes. The problem of complexity with respect to fractional labelling models has been solved elegantly by the introduction of atom mapping matrices (AMMs) (Zupke and Stephanopoulos, 1994), which hold the information on the carbon atom transfers in enzymatic reactions from reactant to product molecules. For example, $AMM_{Pyr>AcCoA}$, which represents the conversion from pyruvate into acetyl coenzyme A (AcCoA) that is catalysed by pyruvate dehydrogenase, results in the following matrix:

$$AMM_{Pyr>AcCoA} = \begin{matrix} 0 & 1 & 0 \\ 0 & 0 & 1 \end{matrix} \quad (3.6)$$

If AcCoA were derived only from pyruvate, its fractional enrichment would be given by:

$$AMM_{Pyr>AcCoA} \times \begin{matrix} Pyr(1) \\ Pyr(2) \\ Pyr(3) \end{matrix}$$

$$= \begin{matrix} Pyr(2) \\ Pyr(3) \end{matrix} = \begin{matrix} AcCoa(1) \\ AcCoa(2) \end{matrix} \quad (3.7)$$

where Pyr(n) and AcCoA(n) represent degrees of ^{13}C enrichment of the carbon atoms of each metabolite in vector form. From the model developed through the introduction of AMMs and ^{1}H-NMR analyses of 10 types of amino acids (Gly, Ser, Ala, Val, Ile, Glu Asp, Thr, Lys, and Phe) for ^{13}C enrichments, the exchange fluxes of glycolysis, pentose phosphate pathway, and the TCA cycle (oxaloacetate to fumarate) were estimated (Marx *et al.*, 1996).

When the lysine production yield is relatively low (<0.2 mol Lys/mol glucose), there exist reversible reactions with extremely high exchange fluxes in glycolysis and the pentose phosphate pathway. Moreover, during the lysine production phase, the fluxes of reactions catalysed by malic enzyme or PEPck in anaplerotic pathways are relatively high (45% of forward flux), even though it is thought that these fluxes have negative effects on lysine production. By introducing *Peptostreptococcus asaccharolyticus*-derived NADH-dependent glutamate dehydrogenase (GDH) into *C. glutamicum* (GDH of *C. glutamicum* is NADPH dependent), responses in intracellular fluxes were analysed (Marx *et al.*, 1999). The flux for the pentose phosphate pathway during lysine production in *C. glutamicum* with NADH-dependent GDH expressed was reduced to one-third of that of the wild type, which strongly suggests that in the wild type, the pentose phosphate pathway plays an important role in compensating NADPH levels. In addition, a positive relationship was discovered between NADPH supply through the pentose phosphate pathway and the lysine production flux by using a glucose-6-phosphate isomerase mutant (Marx *et al.*, 2003). From the ^{1}H-NMR analysis of ^{13}C enrichments in amino acids, the intracellular metabolic flux was analysed when glucose and/or acetate was used as a substrate (Wendisch *et al.*, 2000). The analysis revealed the function of the glyoxylate shunt within the cell through the flux levels.

The two reactions in the pentose phosphate pathway (Xu5p + R5P ⇔ S7P + GAP, Xu5P + E4P ⇔ F6P + GAP) (Xu5P, xylulose-5-phosphate; R5P, ribose-5-phosphate; S7P, sedoheptulose-7-phosphate; GAP, glyceraldehyde-3-phosphate; E4P, erythrose-4-phosphate; F6P, fructose-6-phosphate) are catalysed by one enzyme (transketolase) (Sugimoto and Shiio, 1989). However, ^{13}C-MFA revealed a large discrepancy in their exchange rates (Marx *et al.*, 1996, 1999, 1997). To resolve this discrepancy without the non-accessible biochemical constraints that govern the exchange rates (i.e. *in vivo* velocity, net fluxes, pool sizes of reactants, and thermodynamic equilibrium), Marx *et al.* obtained ^{13}C enrichments of guanosine derivatives, which are derived from R5P (Marx *et al.*, 1997). This showed that the original model was not determined completely. In other words, the accurate estimation of exchange fluxes in the pentose phosphate pathway is difficult when using only ^{1}H-NMR analysis of the fractional ^{13}C enrichment of Phe, as indicated by using a multiple simulation approach (Follstad and Stephanopoulos, 1998; Wiechert and de Graaf, 1997; Wiechert *et al.*, 1997).

Improvement in metabolic flux analysis precision

In order to perform a detailed analysis of flux including intracellular reversible reactions, a greater amount of ^{13}C labelling information on intracellular metabolites is necessary. Moreover, obtaining precise ^{13}C enrichment data for amino acids by using ^{1}H-NMR analysis requires each amino acid to be purified by methods such as ion-exchange chromatography (Marx *et al.*, 1996; Sonntag *et al.*, 1993). However, these obstacles can be overcome by using ^{13}C labelling information obtained through two-dimensional NMR (2D-NMR) (Schmidt *et al.*, 1999a,b;

Szyperski, 1995; Szyperski *et al.*, 1996; Yang *et al.*, 2003). Measurements of the degree of coupling between adjacent ^{13}C nuclei also represent linear constraints on isotopomer distribution. They represent the ratio of isotopomers in which the observed ^{13}C nucleus is covalently bound to another ^{13}C nucleus relative to the mole fraction of isotopomers in which the observed ^{13}C nucleus is attached to non-labelled carbons (Fig. 3.3). This analytical method was applied to determine the fluxes of anaplerotic pathways in *C. glutamicum* in detail (Petersen *et al.*, 2000). Conventionally, metabolic models required that PEP and pyruvate (Pyr), and oxaloacetate and malate be treated as one metabolite pool when estimating the fluxes of anaplerotic pathways (Marx *et al.*, 1996, 1997; Wiechert and de Graaf, 1997; Wiechert *et al.*, 1997). Petersen *et al.* (2000) used [3-^{13}C] lactate and ^{13}C-labelled glucose at the same time. ^{13}C-labelled lactate information can be used directly as those for pyruvate. Thus, ^{13}C-labelled PEP information from glycolysis and pyruvate can be considered independently. From this unique ^{13}C labelling experiment and ^{13}C labelling information for each amino acid obtained by 2D-NMR, detail fluxes of reactions catalysed by enzymes of the anaplerotic pathways such as PEPc, Pc, PEPck, and malic enzyme were revealed (Fig. 3.4). In addition, ^{13}C-MFA of *C. glutamicum* lacking PEPck (decarboxylation of oxaloacetate into PEP) showed that PEPck loss greatly reduces fluxes in Pc and PEPc (carboxylation of pyruvate or PEP into oxaloacetate) (–60% and –80%, respectively) (Petersen *et al.*, 2001).

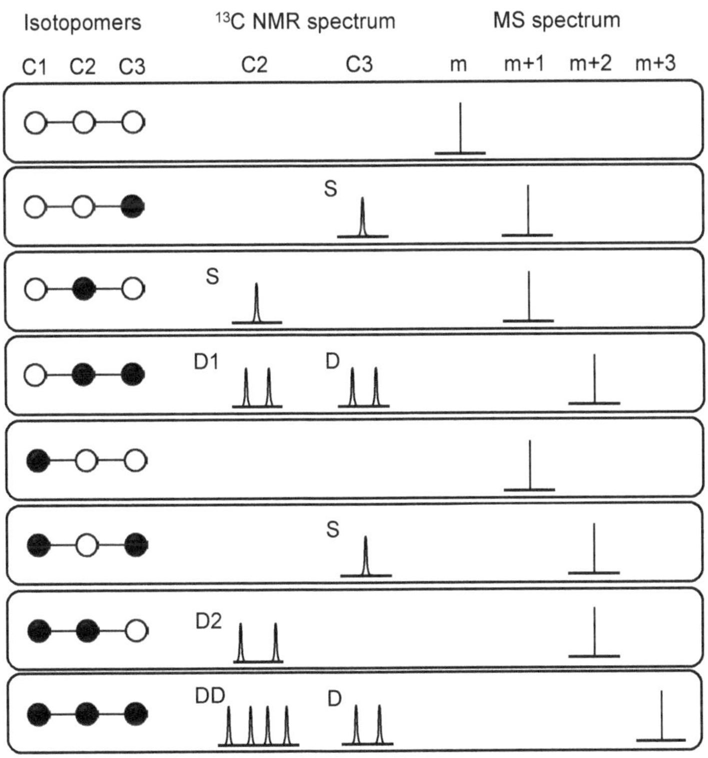

Figure 3.3 The isotope patterns of a molecule with three carbon atoms and the corresponding ^{13}C nuclear magnetic resonance (NMR) and mass spectrometry (MS) spectra. Coupling between adjacent ^{13}C nuclei gives rise to multiplet signals in the form of singlet (S), doublets (D1, D2), and double-doublets (DD) in ^{13}C NMR spectra. Mixtures of isotopomers are obtained as multiplet patterns which are weighted superpositions of the S, D1, D2, and DD signals. Terminal carbon atoms can only give rise to singlet and doublet signals. In the case of equal coupling constants D1 and D2 cannot be distinguished, and double-doublets become triplets. The MS spectrum as analysed by gas chromatography (GC)–MS separate isotopomers with different molecular mass: *m*, *m*+1, *m*+2, *m*+3.

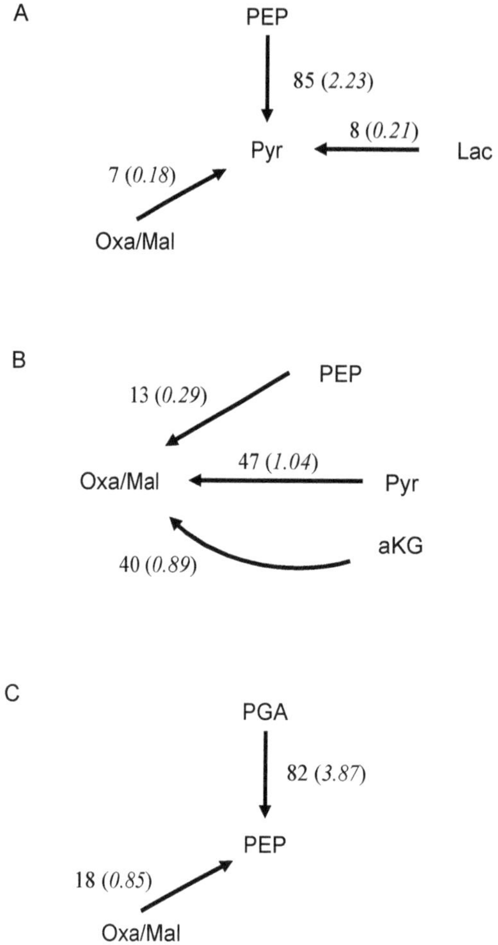

Figure 3.4 Anaplerotic flux distribution in metabolic nodes: pyruvate (A), oxaloacetate/malate (B), phosphoenolpyruvate (C). Numbers in parentheses represent flux rate estimates given in mmol/g-dry weight (italic numbers). Data from Petersen *et al.* (2000).

Solution algorithms of non-linear problems are generally used to estimate fluxes by using ^{13}C labelling information. These include the Levenberg–Marquardt algorithm (Shirai *et al.*, 2005; Wiechert and de Graaf, 1997; Wiechert *et al.*, 1997), a Monte Carlo approach (Kadirkamanathan *et al.*, 2006; Schmidt *et al.*, 1999b), and a hybrid algorithm that combines a genetic algorithm with the Levenberg–Marquardt algorithm (Zhao and Shimizu, 2003). Moreover, universal software frameworks that can design ^{13}C labelling experiments and metabolic models that are necessary for ^{13}C-MFA have been developed (Wiechert *et al.*, 2001), and this framework has been extensively tested by using a variety of experiments and organisms (de Graaf *et al.*, 1999, 2000; Petersen *et al.*, 2000).

Because there are 2^n possible labelling patterns in a molecule of n carbon atoms, each metabolite will account for 2^n model states in an isotopomer model (in fractional labelling models the number of model states per metabolite pool is only n). Therefore, isotopomer models are inherently complex and can easily build up to several hundred or thousands of model states. A general methodology to handle more complex isotopomer models has been developed (Schmidt *et al.*, 1997). This methodology employs isotopomer distribution vectors (IDVs) and isotopomer mapping matrices (IMMs). IDVs contain mole fractions of individual isotopomers. For example, the IDV of pyruvate and AcCoA have $2^3 = 8$ and $2^2 = 4$ elements that are given as:

$$\mathrm{IDV}_{\mathrm{Pyr}} = \begin{matrix} 000 \\ 001 \\ 010 \\ 011 \\ 100 \\ 101 \\ 110 \\ 111 \end{matrix} \tag{3.8}$$

and

$$\mathrm{IDV}_{\mathrm{AcCoA}} = \begin{matrix} 00 \\ 01 \\ 10 \\ 11 \end{matrix} \tag{3.9}$$

respectively. IMMs are defined for each pair of reactant and product molecules in a biochemical carbon exchange reaction. $\mathrm{IMM}_{\mathrm{Pyr>AcCoA}}$ is given as:

$$\mathrm{IMM}_{\mathrm{Pyr>AcCoA}} = \begin{matrix} 1 & 0 & 0 & 0 & 1 & 0 & 0 & 0 \\ 0 & 1 & 0 & 0 & 0 & 1 & 0 & 0 \\ 0 & 0 & 1 & 0 & 0 & 0 & 1 & 0 \\ 0 & 0 & 0 & 1 & 0 & 0 & 0 & 1 \end{matrix} \tag{3.10}$$

By using the $IMM_{Pyr>AcCoA}$, the IDV_{AcCoA} of the product molecules can be calculated from the IDV_{Pyr} of the reactant isotopomers as follows:

$$IDV_{AcCoA} = IMM_{Pyr>AcCoA} \times IDV_{Pyr} \qquad (3.11)$$

This isotopomer model allows flux estimations by using not only ^{13}C labelling information by NMR analysis but also ^{13}C labelling information obtained by other analytical tools. In particular, for NMR analysis, due to the fact that the superposition of signals is caused by a limited number of linear constraints of the different isotopomers, the entire pool of isotopomers cannot always be calculated. It was proposed recently that the labelling patterns of intermediary metabolites that usually occur at low concentrations are difficult to obtain directly by employing NMR (Follstad and Stephanopoulos, 1998). MS has been regarded as being an interesting tool for quantifying metabolic fluxes (Wittmann and Heinzle, 1999). Owing to its high sensitivity, it is applicable to batch processes, which are of more industrial relevance than experiments that entail continuous culturing. In contrast to NMR, the main fraction of metabolites, which appear as being non-labelled, is visible to MS and can be included in the analysis (Fig. 3.3). Concerning the quantification of labelling patterns by MS, several authors demonstrated the use of gas chromatography-MS (GC-MS) in tracer experiments. This also enables mixtures to be analysed without prior separation of the components.

MS can only separate isotopomers with different molecular masses (mass isotopomers) (Lee *et al.*, 1991). Thus, a metabolite with n carbon atoms essentially enables $n+1$ mass isotopomers to be measured. Moreover, the isotopomers of compound fragments that are produced by thermal cleavage in an MS instrument can also be detected (Christensen and Nielsen, 1999; MacKenzie *et al.*, 1987; Mawhinney *et al.*, 1986; Strelkov *et al.*, 2004; Yang *et al.*, 2002). The studies included different organisms such as bacteria (Dauner and Sauer, 2000; Park *et al.*, 1997a) and fungi (Christensen and Nielsen, 1999). In addition, matrix-assisted laser desorption ionization-time of flight (MALDI-TOF)-MS or GC–combustion-isotope ratio (C-IR) MS is applicable for quantifying mass distributions of low molecular weight metabolites (Heinzle *et al.*, 2008; Wittmann and Heinzle, 2001a; Yuan *et al.*, 2010).

In order to precisely estimate the intracellular metabolism flux of lysine-producing *C. glutamicum* by using MS, the appropriate combination of models and ^{13}C-labelled substrates that should be used in ^{13}C labelling experiments was investigated (Wittmann and Heinzle, 2001b). *C. glutamicum* genealogy was investigated from precisely designed from ^{13}C-MFA (Wittmann and Heinzle, 2002). ^{13}C-MFA of five types of *C. glutamicum* (ATCC 13032, ATCC 13287, ATCC 21253, ATCC 21526, and ATCC 21543) revealed that the flux in the pentose phosphate pathway increased concurrently with high lysine yield, while the flux in the TCA cycle decreased. Concurrent with increased lysine yield, the net flux of anaplerotic pathways increased. This net flux result was obtained through the regulation of carboxylation and decarboxylation within the anaplerotic pathways. Obtaining mass isotopomer information on amino acids by GC-MS analysis enabled a deeper understanding of the intracellular physiology (Klapa *et al.*, 2003). Fragmented Gly, Phe, and Val mass isotopomers were analysed by using GC-MS analysis. Representing each of their precursors were 3-phosphoglycerate, PEP, and pyruvate, respectively. The results strongly suggest the existence of oxaloacetate decarboxylase activity as a part of the anaplerotic pathways.

Improvements in GC-MS analytical techniques and the development of flux analysis techniques have contributed to a greater understanding of intracellular metabolisms in *C. glutamicum* under a variety of conditions: ^{13}C-MFA by using substrates such as ^{13}C-labelled fructose and/or sucrose (Becker *et al.*, 2005; Kiefer *et al.*, 2004; Wittmann *et al.*, 2004), ^{13}C-MFA when a gene that encodes a part of the central pathway is deleted (Becker *et al.*, 2008), and that when transcriptional regulators are deleted (Kromer *et al.*, 2008). Exchange fluxes in anaplerotic pathways and by glucose-6-phosphate isomerase have been estimated precisely by using respiratory $^{13}CO_2$ information obtained by performing parallel ^{13}C labelling experiments with different types of ^{13}C-labelled substrates (Hoon Yang *et al.*, 2006; Yang *et al.*, 2005; Yang *et al.*, 2006). ^{13}C labelling information on His obtained from GC-MS analysis allowed for the precise estimation of exchange fluxes in the pentose phosphate pathway, and how much the estimated fluxes can change with the experimental errors from GC-MS analysis was investigated (Shirai *et al.*, 2006).

The development of ^{13}C-MFA techniques has led to a greater understanding of the many layers of intracellular physiology combined with transcriptome and metabolome analysis. Relationships between changes in gene expression and intracellular metabolites, and changes in intracellular metabolic fluxes in *C. glutamicum* from growth to lysine production, have been investigated (Kromer *et al.*, 2004). Furthermore, the understanding of the physiology obtained using ^{13}C-MFA has led to the successful targeting of genes for modification and improvements in lysine productivity (Becker *et al.*, 2007, 2009; Kind *et al.*, 2013).

In order to keep the intracellular metabolic state constant, continuous cultures were required for ^{13}C-MFA. However, batch cultures or fed batch cultures are used for the production of industrial substances. Glutamate production in *C. glutamicum* can occur in batch cultures (Sato *et al.*, 2008; Shimizu *et al.*, 2003; Shirai *et al.*, 2005). ^{13}C-MFA by pursuing the time courses of ^{13}C labelling information introduced to proteinogenic amino acids revealed a positive correlation between glutamate production and the flux catalysed by Pc (Shirai *et al.*, 2007) when glutamate production was induced by Tween40 addition. Moreover, a novel ^{13}C-MFA model with the addition of two dilution parameters has been proposed. These parameters, D and G, correspond to the dilution of the tracer (i.e. glucose) and dilution of the product (i.e. biomass amino acids) (Antoniewicz *et al.*, 2007), respectively. The use of this method led to the successful analysis of changes in intracellular metabolic flux during transglutaminase production in *C. glutamicum* batch cultures (Umakoshi *et al.*, 2011).

Recently, by using advances in analytical techniques such as capillary electrophoresis-MS (CE-MS) and LC-MS/MS, ^{13}C labelling information on intracellular intermediate metabolites of central carbon metabolism and free amino acids was employed for performing precise ^{13}C-MFA (Iwatani *et al.*, 2007; Noack *et al.*, 2011; Noh *et al.*, 2007; Schaub *et al.*, 2008; Toya *et al.*, 2007, 2010). Furthermore, *in silico* genome-scale metabolic models of *C. glutamicum* have been constructed on the basis of genome sequence annotation and physiological data (Kjeldsen and Nielsen, 2009; Shinfuku *et al.*, 2009), and the intracellular metabolic fluxes can be predicted. The combination of the *in silico* design predictive result for maximum production of target and the precise MFA for interpretation of intracellular metabolisms is believed as a powerful tool to improve the production of useful compounds by using *C. glutamicum*.

Conclusion

C. glutamicum has been used as a model organism to improve MFA precisions. High-precision MFA requires ^{13}C trace experiments and ^{13}C labelling information that is obtained using analytical techniques such as NMR and GC-MS, in addition to extracellular measurement data. By using these data and computer calculations to obtain a mathematically optimized solution, a highly precise estimation of fluxes in intracellular branching and reversible reactions can be realized. Highly precise MFA results lead to better understanding of *C. glutamicum* physiology, and at the same time enable the identification of modification targets to improve the production of valuable compounds.

References

Antoniewicz, M. R., Kraynie, D. F., Laffend, L. A., Gonzalez-Lergier, J., Kelleher, J. K., and Stephanopoulos, G. (2007) Metabolic flux analysis in a nonstationary system: fed-batch fermentation of a high yielding strain of *E. coli* producing 1,3-propanediol. *Metab. Eng.* **9**, 277–292.

Becker, J., Klopprogge, C., Zelder, O., Heinzle, E., and Wittmann, C. (2005) Amplified expression of fructose 1,6-bisphosphatase in *Corynebacterium glutamicum* increases *in vivo* flux through the pentose phosphate pathway and lysine production on different carbon sources. *Appl. Environ. Microbiol.* **71**, 8587–8596.

Becker, J., Klopprogge, C., Herold, A., Zelder, O., Bolten, C. J., and Wittmann, C. (2007) Metabolic flux engineering of L-lysine production in *Corynebacterium glutamicum* – over expression and modification of G6P dehydrogenase. *J. Biotechnol.* **132**, 99–109.

Becker, J., Klopprogge, C., and Wittmann, C. (2008) Metabolic responses to pyruvate kinase deletion in lysine producing *Corynebacterium glutamicum*. *Microb. Cell Fact.* 7, 8.

Becker, J., Klopprogge, C., Schroder, H., and Wittmann, C. (2009) Metabolic engineering of the tricarboxylic acid cycle for improved lysine production by *Corynebacterium glutamicum*. *Appl. Environ. Microbiol.* **75**, 7866–7869.

Blombach, B., Riester, T., Wieschalka, S., Ziert, C., Youn, J. W., Wendisch, V. F., and Eikmanns, B. J. (2011) *Corynebacterium glutamicum* tailored for efficient isobutanol production. *Appl. Environ. Microbiol.* **77**, 3300–3310.

Christensen, B., and Nielsen, J. (1999) Isotopomer analysis using GC-MS. *Metab. Eng.* **1**, 282–290.

Dauner, M., and Sauer, U. (2000) GC-MS analysis of amino acids rapidly provides rich information for isotopomer balancing. *Biotechnol. Prog.* **16**, 642–649.

Delaunay, S., Lapujade, P., Engasser, J. M., and Goergen, J. L. (2002) Flexibility of the metabolism of *Corynebacterium glutamicum* 2262, a glutamic acid-producing bacterium, in response to temperature upshocks. *J. Ind. Microbiol. Biotechnol.* **28**, 333–337.

Dominguez, H., Rollin, C., Guyonvarch, A., Guerquin-Kern, J. L., Cocaign-Bousquet, M., and Lindley, N. D. (1998) Carbon-flux distribution in the central metabolic pathways of *Corynebacterium glutamicum* during growth on fructose. *Eur. J. Biochem./FEBS* **254**, 96–102.

Eikmanns, B. J., Follettie, M. T., Griot, M. U., and Sinskey, A. J. (1989) The phosphoenolpyruvate carboxylase gene of *Corynebacterium glutamicum:* molecular cloning, nucleotide sequence, and expression. *Mol. Gen. Genet.* **218**, 330–339.

Follstad, B. D., and Stephanopoulos, G. (1998) Effect of reversible reactions on isotope label redistribution – analysis of the pentose phosphate pathway. *Eur. J. Biochem./FEBS* **252**, 360–371.

de Graaf, A. A., Striegel, K., Wittig, R. M., Laufer, B., Schmitz, G., Wiechert, W., Sprenger, G. A., and Sahm, H. (1999) Metabolic state of Zymomonas mobilis in glucose-, fructose-, and xylose-fed continuous cultures as analysed by ^{13}C- and ^{31}P-NMR spectroscopy. *Arch. Microbiol.* **171**, 371–385.

de Graaf, A. A., Mahle, M., Mollney, M., Wiechert, W., Stahmann, P., and Sahm, H. (2000) Determination of full ^{13}C isotopomer distributions for metabolic flux analysis using heteronuclear spin echo difference NMR spectroscopy. *J. Biotechnol.* **77**, 25–35.

Gubler, M., Jetten, M., Lee, S. H., and Sinskey, A. J. (1994) Cloning of the pyruvate kinase gene (*pyk*) of *Corynebacterium glutamicum* and site-specific inactivation of pyk in a lysine-producing *Corynebacterium lactofermentum* strain. Appl. Environ. Microbiol. **60**, 2494–2500.

Heinzle, E., Yuan, Y., Kumar, S., Wittmann, C., Gehre, M., Richnow, H. H., Wehrung, P., Adam, P., and Albrecht, P. (2008) Analysis of ^{13}C labeling enrichment in microbial culture applying metabolic tracer experiments using gas chromatography–combustion-isotope ratio mass spectrometry. *Anal. Biochem.* **380**, 202–210.

Hoon Yang, T., Wittmann, C., and Heinzle, E. (2006) Respirometric ^{13}C flux analysis – Part II: *in vivo* flux estimation of lysine-producing *Corynebacterium glutamicum*. *Metab. Eng.* **8**, 432–446.

Inui, M., Kawaguchi, H., Murakami, S., Vertes, A. A., and Yukawa, H. (2004a) Metabolic engineering of *Corynebacterium glutamicum* for fuel ethanol production under oxygen-deprivation conditions. *J. Mol. Microbiol. Biotechnol.* **8**, 243–254.

Inui, M., Murakami, S., Okino, S., Kawaguchi, H., Vertes, A. A., and Yukawa, H. (2004b) Metabolic analysis of *Corynebacterium glutamicum* during lactate and succinate productions under oxygen deprivation conditions. *J. Mol. Microbiol. Biotechnol.* **7**, 182–196.

Iwatani, S., Van Dien, S., Shimbo, K., Kubota, K., Kageyama, N., Iwahata, D., Miyano, H., Hirayama, K., Usuda, Y., Shimizu, K., *et al.* (2007) Determination of metabolic flux changes during fed-batch cultivation from measurements of intracellular amino acids by LC-MS/MS. *J. Biotechnol.* **128**, 93–111.

Jetten, M. S. M., and Sinskey, A. J. (1993) Characterization of phosphoenolpyruvate carboxykinase from *Corynebacterium glutamicum*. *FEMS Microbiol. Lett.* **111**, 183–188.

Jo, S. J., Maeda, M., Ooi, T., and Taguchi, S. (2006) Production system for biodegradable polyester polyhydroxybutyrate by *Corynebacterium glutamicum*. *J. Biosci. Bioeng.* **102**, 233–236.

Kacser, H., and Burns, J. A. (1973) The control of flux. *Symp. Soc. Exp. Biol.* **27**, 65–104.

Kadirkamanathan, V., Yang, J., Billings, S. A., and Wright, P. C. (2006) Markov Chain Monte Carlo Algorithm based metabolic flux distribution analysis on *Corynebacterium glutamicum*. *Bioinformatics* **22**, 2681–2687.

Kiefer, P., Heinzle, E., Zelder, O., and Wittmann, C. (2004) Comparative metabolic flux analysis of lysine-producing *Corynebacterium glutamicum* cultured on glucose or fructose. *Appl. Environ. Microbiol.* **70**, 229–239.

Kind, S., Jeong, W. K., Schroder, H., and Wittmann, C. (2010) Systems-wide metabolic pathway engineering in *Corynebacterium glutamicum* for bio-based production of diaminopentane. *Metab. Eng.* **12**, 341–351.

Kind, S., Becker, J., and Wittmann, C. (2013) Increased lysine production by flux coupling of the tricarboxylic acid cycle and the lysine biosynthetic pathway-Metabolic engineering of the availability of succinyl-CoA in *Corynebacterium glutamicum*. *Metab. Eng.* **15**, 184–195.

Kinoshita, S., Udaka, S., and Shimono, M. (1957) Studies on the amino acid fermentation. Part 1. Production of L-glutamic acid by various microorganisms. *J. Gen. Appl. Microbiol. Japan* **3**, 193–205.

Kiss, R. D., and Stephanopoulos, G. (1992) Metabolic characterization of a L-lysine-producing strain by continuous culture. *Biotechnol. Bioeng.* **39**, 565–574.

Kjeldsen, K. R., and Nielsen, J. (2009) *In silico* genome-scale reconstruction and validation of the *Corynebacterium glutamicum* metabolic network. *Biotechnol. Bioeng.* **102**, 583–597.

Klapa, M. I., Aon, J. C., and Stephanopoulos, G. (2003) Systematic quantification of complex metabolic flux networks using stable isotopes and mass spectrometry. *Eur. J. Biochem./FEBS* **270**, 3525–3542.

Kromer, J. O., Sorgenfrei, O., Klopprogge, K., Heinzle, E., and Wittmann, C. (2004) In-depth profiling of lysine-producing *Corynebacterium glutamicum* by combined analysis of the transcriptome, metabolome, and fluxome. *J. Bacteriol.* **186**, 1769–1784.

Kromer, J. O., Bolten, C. J., Heinzle, E., Schroder, H., and Wittmann, C. (2008) Physiological response of *Corynebacterium glutamicum* to oxidative stress induced by deletion of the transcriptional repressor McbR. *Microbiology* **154**, 3917–3930.

Lawson, C. L., and Hanson, R. J. (1995) Solving Least Squares Problems, Vol. 15. Society for Industrial and Applied Mathematics, Philadelphia, PA, USA.

Lee, W. N., Byerley, L. O., Bergner, E. A., and Edmond, J. (1991) Mass isotopomer analysis: theoretical and practical considerations. *Biol. Mass Spectrom.* **20**, 451–458.

MacKenzie, S. L., Tenaschuk, D., and Fortier, G. (1987) Analysis of amino acids by gas–liquid chromatography as tert.-butyldimethylsilyl derivatives. Preparation of derivatives in a single reaction. *J. Chromatogr.* **387**, 241–253.

Marx, A., de Graaf, A. A., Wiechert, W., Eggeling, L., and Sahm, H. (1996) Determination of the fluxes in the central metabolism of *Corynebacterium glutamicum* by nuclear magnetic resonance spectroscopy combined with metabolite balancing. *Biotechnol. Bioeng.* **49**, 111–129.

Marx, A., Striegel, K., de Graaf, A. A., Sahm, H., and Eggeling, L. (1997) Response of the central metabolism of *Corynebacterium glutamicum* to different flux burdens. *Biotechnol. Bioeng.* **56**, 168–180.

Marx, A., Eikmanns, B. J., Sahm, H., de Graaf, A. A., and Eggeling, L. (1999) Response of the central metabolism in *Corynebacterium glutamicum* to the use of an NADH-dependent glutamate dehydrogenase. *Metab. Eng.* **1**, 35–48.

Marx, A., Hans, S., Mockel, B., Bathe, B., de Graaf, A. A., McCormack, A. C., Stapleton, C., Burke, K., O'Donohue, M., and Dunican, L. K. (2003) Metabolic phenotype of phosphoglucose isomerase mutants of *Corynebacterium glutamicum*. *J. Biotechnol.* **104**, 185–197.

Mawhinney, T. P., Robinett, R. S., Atalay, A., and Madson, M. A. (1986) Analysis of amino acids as their tert.-butyldimethylsilyl derivatives by gas–liquid chromatography and mass spectrometry. *J. Chromatogr.* **358**, 231–242.

Neijssel, O. M., and Teixeira de Mattos, M. J. (1994) The energetics of bacterial growth: a reassessment. *Mol. Microbiol.* **13**, 172–182.

Noack, S., Noh, K., Moch, M., Oldiges, M., and Wiechert, W. (2011) Stationary versus non-stationary (13) C-MFA: a comparison using a consistent dataset. *J. Biotechnol.* **154**, 179–190.

Noh, K., Gronke, K., Luo, B., Takors, R., Oldiges, M., and Wiechert, W. (2007) Metabolic flux analysis at ultra short time scale: isotopically non-stationary ^{13}C labeling experiments. *J. Biotechnol.* **129**, 249–267.

Nunheimer, T. D., Birnbaum, J., Ihnen, E. D., and Demain, A. L. (1970) Product inhibition of the fermentative formation of glutamic acid. *Appl. Microbiol.* **20**, 215–217.

Park, S. M., Shaw-Reid, C., Sinskey, A. J., and Stephanopoulos, G. (1997a) Elucidation of anaplerotic pathways in *Corynebacterium glutamicum* via ^{13}C-NMR spectroscopy and GC-MS. *Appl. Microbiol. Biotechnol.* **47**, 430–440.

Park, S. M., Sinskey, A. J., and Stephanopoulos, G. (1997b) Metabolic and physiological studies of *Corynebacterium glutamicum* mutants. *Biotechnol. Bioeng.* **55**, 864–879.

Peters-Wendisch, P. G., Eikmanns, B. J., Thierbach, G., Bachmann, B., and Sahm, H. (1993) Phosphoenolpyruvate carboxylase in *Corynebacterium glutamicum* is dispensable for growth and lysine production. *FEMS Microbiol. Lett.* **112**, 269–274.

Peters-Wendisch, P. G., Kreutzer, C., Kalinowski, J., Patek, M., Sahm, H., and Eikmanns, B. J. (1998) Pyruvate carboxylase from *Corynebacterium glutamicum*: characterization, expression and inactivation of the pyc gene. *Microbiology* **144**, 915–927.

Petersen, S., de Graaf, A. A., Eggeling, L., Mollney, M., Wiechert, W., and Sahm, H. (2000) *In vivo* quantification of parallel and bidirectional fluxes in the anaplerosis of *Corynebacterium glutamicum*. *J. Biol. Chem.* **275**, 35932–35941.

Petersen, S., Mack, C., de Graaf, A. A., Riedel, C., Eikmanns, B. J., and Sahm, H. (2001) Metabolic consequences of altered phosphoenolpyruvate carboxykinase activity in *Corynebacterium glutamicum* reveal anaplerotic regulation mechanisms *in vivo*. *Metab. Eng.* **3**, 344–361.

Rollin, C., Morgant, V., Guyonvarch, A., and Guerquin-Kern, J. L. (1995) ^{13}C-NMR studies of *Corynebacterium melassecola* metabolic pathways. *Eur. J. Biochem./FEBS* **227**, 488–493.

Russell, J. B., and Cook, G. M. (1995) Energetics of bacterial growth: balance of anabolic and catabolic reactions. *Microbiol. Rev.* **59**, 48–62.

Sato, H., Orishimo, K., Shirai, T., Hirasawa, T., Nagahisa, K., Shimizu, H., and Wachi, M. (2008) Distinct roles of two anaplerotic pathways in glutamate production induced by biotin limitation in *Corynebacterium glutamicum*. *J. Biosci. Bioeng.* **106**, 51–58.

Schaub, J., Mauch, K., and Reuss, M. (2008) Metabolic flux analysis in *Escherichia coli* by integrating isotopic dynamic and isotopic stationary ^{13}C labeling data. *Biotechnol. Bioeng.* **99**, 1170–1185.

Schmidt, K., Carlsen, M., Nielsen, J., and Villadsen, J. (1997) Modeling isotopomer distributions in biochemical networks using isotopomer mapping matrices. *Biotechnol. Bioeng.* **55**, 831–840.

Schmidt, K., Nielsen, J., and Villadsen, J. (1999a) Quantitative analysis of metabolic fluxes in *Escherichia coli*, using two-dimensional NMR spectroscopy and complete isotopomer models. *J. Biotechnol.* **71**, 175–189.

Schmidt, K., Norregaard, L. C., Pedersen, B., Meissner, A., Duus, J. O., Nielsen, J. O., and Villadsen, J. (1999b) Quantification of intracellular metabolic fluxes from fractional enrichment and ^{13}C-^{13}C coupling constraints on the isotopomer distribution in labeled biomass components. *Metab. Eng.* **1**, 166–179.

Schrumpf, B., Schwarzer, A., Kalinowski, J., Puhler, A., Eggeling, L., and Sahm, H. (1991) A functionally split pathway for lysine synthesis in *Corynebacterium glutamicium. J. Bacteriol.* **173**, 4510–4516.

Shiio, I., Miyajima, R., and Sano, K. (1970) Genetically desensitized aspartate kinase to the concerted feedback inhibition in Brevibacterium flavum. *J. Biochem.* **68**, 701–710.

Shiio, I., Otsuka, S. I., and Katsuya, N. (1962a) Effect of biotin on the bacterial formation of glutamic acid. II. Metabolism of glucose. *J. Biochem.* **52**, 108–116.

Shiio, I., Otsuka, S. I., and Takahashi, M. (1962b) Effect of biotin on the bacterial formation of glutamic acid. I. Glutamate formation and cellular premeability of amino acids. *J. Biochem.* **51**, 56–62.

Shimizu, H., Tanaka, H., Nakato, A., Nagahisa, K., Kimura, E., and Shioya, S. (2003) Effects of the changes in enzyme activities on metabolic flux redistribution around the 2-oxoglutarate branch in glutamate production by *Corynebacterium glutamicum. Bioprocess Biosyst. Eng.* **25**, 291–298.

Shinfuku, Y., Sorpitiporn, N., Sono, M., Furusawa, C., Hirasawa, T., and Shimizu, H. (2009) Development and experimental verification of a genome-scale metabolic model for *Corynebacterium glutamicum. Microb. Cell Fact.* **8**, 43.

Shirai, T., Nakato, A., Izutani, N., Nagahisa, K., Shioya, S., Kimura, E., Kawarabayasi, Y., Yamagishi, A., Gojobori, T., and Shimizu, H. (2005) Comparative study of flux redistribution of metabolic pathway in glutamate production by two coryneform bacteria. *Metab. Eng.* **7**, 59–69.

Shirai, T., Matsuzaki, K., Kuzumoto, M., Nagahisa, K., Furusawa, C., Shioya, S., and Shimizu, H. (2006) Precise metabolic flux analysis of coryneform bacteria by gas chromatography–mass spectrometry and verification by nuclear magnetic resonance. *J. Biosci. Bioeng.* **102**, 413–424.

Shirai, T., Fujimura, K., Furusawa, C., Nagahisa, K., Shioya, S., and Shimizu, H. (2007) Study on roles of anaplerotic pathways in glutamate overproduction of *Corynebacterium glutamicum* by metabolic flux analysis. *Microb. Cell Fact.* **6**, 19.

Smith, K. M., Cho, K. M., and Liao, J. C. (2010) Engineering *Corynebacterium glutamicum* for isobutanol production. *Appl. Microbiol. Biotechnol.* **87**, 1045–1055.

Sonntag, K., Eggeling, L., De Graaf, A. A., and Sahm, H. (1993) Flux partitioning in the split pathway of lysine synthesis in *Corynebacterium glutamicum.* Quantification by ^{13}C- and 1H-NMR spectroscopy. *Eur. J. Biochem./FEBS* **213**, 1325–1331.

Stephanopoulos, G., Aristidou, A. A., Nielsen, J. H., and Nielsen, J. (1998) Metabolic Engineering: Principles and Methodologies. Academic Press, San Diego, CA, USA.

Strelkov, S., von Elstermann, M., and Schomburg, D. (2004) Comprehensive analysis of metabolites in *Corynebacterium glutamicum* by gas chromatography/mass spectrometry. *Biol. Chem.* **385**, 853–861.

Sugimoto, S.-i., and Shiio, I. (1989) Regulation of enzymes for erythrose 4-phosphate synthesis in *Brevibacterium flavum. Agric. Biol. Chem.* **53**, 2081–2087.

Szyperski, T. (1995) Biosynthetically directed fractional ^{13}C-labeling of proteinogenic amino acids. An efficient analytical tool to investigate intermediary metabolism. *Eur. J. Biochem/FEBS* **232**, 433–448.

Szyperski, T., Bailey, J. E., and Wüthrich, K. (1996) Detecting and dissecting metabolic fluxes using biosynthetic fractional ^{13}C labeling and two-dimensional NMR spectroscopy. *Trends Biotechnol.* **14**, 453–459.

Takiguchi, N., Shimizu, H., and Shioya, S. (1997) An on-line physiological state recognition system for the lysine fermentation process based on a metabolic reaction model. *Biotechnol. Bioeng.* **55**, 170–181.

Takinami, K., Yamada, Y., and Okada, H. (1966) Biochemical effects of fatty acid and its derivatives on L-glutamic acid fermentation. 4. Biotin content of growing cells of *Brevibacterium lactofermentum. Agric. Biol. Chem.* **30**, 674–680.

Tateno, T., Okada, Y., Tsuchidate, T., Tanaka, T., Fukuda, H., and Kondo, A. (2009) Direct production of cadaverine from soluble starch using *Corynebacterium glutamicum* coexpressing alpha-amylase and lysine decarboxylase. *Appl. Microbiol. Biotechnol.* **82**, 115–121.

Toya, Y., Ishii, N., Hirasawa, T., Naba, M., Hirai, K., Sugawara, K., Igarashi, S., Shimizu, K., Tomita, M., and Soga, T. (2007) Direct measurement of isotopomer of intracellular metabolites using capillary electrophoresis time-of-flight mass spectrometry for efficient metabolic flux analysis. *J. Chromatogr. A* **1159**, 134–141.

Toya, Y., Ishii, N., Nakahigashi, K., Hirasawa, T., Soga, T., Tomita, M., and Shimizu, K. (2010) ^{13}C-metabolic flux analysis for batch culture of *Escherichia coli* and its *Pyk* and *Pgi* gene knockout mutants based on mass isotopomer distribution of intracellular metabolites. *Biotechnol. Prog.* **26**, 975–992.

Umakoshi, M., Hirasawa, T., Furusawa, C., Takenaka, Y., Kikuchi, Y., and Shimizu, H. (2011) Improving protein secretion of a transglutaminase-secreting *Corynebacterium glutamicum* recombinant strain on the basis of ^{13}C metabolic flux analysis. *J. Biosci. Bioeng.* **112**, 595–601.

Uy, D., Delaunay, S., Goergen, J. L., and Engasser, J. M. (2005) Dynamics of glutamate synthesis and excretion fluxes in batch and continuous cultures of temperature-triggered *Corynebacterium glutamicum. Bioprocess Biosyst. Eng.* **27**, 153–162.

Vallino, J. J., and Stephanopoulos, G. (1993) Metabolic flux distributions in *Corynebacterium glutamicum* during growth and lysine overproduction. *Biotechnol. Bioeng.* **41**, 633–646.

Wendisch, V. F., de Graaf, A. A., and Sahm, H. (1997) Accurate determination of ^{13}C enrichments in nonprotonated carbon atoms of isotopically enriched amino acids by 1H nuclear magnetic resonance. *Anal. Biochem.* **245**, 196–202.

Wendisch, V. F., de Graaf, A. A., Sahm, H., and Eikmanns, B. J. (2000) Quantitative determination of metabolic fluxes during coutilization of two carbon sources: comparative analyses with *Corynebacterium glutamicum* during growth on acetate and/or glucose. *J. Bacteriol.* **182**, 3088–3096.

Wiechert, W., and de Graaf, A. A. (1997) Bidirectional reaction steps in metabolic networks: I. Modeling and simulation of carbon isotope labeling experiments. *Biotechnol. Bioeng.* **55**, 101–117.

Wiechert, W., Siefke, C., de Graaf, A. A., and Marx, A. (1997) Bidirectional reaction steps in metabolic networks: II. Flux estimation and statistical analysis. *Biotechnol. Bioeng.* **55**, 118–135.

Wiechert, W., Mollney, M., Petersen, S., and de Graaf, A. A. (2001) A universal framework for ^{13}C metabolic flux analysis. *Metab. Eng.* **3**, 265–283.

Wittmann, C., and Heinzle, E. (1999) Mass spectrometry for metabolic flux analysis. *Biotechnol. Bioeng.* **62**, 739–750.

Wittmann, C., and Heinzle, E. (2001a) MALDI-TOF MS for quantification of substrates and products in cultivations of *Corynebacterium glutamicum*. *Biotechnol. Bioeng.* **72**, 642–647.

Wittmann, C., and Heinzle, E. (2001b) Modeling and experimental design for metabolic flux analysis of lysine-producing Corynebacteria by mass spectrometry. *Metab. Eng.* **3**, 173–191.

Wittmann, C., and Heinzle, E. (2002) Genealogy profiling through strain improvement by using metabolic network analysis: metabolic flux genealogy of several generations of lysine-producing corynebacteria. *Appl. Environ. Microbiol.* **68**, 5843–5859.

Wittmann, C., Kiefer, P., and Zelder, O. (2004) Metabolic fluxes in *Corynebacterium glutamicum* during lysine production with sucrose as carbon source. *Appl. Environ. Microbiol.* **70**, 7277–7287.

Yang, C., Hua, Q., and Shimizu, K. (2002) Quantitative analysis of intracellular metabolic fluxes using GC-MS and two-dimensional NMR spectroscopy. *J. Biosci. Bioeng.* **93**, 78–87.

Yang, C., Hua, Q., Baba, T., Mori, H., and Shimizu, K. (2003) Analysis of *Escherichia coli* anaplerotic metabolism and its regulation mechanisms from the metabolic responses to altered dilution rates and phosphoenolpyruvate carboxykinase knockout. *Biotechnol. Bioeng.* **84**, 129–144.

Yang, T. H., Heinzle, E., and Wittmann, C. (2005) Theoretical aspects of ^{13}C metabolic flux analysis with sole quantification of carbon dioxide labeling. *Comput. Biol. Chem.* **29**, 121–133.

Yang, T. H., Wittmann, C., and Heinzle, E. (2006) Respirometric ^{13}C flux analysis, Part I: design, construction and validation of a novel multiple reactor system using on-line membrane inlet mass spectrometry. *Metab. Eng.* **8**, 417–431.

Yuan, Y., Yang, T. H., and Heinzle, E. (2010) ^{13}C metabolic flux analysis for larger scale cultivation using gas chromatography–combustion-isotope ratio mass spectrometry. *Metab. Eng.* **12**, 392–400.

Zhao, J., and Shimizu, K. (2003) Metabolic flux analysis of *Escherichia coli* K12 grown on ^{13}C-labeled acetate and glucose using GC-MS and powerful flux calculation method. *J. Biotechnol.* **101**, 101–117.

Zupke, C., and Stephanopoulos, G. (1994) Modeling of Isotope Distributions and Intracellular Fluxes in Metabolic Networks Using Atom Mapping Matrices. *Biotechnol. Prog.* **10**, 489–498.

Growth and Production Capabilities of *Corynebacterium glutamicum*: Interrogating a Genome-scale Metabolic Network Model

4

Elisabeth Zelle, Katharina Nöh and Wolfgang Wiechert

Abstract

In recent years, the assembly of genome-scale metabolic networks has been established as a powerful tool for obtaining a comprehensive understanding of microbial production hosts and for exploring their industrial potential. Network modelling provides a versatile framework for analysing, predicting and, eventually, optimizing cellular processes. This book chapter reviews the current state of stoichiometric modelling for *Corynebacterium glutamicum* and presents a new curated network containing 475 reaction steps and 408 metabolites. This network is interrogated to obtain answers for several fundamental questions on quantitative growth and production capabilities of the organism. Using flux balance analysis the model is validated with experimental data, maximal yields for all 20 amino acids are exemplarily computed and the impact of biomass composition data on model prediction is systematically investigated. As an extension of stoichiometry, the structural prerequisites for metabolic flux analysis are discussed. For the purpose of isotope-based flux analysis, the central metabolic network is extended by carbon transition information required for tracer studies. The genome-scale metabolic network presented in this chapter is supplied via a web server.

Introduction

Since its discovery roughly 60 years ago, *Corynebacterium glutamicum* has emerged as one of the major production hosts in industrial biotechnology. Particularly, the production of amino acids paved the way for the success of this Gram-positive soil bacterium as a biotechnological workhorse. For example, more than 2 million tons of L-lysine per year originate from *C. glutamicum*. However, the inherent complexity of a living organism makes further strain improvement more and more difficult and, particularly, improvements achievable by conventional mutation and screening techniques approach a saturation phase. This finding motivates a complementation of classical techniques with a systems biology approach to obtain rational model-based routes towards a streamlined *C. glutamicum* cell factory with improved capabilities (Wendisch *et al.*, 2006). The ultimate aim is to have a predictive tool at hand that generates testable hypotheses for metabolic engineering targets on demand.

For metabolic network modelling the living cell is considered as a complex biochemical reactor in which thousands of metabolites are converted by enzymatic reactions in a highly cross-linked manner. The structure of these interactions is captured in the reaction stoichiometry which is the basic information contained in metabolic network models. If some governing supplementary data are given, stoichiometric network models provide a versatile framework for exploring, analysing, predicting and, eventually, optimizing cellular processes.

This review is concerned with stoichiometry-based methods to analyse *C. glutamicum*. The current state of genome-scale modelling for *C. glutamicum* is overviewed and a new curated network is presented. Stoichiometry-based tools are then applied to validate the network and some exemplary applications demonstrate the added value of such a model. In particular, some new insights on the energy metabolism are obtained that are otherwise impossible to derive. A special focus is placed on metabolic engineering objectives. In the final sections the structural modelling fundamentals for ^{13}C-based metabolic flux analysis in *C. glutamicum* are laid. They are given by the carbon transition network which complements stoichiometry and allows central questions on isotope tracing to be answered.

The metabolic network of *Corynebacterium glutamicum*

Genome-wide metabolic networks

Metabolic network analysis contributed much to the current knowledge on the *in vivo* distribution of intracellular carbon flows in microbial production hosts. The general methodology of stoichiometry-based methods is well described (see Palsson, 2006) and, thus, only briefly reviewed here. Traditionally, genome-wide network representations are constructed manually from historically earlier models augmented with information on reactions newly available from experts or databases successively refined and updated. Paramount examples for this process are the consensus network reconstructions of the model organism *Escherichia coli* (Orth *et al.*, 2011; McCloskey *et al.*, 2013), *Saccharomyces cerevisiae* (Mo *et al.*, 2009) or the human metabolism (Duarte *et al.*, 2007; Thiele *et al.*, 2013).

As one important information source DNA sequencing has nowadays become an indispensable tool. Recent advanced in tools for annotating the sequenced genome (Kawaji and Hayashizaki, 2008) simplify the reconstruction of metabolic network models although they are still far from operating without errors (Thiele and Palsson, 2010). A continuously updated list of metabolic reconstructions can be found at gcrg.ucsd.edu/InSilicoOrganisms/OtherOrganisms. Currently, the genome-scale metabolic reconstruction of *E. coli* is regarded as the largest and best verified network containing 2251 reactions and 1136 metabolites organized in 11 functional categories. The assembly took several man-years of ongoing effort (Orth *et al.*, 2011), mainly because manual curation remains an indispensable step in the reconstruction pipeline (Thiele and Palsson, 2010). The following section summarizes the, compared with the mentioned organisms, brief history of the *C. glutamicum* stoichiometric network model.

Network reconstructions of *Corynebacterium glutamicum*

The determination of the whole genome sequence of the representative *C. glutamicum* wild-type strain ATCC 13032 marked an important breakthrough for this organism. Two competing genome projects led to the publication of two complete genome sequences (Ikeda and Nakagawa, 2003; Kalinowski *et al.*, 2003). Both sequences differ by the presence or absence of insertion element copies and a prophage (Wendisch *et al.*, 2006; Frunzke *et al.*, 2008; Baumgart *et al.*, 2013). The *C. glutamicum* genome as described in (Kalinowski *et al.*, 2003) consists of a single circular chromosome comprising over 3 million base pairs (Kalinowski, 2004). Furthermore, *C. glutamicum* is characterized by an, at least for Gram-positive bacteria, high G + C content of the chromosomal DNA (53.8%) (Bott and Niebisch, 2003).

In 2009, two genome-scale reconstructions were published almost simultaneously for *C. glutamicum* by Kjeldsen and Nielsen (2009) as well as Shinfuku *et al.* (2009). The network developed by Kjeldsen and Nielsen (herein named *iKK446*) is based on the genome annotation of Kalinowski *et al.* (2003). It contains 411 metabolites and 446 reactions. The network was verified with various experimental data sets. The reconstructed network presented by Shinfuku *et al.* (herein named *iYS502*) contains 423 metabolites and 502 reactions.

In both cases the reactions were assembled from databases and literature. Nevertheless, there are some differences between the networks. For example, the respiratory chain included in *iYS502* does not represent the strain-specific properties of *C. glutamicum* respiration that were given in (Bott and Niebisch, 2003; Nantapong *et al.*, 2005; Follmann *et al.*, 2009). On the other hand, the Shinfuku network *iYS502* accounts for the balancing of protons and water for all reactions, while *iKK446* is restricted to the protons translocated by the respiratory chain.

The model extensions shown in the following start from the *iKK446* network. The reconstruction *iKK446* already covers the whole central carbon metabolism including glycolysis (Embden–Meyerhof–Parnas pathway; EMP), gluconeogenesis, the pentose phosphate pathway (PPP), the citric acid cycle (TCA cycle) as well as the anaplerotic reactions. Furthermore biomass synthesis is taken into account by amino acids, nucleotide lipids, peptidoglycan precursors and cofactor synthesis. These cellular building blocks are assembled in a biomass synthesis equation. Finally, the reconstruction incorporates various transport reactions and maintenance requirements.

A new extended and curated *Corynebacterium glutamicum* network

For this chapter the *iKK446* network was further extended and revised (Fig. 4.1). This revision – denoted *iEZ475* – partly follows the available *E. coli* reconstructions because they are highly developed. As a first step, the reactions were augmented with the missing balances for protons and water by a direct comparison with a genome-scale *E. coli* network (Feist *et al.*, 2007). Furthermore, the notation of the metabolites was changed to that used in the *E. coli* reconstruction. This adaption discriminates between cytosolic and extracellular compounds. Extracellular protons are denoted by h[e] while cytosolic protons are named h[c].

In a next step 'dead-end' metabolites (and the corresponding reactions) were deleted, e.g. folate. Otherwise, the network reconstruction was extended by four reactions that describe the glycogen metabolism (Woo *et al.*, 2010). Likewise, reactions for the uptake and conversion of various alternative substrates, namely glucosamine, arabitol (arabinitol), maltose, and protocatechuate (protocatechuic acid) are added. These steps are only active if the culture medium contains these particular substrates.

Based on literature the cofactor usage was checked and adapted if required, e.g. the imidazole glycerol phosphate synthase (*hisHF*) that is one step of L-histidine synthesis uses NH_4^+ instead of glutamine as nitrogen donor (Kulis-Horn *et al.*, 2014). Besides that the CoA transferase (acetate:succinate CoA transferase; *actA*) reaction, described in (Yasuda *et al.*, 2007), is added. This *actA* reaction represents an alternative to the succinyl-CoA synthetase (*sucD*) reaction, which contributes to ATP formation by substrate-level phosphorylation. Moreover, the reversibility specification was changed according to literature. This means that reaction directionalities are assigned to reflect available knowledge on reaction thermodynamics (Lewis *et al.*, 2012). Finally, the maintenance requirements are adjusted to strain specific values determined in (Koch-Koerfges *et al.*, 2013).

The resulting extended and curated network *iEZ475* contains 475 reactions and 408 metabolites (Fig. 4.1). The metabolites comprise 68 extracellular and 340 intracellular pools. The reactions are grouped in nine major pathway families (Table 4.1). Compared with *E. coli* reconstructions it becomes clear that the genome-scale network of *C. glutamicum* is still in its infancy. However, in the context of available genome-scale metabolic network reconstructions in general it is already a valuable tool (Feist *et al.*, 2009).

Stoichiometric modelling fundamentals

Stoichiometric equations

Stoichiometric methods in general are used to address a broad spectrum of basic questions and practical applications (Feist and Palsson, 2008). The primary input for any such method is a metabolic (or biochemical reaction) network defined by a set of reaction steps interconverting the metabolites. In order to apply mathematical tools to the network, it has first to be converted into a matrix notation containing essentially the same information (Palsson, 2006). The resulting stoichiometric matrix $\mathbf{S}$ has m rows representing the metabolites and n columns representing the reactions. The entry ($\mathbf{S}_{i,j}$) in row i and column j gives the stoichiometric coefficient of the ith metabolite in the jth reaction step. The stoichiometric matrix is sparsely populated (i.e. most entries are zero)

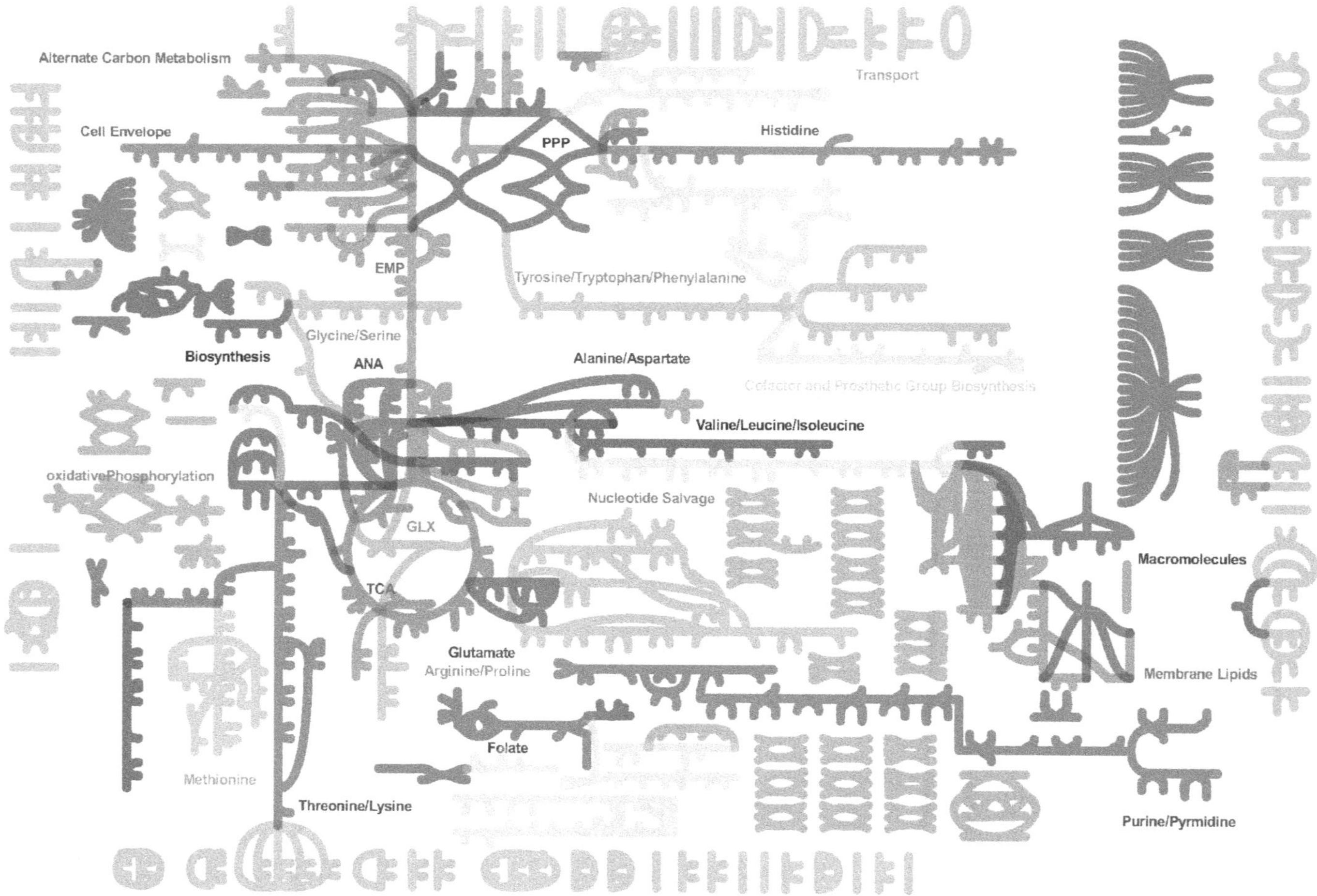

Figure 4.1 Graphical representation of the new *Corynebacterium glutamicum* genome-scale network *iEZ475*. The network comprises 475 reactions interconverting 408 metabolites. The reactions are classified in 25 pathways (see Table 4.1 for statistical data). Details of the reactions and metabolites are visually suppressed due to the high complexity but can be interactively explored using the Omix® network visualization tool. See the section 'Web resources' for availability. A colour version of this figure is available in the plate section at the back of the book.

Table 4.1 Statistics of pathway families covered by the genome-scale *Corynebacterium glutamicum* network *iEZ475*

Pathway	No. of reactions
Central carbon metabolism	43
Amino acid synthesis	110
Oxidative phosphorylation	13
Membrane lipid metabolism	20
Nucleotide salvage pathway	48
Cofactor and prosthetic group biosynthesis	61
Biomass formation	52
Alternate carbon metabolism	35
Transport reactions	93

because biochemical reactions normally involve only a few different metabolites.

In this book chapter we are primarily interested in fluxes (i.e. reaction rates) through the metabolic network. In total these fluxes define the physiological state of an organism. Typically, under certain external conditions only a subset of the reaction steps is active (i.e. non-zero). The physiological state of a cell is described by a flux distribution vector $\boldsymbol{v}$ which assigns a value to every single reaction step.

One fundamental assumption underlying all stoichiometric methods used in this chapter is the presence of an intra-cellular steady state. This means that all intracellular metabolite levels stay constant over time or at least during the considered growth phase. Without discussing this common assumption (Clarke, 1981; Schilling *et al.*, 2000), the steady state assumption is equivalent with the general stoichiometric equation:

$$\mathbf{S}\cdot\boldsymbol{v} = 0 \tag{4.1}$$

which says that for each intermediate metabolite pool the sum of ingoing fluxes equals the sum of the outgoing fluxes.

In genome-scale network models the number of metabolites is far less than the number of reactions (i.e. $m << n$). Thus, stoichiometry in combination with the steady state assumption leads to an underdetermined system. This leaves many degrees of freedom to adjust metabolic fluxes. In other words, there are many different possible flux distributions over the network fulfilling the stoichiometric equation.

Flux balance analysis

The constraint-based reconstruction and analysis (COBRA) approach is a well-established stoichiometry-based modelling method introduced in 1984 and popularized by the Palsson group and a wide variety of COBRA methods have been developed over the years (Lewis *et al.*, 2012). The idea of constraint-based modelling is to further narrow the space of possible steady state flux distributions over the metabolic network by adding additional assumptions to stoichiometry (Equation 4.1).

Therefore, additional experimentally derived constraints are introduced in the form of flux inequality constraints (Reed, 2012). For example, inequalities can be used to express experimental restrictions or bounds on the values of individual fluxes ($v_{i,\min} \leq v_i \leq v_{i,\max}$) or to constrain the direction of a reaction ($v_i \leq 0$) reflecting thermodynamic knowledge. A set of inequalities can be formally represented in matrix notation:

$$\mathbf{C}\cdot\boldsymbol{v} \leq \boldsymbol{b} \tag{4.2}$$

Here, $\mathbf{C}$ is the constraint matrix and $\boldsymbol{b}$ the flux capacity vector. A comprehensive list of different types of constraints can be found in (Price *et al.*, 2004; Gianchandani *et al.*, 2010).

Although, the introduction of constraints reduces the space of possible flux distributions, and hence, the potential of the organism, there usually still remain many possible flux distributions fulfilling both (Equation 4.1) and (Equation 4.2). With the aim of computing useful predictions it is desirable to select one single flux distribution from this variety by imposing a reasonable selection criterion. This approach is called flux balance analysis (FBA) (Orth *et al.*, 2010) which is the most widely applied stoichiometry-based computational tool to date. In FBA a selection criterion must be chosen such as cell growth (Feist and Palsson, 2010), redox-potential or ATP production (Nolan *et al.*, 2006; Schuetz *et al.*, 2007). The biological significance of these objective functions is subject of ongoing debate (Schuetz *et al.*, 2007; Schuster *et al.*, 2008).

Mathematically, the selection criterion is usually expressed as a linear objective function. Comprising the corresponding weight coefficients for the network fluxes in one vector *f*, FBA turns out to be a linear programming (LP) problem:

$$\max_{v} \boldsymbol{f}^{T}{\cdot}v \text{ s.t. } \mathbf{S}{\cdot}\boldsymbol{v}=0,\ \mathbf{C}{\cdot}\boldsymbol{v}\leq b \qquad (4.3)$$

The LP solution selects one flux distribution which fulfils stoichiometry and all inequality constraints while, at the same time, it maximizes the criterion. If for example, the growth rate is maximized, the result is a physiological steady state with highest possible growth rate which is, simultaneously, in accordance with the given constraints.

Linear programming is a classical optimization problem and lots of software tools exist to deal with it. Particularly, there are several extensions as well as software systems related to FBA (Lewis *et al.*, 2012), e.g. the MATLAB™-based COBRA toolbox (Becker *et al.*, 2007; Schellenberger *et al.*, 2011) the R package sybil (Gelius-Dietrich *et al.*, 2013) or the commercially available SimPheny® modelling software (Genomatica Inc., San Diego, CA, USA).

Model validation

Comparing model predictions with experimental data

By applying FBA, Kjeldsen and Nielsen (2009) validated their reconstruction *iKK446* against published fluxome data (Vallino and Stephanopoulos, 1993; Wittmann and Heinzle, 2001, 2002; Sorgenfrei *et al.*, 2004) for different strains and growth conditions. Thereby, the focus was placed on L-lysine-producing strains. Using the FBA approach, it was shown that the *iKK446* model is able to reproduce metabolic fluxes during L-lysine production and growth under various conditions. Moreover, it was confirmed that the model correctly predicts main physiological properties, such as the ability to synthesize biomass precursors (such as amino acids, nucleotides triphosphates and lipids).

The revised reconstruction *iEZ475* was further validated against experimental observations and it correctly predicts major physiological properties of different *C. glutamicum* strains, such as aerobic and anaerobic growth (not shown here). Owing to the added proton balances the validation especially focuses on energetic properties, i.e. cellular respiration (substrate level phosphorylation SLP and electron transport phosphorylation ETP) and the phosphate/oxygen ratio (P/O ratio) (Hinkle, 2005; Ferguson, 2010). The P/O ratio refers to the amount of ATP produced per oxygen atom reduced by the respiratory chain and, thus, is crucial for energetic efficiency. The exact P/O ratios are still in question (Hinkle, 2005).

Studying the branched respiratory chain of *Corynebacterium glutamicum*

C. glutamicum exhibits a branched respiratory chain. There are two terminal oxidases with proton coefficients ($H^+/2e^-$) of 2 (*bd* branch) and 6 (bc_1-aa_3 branch). Thus, the P/O ratio may vary between 0.5 and 2 (Bott and Niebisch, 2003; Kabus *et al.*, 2007). Clearly, a variable P/O-ratio cannot be reproduced with genome-scale networks, when maximizing for growth, because FBA will necessarily calculate flux distributions that exclusively use the more efficient bc_1-aa_3 branch of oxidative phosphorylation (Kjeldsen and Nielsen, 2009).

For investigating the impact of varying usage of the two terminal oxidases the $H^+/2e^-$ ratio is varied in the possible range between 2 (100% *bd*) and 6 (100% bc_1-aa_3). For each $H^+/2e^-$ ratio the amount of protons needed to synthesize one ATP (H^+/ATP) is varied between 3 and 10 H^+/ATP. Here, the lower bound of 3H^+/ATP is based on the assumption that the proton motive force (*pmf*) is exclusively used for ATP production, whereas the maximal value of 10 H^+/ATP accounts for extracellular protons required to operate transport processes (Koch-Koerfges *et al.*, 2012). For all value pairs of $H^+/2e^-$ and H^+/ATP ratios the FBA problem is solved for the objective 'maximize for growth'. Here, the glucose uptake rate is fixed to the experimentally determined value of 5.4 mmol/g_{CDW}/h (Koch-Koerfges *et al.*, 2012). The result is shown in Fig. 4.2.

Two main conclusions are drawn from the *in silico* analysis. First, the optimal growth

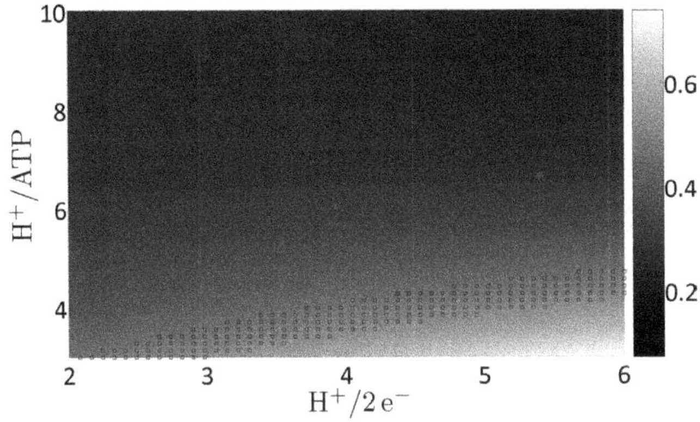

Figure 4.2 Model prediction of the optimal growth rate [μ (h^{-1}), grey shades] as function of the H^+/ATP and the $H^+/2e^-$ ratios, respectively, i.e. the ratios by which the two terminal oxidases are used. The marked area represents the experimentally measured growth rate $\mu_{exp} = 0.4 \pm 0.01$.

rate directly correlates with the P/O ratio. As expected, a smaller P/O ratio is caused by a decreased $H^+/2e^-$ or an increased H^+/ATP, which both result in less generated ATP and, consequently, smaller growth rates. Second, assuming that four protons are needed for the production of one ATP, the experimentally observed growth rate $\mu_{exp} = 0.4 \pm 0.01$/h (Koch-Koerfges *et al.*, 2012) is obtained when both branches are equally used. An equal usage of both oxidases is also predicted in (Kabashima *et al.*, 2009). Following this further, the model will slightly overestimate the growth rate, when not enforcing a specific split-ratio of the two oxidases ($\mu_{sim} = 0.48$ compared with $\mu_{exp} = 0.4 \pm 0.01$), because in that case only the bc_1-aa_3 branch is used.

Predictions for oxidative phosphorylation mutants

Besides examining the influence of the P/O ratio, it is analysed whether the network *iEZ475* can reproduce the growth phenotypes of oxidative phosphorylation mutants described in (Koch-Koerfges *et al.*, 2012, 2013). All mutants are deletion mutants lacking either the cytochrome *bd* branch (*C. glutamicum* Δ*cydAB*), the cytochrome bc_1-aa_3 branch (*C. glutamicum* Δ*qcr*), or the ATP synthase (*C. glutamicum* ΔF_1F_0). In a first step it was confirmed that the model can reproduce the theoretical maximal ATP production of each mutant strain obtained before by elementary considerations (Bott and Niebisch, 2003).

Secondly, the growth behaviour of the mutants was investigated. Therefore, the extracellular rates (i.e. the glucose uptake rate as well as by-product production rates) were set to the measured values given in (Koch-Koerfges *et al.*, 2012, 2013). In accordance with the experimental observation the simulated optimal growth behaviour of all three mutants still show a growth-phenotype, which, however, is different to the wild type. Thus, deleting one of the two oxidases or the ATP synthase is non-lethal for growth with glucose as sole carbon and energy source. Furthermore, the growth rate of the Δ*cydAB* mutant is as large as the growth rate of the wild type, whereas the growth rate of the Δ*qcr* mutant as well as that of the ΔF_1F_0 mutant was 50% lower than the wild type. In conclusion, the simulated growth phenotypes agree very well with the experimental observations. Table 4.2 shows an exemplary result.

Predicting production capabilities for amino acids

A major application of genome-wide models is the prediction of production capabilities of industrial host organisms. As an example the genome-scale reconstruction *iEZ475* is now used to investigate the ATP requirements for amino acid synthesis (Fig. 4.3). The energetic cost is calculated as the amount of ATP required for converting a maximum of the substrate to the correspoding amino acid. Clearly, this cost depends on the particular

Table 4.2 Measured (meas) and predicted (pred) growth rates (μ) for *Corynebacterium glutamicum* wild-type strain and the ΔF_1F_0 mutant lacking the ATP synthase. Glucose uptake rate and oxygen uptake rate were taken from experimental data. Low amounts of pyruvate, acetate, succinate and lactate by-production were fixed to amounts observed in Koch-Koerfges *et al.* (2012)

	Wild type	ΔF_1F_0
Glucose uptake rate	5.4 mmol g_{CDW}/h	19.7 mmol g_{CDW}/h
Oxygen uptake rate	4.1 mmol g_{CDW}/h	11.1 mmol g_{CDW}/h
μ_{meas}	0.40/h	0.19 h^{-1}
μ_{pred}	0.41/h	0.23 h^{-1}

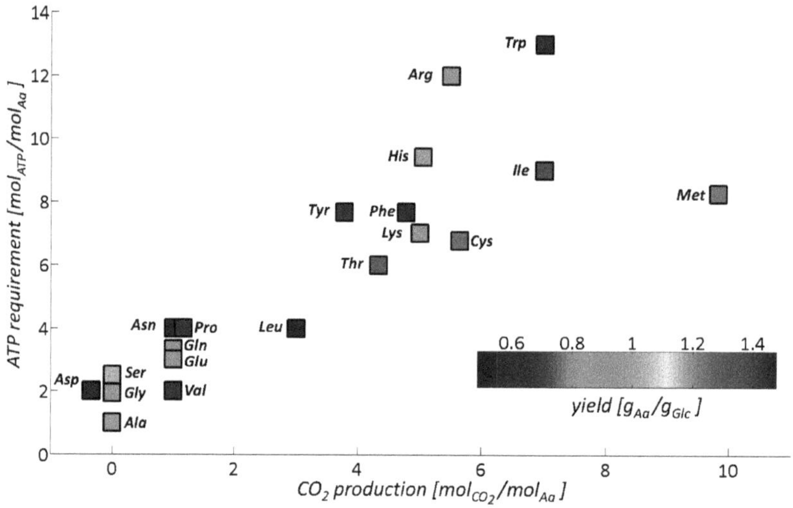

Figure 4.3 ATP requirement versus CO_2 production and yield (colour coded) of *in silico* amino acid production determined by FBA with the genome-scale *Corynebacterium glutamicum* network *iEZ475* using glucose as sole carbon source. Each square corresponds to a single FBA study that maximizes the associated amino acid export flux. A colour version of this figure is available in the plate section at the back of the book.

carbon source (Akashi and Gojobori, 2002). Exemplarily, the energetic costs shown here are based on glucose as sole carbon source.

In general, it can be expected that amino acids derived from chemical intermediates occurring in 'early' fuelling reaction pathways (i.e. the aromatic amino acids and histidine) are more costly than those derived from intermediates downstream in glycolysis (the serine family) and the tricarboxylic acid cycle (the aspartate and glutamate families) (Akashi and Gojobori, 2002). The metabolic cost of synthesizing the 20 amino acids used in proteins, as calculated with the metabolic reconstruction *iEZ475* of *C. glutamicum*, are shown in Fig. 4.3.

It turns out that alanine, glycine, aspartate, valine and serine have the lowest energetic cost, which is below 3 mol_{ATP} per mol of the corresponding amino acid. This low ATP requirement can partly be explained by the tight connection of these amino acids to the central carbon metabolism, i.e. the synthesis of these amino acids is an unbranched pathway with only a few intermediate reactions. Moreover, amino acids that are energetically cheap rather contain low numbers of carbon atoms. In contrast, carbon-rich amino acids (i.e. L-tyrosine, L-phenylalanine, and L-tryptophan) tend to be energetically costly. In accordance with the findings of Akashi and Gojobori (2002) for *E. coli*, the aromatic amino acids and L-histidine have the highest ATP requirements. The energetically most expansive amino acid is L-tryptophan:

$$13.0\,\text{mol ATP mol}_{Trp}^{-1}$$

followed by L-arginine:

$$12.0\,\text{mol ATP mol}^{-1}_{Arg}$$

and L-histidine:

$$9.4\,\text{mol ATP mol}^{-1}_{His}$$

Of special interest is the finding that L-histidine biosynthesis requires 9.4 ATP molecules per L-histidine molecule. This result is dramatically lower than published values (Kulis-Horn *et al.*, 2014). Brenner and Ames (1971) calculated a demand of 41 ATP equivalents for the synthesis of one L-histidine molecule in *Salmonella* Typhimurium. This value was carried over to the L-histidine synthesis in general (Alifano *et al.*, 1996; Stepansky and Leustek, 2006; Kulis-Horn *et al.*, 2014), but cannot be confirmed for *C. glutamicum*.

Uncertainties in metabolic network models

Uncertain stoichiometric coefficients

For newcomers it is surprising that genome-wide stoichiometric models contain several non-integer coefficients. This is because the stoichiometric relation between precursor molecules, macromolecular building blocks and cell mass can only be described by phenomenological coefficients. For example, in the *iEZ475* model the following pseudo reaction equations for biomass composition can be found. Here, damp, dtmp, dcmp, dgmp are the four monophosphates forming DNA, ala-L, ..., val-L denote the 20 amino acids forming cell protein and DNA, ..., PROTEIN are macromolecular building blocks of cell mass:

$$0.75\ \text{damp} + 0.75\ \text{dtmp} + 0.87\ \text{dcmp} + 0.87\ \text{dgmp} + 4.4\ \text{ATP} \rightarrow \text{DNA} + 4.4\ \text{ADP}$$

$$0.67\ \text{ala-L} + 0.19\ \text{arg-L} + \ldots + 0.27\ \text{val-L} + 19.7\ \text{ATP} \rightarrow \text{PROTEIN} + 19.7\ \text{ADP}$$

$$\text{DNA} + \ldots + \text{PROTEIN} + 47.7\ \text{ATP} \rightarrow \text{BIOMASS} + 47.7\ \text{ADP}$$

Large stoichiometric models use the macromolecular composition to introduce a growth equation (Thiele and Palsson, 2010). The biomass synthesis equation is then designed so that the summed mole fractions of all precursors yield one gram cell dry weight. Hence, the unit of the biomass reaction is h^{-1} (Orth *et al.*, 2010) and, thus, is a proxy of the specific growth rate μ.

The non-integer coefficients are experimentally determined but usually assumed to be constant for all physiological states. Moreover, although it is recommended that a detailed biomass composition of the target organism has to be determined (Thiele and Palsson, 2010) the required experiments are rarely performed due to the high workload of the experimental procedure for analysing the biomass. Instead, the biomass composition incorporated in network models is often received by collecting information from various literature sources. For example, data from phylogenetically related organisms are used (Thiele and Palsson, 2010).

As a consequence, most non-integer stoichiometric coefficients in genome-scale networks are error-prone. For this reason, the biomass synthesis equation incorporated in metabolic network models must be carefully analysed (Benyamini *et al.*, 2010). Notably, the biomass formation of the genome-scale network reconstruction *iKK446*, which forms the basis for the work presented here, is collected from various sources. For example, the energetic requirements for polymerization of the macromolecules were approximated from *E. coli*. The macromolecular composition itself was taken from Cocaign-Bousquet *et al.* (1996), whereas the composition of the macromolecules was extracted from different references (Kjeldsen and Nielsen, 2009).

Sampling-based flux balance analysis

Monte Carlo sampling methods have already been used earlier in the context of FBA to sample the feasible flux space (Wiback *et al.*, 2004; Schellenberger and Palsson, 2009). In this section another field of application for Monte Carlo sampling is proposed. The aim is to investigate how errors in non-integer stoichiometric coefficients influence the optimal flux distribution. The idea is to randomly sample the coefficients in a range defined by the experimentally measured error (in the following called uncertainty level δ). Based on the generated samples repeated FBAs are performed and the variation in the resulting flux distributions is analysed. The algorithm referred to as sampling-based FBA repeats the following procedure *N* times and, finally, collects and evaluates all results:

1 Draw a random sample from an assumed error distribution P_δ for each uncertain coefficient.
2 Create the corresponding perturbed stoichiometric matrix.
3 Solve the FBA problem for a given criterion (Equation 4.5) using the perturbed stoichiometry.

Impact of biomass composition data on model prediction

An uncertainty level of δ = 10% is regarded as a typical measurement error. The cell wall composition is very complex hampering the formulation of its synthesis pathways from precursors. Thus, the part of the biomass equation covering the cell wall synthesis demand is rather uncertain. An even higher uncertainty level of 50% might have to be considered when various effects contribute to an error-prone biomass composition. To keep things simple, the biomass equation coefficients are independently varied at uncertainty levels δ = 10%, 30%, and 50%. In the biomass synthesis equation the coefficients are perturbed based on a truncated normal distribution and 100,000 samples.

Fig. 4.4 shows the histogram of the calculated optimal growth rates for all three uncertainty levels. It turns out that the range of predicted growth rates broadens asymmetrically with increasing uncertainty level. More precisely, for an uncertainty level of 10% the minimal growth rate is 8% smaller than the nominal one, and the maximal growth rate is 10% larger. When perturbing the coefficients by 50% a 32% smaller as well as an 80% larger optimal growth rate is predicted. This shows that a realistic 10% variation in the biomass composition already leads to a 10% variation in the growth rate prediction. This should be taken into account when interpreting and comparing FBA results.

Metabolic flux analysis

Flux balance analysis is a speculative method

In the previous sections several illustrative examples have been given to demonstrate the benefits of a genome-scale metabolic network model. It has been shown how this network can be used for predicting physiological phenotypes if some

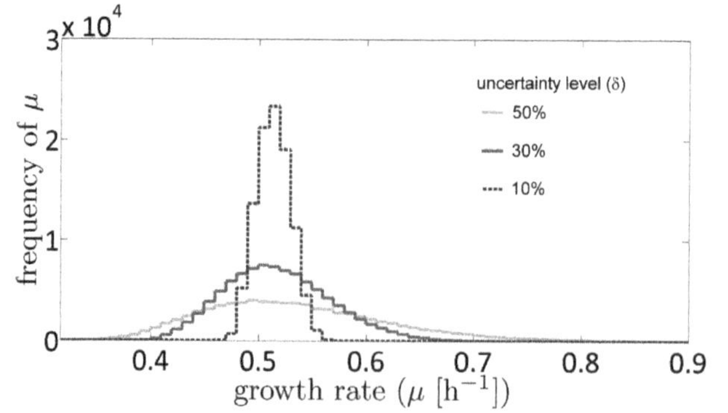

Figure 4.4 Results of a sampling-based FBA study performed with the *Corynebacterium glutamicum* network. The perturbation of the uncertain non-integer coefficients is based on a truncated normal distribution with three different uncertainty levels δ (10%, 30%, and 50%). For each uncertainty level 100,000 samples are generated. The frequency of the found objective values (maximal growth rate) of the perturbed FBA problems is represented as a histogram.

supplementary information is given. For example, knowledge on reaction thermodynamics or measured extracellular flux rates can be exploited. However, it must be clear that even if extracellular measurement data are available, Equations 4.1 and 4.2 usually still allow for a variety of different possible phenotypes. All these flux distributions are consistent with stoichiometry, assumed constraints and measured data. Summarizing, although being well motivated, FBA is still a kind of educated guessing. The method is not able to fully determine *in vivo* flux distributions based on measured extracellular flux data.

As an example, the studies summarized in Table 4.2 have been performed for model validation: the experimentally determined extracellular data with their measurement uncertainties (as upper and lower bounds) was fed into the model. By performing a growth maximizing FBA it can be found out if there exists an intracellular flux distribution which reproduces the measured data. However, the calculated flux distribution is usually one of many possible flux distributions consistent with the data.

Towards unambiguous *in vivo* flux determination

Historically, the first approach to really determine *in vivo* fluxes solely from measured extracellular rates (growth rate, substrate uptake, product formation, CO_2 evolution, O_2 uptake, etc.) was stoichiometric metabolic flux analysis (S-MFA) (Vallino and Stephanopoulos, 1993). To reach this goal S-MFA makes one additional but very critical assumption. It has to do with the energy metabolites ATP, NADH, NADPH, etc. In modern genome-scale networks the fact that not all sinks for these metabolites, maintenance requirements or even futile cycles are known is taken into account by introducing dummy energy loss reactions of type ATP → ADP + Pi. Omitting these equations means to assume closed energy metabolite balances. Although these balances are certainly not correct (Schmidt *et al.*, 1998; Petersen *et al.*, 2000) this strongly reduces the degrees of freedom for possible flux distributions. This finding was the advent of the S-MFA method. Interestingly, L-lysine production with *C. glutamicum* was the subject of the pioneering paper (Vallino and Stephanopoulos, 1993). S-MFA is still in use but must be considered as strongly biased.

Additional information is needed to generate phenotypic predictions which are based on less assumptions or guesses. One approach is to include information on cellular regulation on the metabolic and genetic level. Regulation is certainly one major reason why the real phenotype usually stays below the optimal performance criteria imposed for FBA. Modelling regulation leads into the field of mechanistic modelling which is a completely different story. Particularly, quite different input data and methods are needed in comparison to purely stoichiometric approaches (Wiechert and Noack, 2011). To date mechanistic modelling of *C. glutamicum* is still in its infancy (Degenring *et al.*, 2004; Dräger *et al.*, 2009).

^{13}C-metabolic flux analysis

^{13}C-based metabolic flux analysis (^{13}C-MFA) finally solved the problem of unambiguous experimental flux determination. Fig. 4.5 shows a typical result given by an *in vivo* flux distribution over the central metabolic pathways of *C. glutamicum*. Notably, the development of the nowadays broadly used ^{13}C-MFA method and its variants is intimately related to the study of *C. glutamicum* as a general platform organism for amino acids production. The pioneering papers on ^{13}C-MFA dealt with L-lysine production by *C. glutamicum* (Marx *et al.*, 1996) and several other ^{13}C-MFA papers on this organism followed (Bartek *et al.*, 2011; Noack *et al.*, 2011; van Ooyen *et al.*, 2012). In recent years the more advanced isotopically non-stationary INST-MFA method (Nöh *et al.*, 2007; Nöh and Wiechert, 2011; Wiechert and Nöh, 2013) was also tested with *C. glutamicum* as a classical ^{13}C-MFA model system.

In ^{13}C-MFA the additional information needed to complement the stoichiometric information (Equation 4.1) is given by measured extracellular fluxes (as in S-MFA) and – most importantly – isotopic tracer data. Similar to all other stoichiometry-based methods in this chapter, a metabolic steady state is a precondition. The fundamentals of ^{13}C-MFA are treated in depth in Chapter 3 and also in several review papers (Wiechert, 2001; Zamboni, 2011; Klein

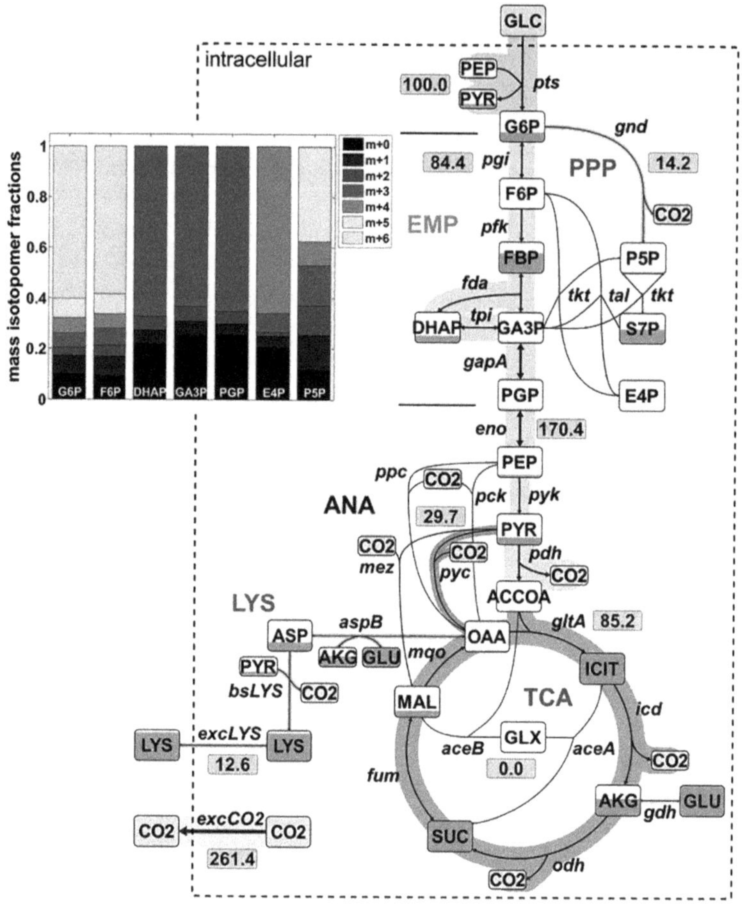

Figure 4.5 Flux map for the L-lysine producer strain *Corynebacterium glutamicum* DM 1730 under industrially relevant fed-batch conditions obtained with the INST-MFA method. The map shows a high activity of glycolysis and citric acid cycle. The flux in pentose phosphate pathway is known to be correlated to lysine production. Flux values are normalized to a glucose uptake rate of 100. Metabolite pool sizes (indicated by the filling heights of the nodes) are normalized to the concentration of glutamate. The inset shows labelling fractions for selected intracellular metabolites. For more details of the data set the reader is referred to Noack *et al.* (2011). A colour version of this figure is available in the plate section at the back of the book.

and Heinzle, 2012). In each case, the structural basis of ^{13}C-MFA is given by the carbon transition network, which is closely linked to the metabolic network. The present section focuses on the structural extensions of the metabolic network needed to trace labelled carbon atoms from the input substrate to the intracellular metabolite pools.

The carbon transition network of *Corynebacterium glutamicum*

In order to link isotopic enrichment measurements to the metabolic network structure additional knowledge on the single reactions on atomic level has to be supplied. For tracing labelled carbon atoms through the metabolic network, the fate of each single carbon atom in each reaction step needs to be defined. This information constitutes the carbon transition network.

The required knowledge to build up a carbon transition network can be taken from biochemical textbooks, databases (Arita *et al.*, 2006; Mu *et al.*, 2007; Pitkänen *et al.*, 2008) and published networks (Suthers *et al.*, 2007; Noack *et al.*, 2011; Ravikirthi *et al.*, 2011). Recently, automated computational approaches also became available (Heinonen *et al.*, 2011; Latendresse *et al.*, 2012).

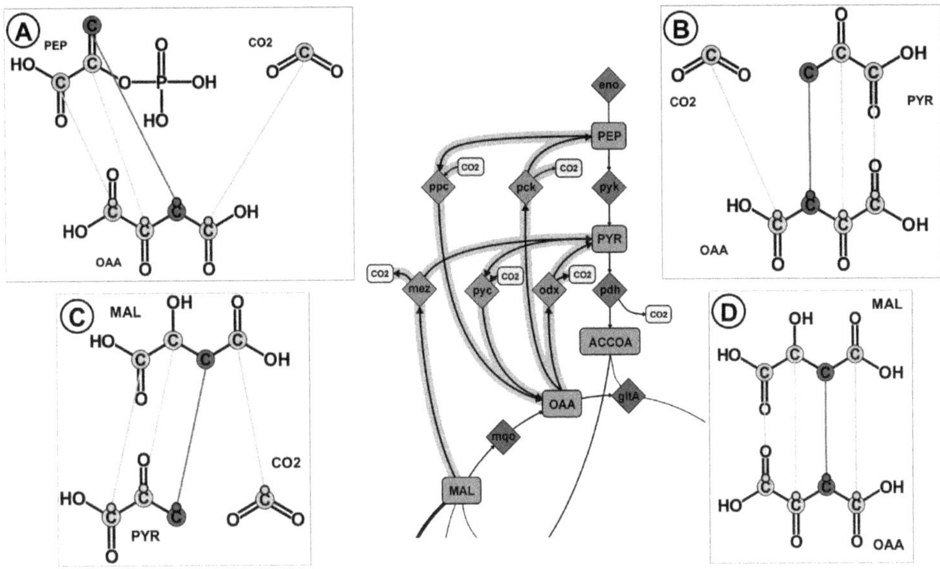

Figure 4.6 Carbon atom tansitions of the anaplerotic section in *C. glutamicum*. (A) PEP carboxylase (*pck*, EC:4.1.1.32)/PEP carboxykinase (*ppc*,EC:4.1.1.49), (B) pyruvate carboxylase (*pyc*, EC:6.4.1.1)/oxaloacetate decarboxylation (*odx*, EC 4.1.1.3), (C) malic enzyme (*mez*, EC:1.1.1.40), and (D) malate dehydrogenase (*mqo*, EC:1.1.5.4). Co-factors and energy equivalents are not shown for clarity. The same colors are used for substrate and target atom pairs. A colour version of this figure is available in the plate section at the back of the book.

However, in any case great care has to be taken to avoid pitfalls. For this reason, available database information must be carefully curated, at best visually (Fig. 4.6).

In particular, homotopic carbons within a molecule introduce alternative mappings that need to be considered. Although recently some advances have been made (Zhou and Nakhleh, 2012), the lack of fully automated approaches for atom network reconstruction makes a manual treatment and careful proofreading by a biochemical expert indispensable. A high number of homotopic atoms eventually leads to a combinatorial explosion of reactants of the same type and, consequently, aggregated atom paths. Another complication of in particular large-scale models is that their compilation from different sources is highly error prone owing to the lack of standards for carbon atom numbering within every metabolite.

Having the complete carbon atom transition network of *C. glutamicum* (published along with (Noack *et al.*, 2011) in the classical letter notation) at hand, the carbon atoms can be traced through the network. This can be realized by interactive visualization tools such as Omix Visualization using the CumoVis plug-in (Droste *et al.*, 2008, 2013).

Tools for ^{13}C-metabolic flux analysis

From the metabolic network, the carbon atom transition network, the given assumptions on flux directions, the measured extracellular fluxes, and the measured intracellular labelling data modern ^{13}C-MFA tools such as 13CFLUX2 (Weitzel *et al.*, 2013) can automatically generate all the balance and measurement equations needed to calculate the intracellular fluxes (Fig. 4.5). Notably, as distinguishing feature, ^{13}C-MFA generates confidence intervals for all reaction rates which originate from the uncertainty in the measurement data. This enables scientists to distinguish between 'black, grey and white spots' of the calculated quantitative metabolic flux map. The details would exceed the scope of this chapter.

Conclusions

It became clear that stoichiometry and isotope-based analysis contributed much to the current knowledge of the metabolism of *C. glutamicum*.

Genome-wide models do not only collect the present knowledge in a comprehensive way but also can guide research projects by calculating testable predictions. This, in turn, helps to distinguish between physiological states that are rather well understood and states that need deeper investigation. In this way, the new extended network model supplied with this book chapter will further help to advance the knowledge on *C. glutamicum*. It is based on several decades of experience with this organism and was extensively tested with experimental data from which only a small part could be exemplarily shown here.

As an extension of stoichiometry, ^{13}C-MFA yields detailed quantitative insights into the pathways of central metabolism and their actual *in vivo* operation mode. On the other hand, FBA yields flux distribution compliant with a putative optimality criterion. With that ^{13}C-MFA and FBA have their own playgrounds and, thus, can be considered as complementary technologies linked by the common structural information coded in the stoichiometry as well as the calibration by extracellular rate measurements. In this context, ^{13}C-MFA can be seen as a framework to directly test the optimality assumption that is arguably valid for all wild-type near strains operating under strong evolutionary pressure, but is critically discussed for highly engineered systems such as production strains (Schuetz *et al.*, 2012).

Acknowledgements

We thank Stephan Noack and Jana Tillack for significant contributions in the curation and documentation of the genome-scale network *C. glutamicum iEZ475*.

Web resources

The reaction and metabolite lists (including flux directions) as well as metabolic network maps and an SBML file for *C. glutamicum iEZ475* can be downloaded from www.13cflux.net. The network diagram (Fig. 4.1) is drawn using the visualization software Omix® (Droste *et al.*, 2013), which can be obtained from www.omix-visualization.com.

References

Akashi, H., and Gojobori, T. (2002) Metabolic efficiency and amino acid composition in the proteomes of *Escherichia coli* and *Bacillus subtilis*. *Proc. Natl. Acad. Sci. U.S.A.* **99**, 3695–3700.

Alifano, P., Fani, R., Liò, P., Lazcano, A., Bazzicalupo, M., Carlomagno, M. S., and Bruni, C. B. (1996) Histidine biosynthetic pathway and genes: structure, regulation, and evolution. *Microbiol. Rev.* **60**, 44–69.

Arita, M., Fujiwara, Y., and Nakanishi, Y. (2006) Map editor for the atomic reconstruction of metabolism (ARM). *Plant Metabolomics* **57**, 129–139.

Bartek, T., Blombach, B., Lang, S., Eikmanns, B. J., Wiechert, W., Oldiges, M., Nöh, K., and Noack, S. (2011) Comparative ^{13}C metabolic flux analysis of pyruvate dehydrogenase complex-deficient, L-valine-producing *Corynebacterium glutamicum*. *Appl. Environ. Microbiol.* **77**, 6644–6652.

Baumgart, M., Unthan, S., Rückert, C., Sivalingam, J., Grünberger, A., Kalinowski, J., Bott, M., Noack, S., and Frunzke, J. (2013) Construction of a prophage-free variant of *Corynebacterium glutamicum* ATCC 13032 for use as a platform strain for basic research and industrial biotechnology. *Appl. Environ. Microbiol.* **79**, 6006–6015.

Becker, S. A., Feist, A. M., Mo, M. L., Hannum, G., Palsson, B. Ø., and Herrgård, M. J. (2007) Quantitative prediction of cellular metabolism with constraint-based models: the COBRA toolbox. *Nat. Protoc.* **2**, 727–738.

Benyamini, T., Folger, O., Ruppin, E., and Shlomi, T. (2010) Flux balance analysis accounting for metabolite dilution. *Genome Biol.* **11**, R43.

Bott, M., and Niebisch, A. (2003) The respiratory chain of *Corynebacterium glutamicum*. *J. Biotechnol.* **104**, 129–153.

Brenner, M., and Ames, B. N. (1971) The histidine operon and its regulation. In: Vogel, H. J. (ed.) Metabolic Regulation. Academic Press, New York, NY, USA, pp. 349–387.

Clarke, B. L. (1981) Complete set of steady states for the general stoichiometric dynamical system. *J. Chem. Phys.* **75**, 4970–4979.

Cocaign-Bousquet, M., Guyonvarch, A., and Lindley, N. D. (1996) Growth rate-dependent modulation of carbon flux through central metabolism and the kinetic consequences for glucose-limited chemostat cultures of *Corynebacterium glutamicum*. *Appl. Environ. Microbiol.* **62**, 429–436.

Degenring, D., Froemel, C., Dikta, G., and Takors, R. (2004) Sensitivity analysis for the reduction of complex metabolism models. *J. Process Control* **14**, 729–745.

Dräger, A., Kronfeld, M., Ziller, M. J., Supper, J., Planatscher, H., Magnus, J. B., Oldiges, M., Kohlbacher, O., and Zell, A. (2009) Modeling metabolic networks in *C. glutamicum*: a comparison of rate laws in combination with various parameter optimization strategies. *BMC Syst. Biol.* **3**, 5.

Droste, P., Weitzel, M., and Wiechert, W. (2008) Visual exploration of isotope labeling networks in 3D. *Bioprocess Biosyst. Eng.* **31**, 227–239.

Droste, P., Nöh, K., and Wiechert, W. (2013) Omix – a visualization tool for metabolic networks with highest usability and customizability in focus. *Chemie-Ing.-Tech.* **85**, 849–862.

Duarte, N. C., Becker, S. A., Jamshidi, N., Thiele, I., Mo, M. L., Vo, T. D., Srivas, R., and Palsson, B. Ø. (2007) Global reconstruction of the human metabolic network based on genomic and bibliomic data. *Proc. Natl. Acad. Sci. U.S.A.* **104**, 1777–1782.

Feist, A. M., and Palsson, B. Ø. (2008) The growing scope of applications of genome-scale metabolic reconstructions using *Escherichia coli*. *Nat. Biotechnol.* **26**, 659–667.

Feist, A. M., and Palsson, B. Ø. (2010) The biomass objective function. *Curr. Opin. Microbiol.* **13**, 344–349.

Feist, A. M., Henry, C. S., Reed, J. L., Krummenacker, M., Joyce, A. R., Karp, P. D., Broadbelt, L. J., Hatzimanikatis, V., and Palsson, B. Ø. (2007) A genome-scale metabolic reconstruction for *Escherichia coli* K-12 MG1655 that accounts for 1260 ORFs and thermodynamic information. *Mol. Syst. Biol.* **3**, 121.

Feist, A. M., Herrgård, M. J., Thiele, I., Reed, J. L., and Palsson, B. Ø. (2009) Reconstruction of biochemical networks in microorganisms. *Nat. Rev. Microbiol.* **7**, 129–143.

Ferguson, S. J. (2010) ATP synthase: from sequence to ring size to the P/O ratio. *Proc. Natl. Acad. Sci. U.S.A.* **107**, 16755–16756.

Follmann, M., Ochrombel, I., Krämer, R., Trötschel, C., Poetsch, A., Rückert, C., Hüser, A., Persicke, M., Seiferling, D., Kalinowski, J., *et al.* (2009) Functional genomics of pH homeostasis in *Corynebacterium glutamicum* revealed novel links between pH response, oxidative stress, iron homeostasis and methionine synthesis. *BMC Genomics* **10**, 621.

Frunzke, J., Bramkamp, M., Schweitzer, J.-E., and Bott, M. (2008) Population heterogeneity in *Corynebacterium glutamicum* ATCC 13032 caused by prophage CGP3. *J. Bacteriol.* **190**, 5111–5119.

Gelius-Dietrich, G., Desouki, A. A., Fritzemeier, C. J., and Lercher, M. J. (2013) Sybil – efficient constraint-based modelling in R. *BMC Syst. Biol.* **7**, 125.

Gianchandani, E. P., Chavali, A. K., and Papin, J. A. (2010) The application of flux balance analysis in systems biology. *Syst. Biol. Med.* **2**, 372–382.

Heinonen, M., Lappalainen, S., Mielikäinen, T., and Rousu, J. (2011) Computing atom mappings for biochemical reactions without subgraph isomorphism. *J. Comput. Biol.* **18**, 43–58.

Hinkle, P. C. (2005) P/O ratios of mitochondrial oxidative phosphorylation. *Biochim. Biophys. Acta* **1706**, 1–11.

Ikeda, M., and Nakagawa, S. (2003) The *Corynebacterium glutamicum* genome: features and impacts on biotechnological processes. *Appl. Microbiol. Biotechnol.* **62**, 99–109.

Kabashima, Y., Kishikawa, J., Kurokawa, T., and Sakamoto, J. (2009) Correlation between proton translocation and growth: genetic analysis of the respiratory chain of *Corynebacterium glutamicum*. *J. Biochem.* **146**, 845–855.

Kabus, A., Niebisch, A., and Bott, M. (2007) Role of cytochrome bd oxidase from *Corynebacterium glutamicum* in growth and lysine production. *Appl. Environ. Microbiol.* **73**, 861–868.

Kalinowski, J. (2004) The genome of amino acid-producing corynebacteria. In: Eggeling, L., and Bott, M. (eds.) Handbook of *Corynebacterium glutamicum*. Taylor & Francis Group, Boca Raton, FL, USA, pp. 37–56.

Kalinowski, J., Bathe, B., Bartels, D., Bischoff, N., Bott, M., Burkovski, A., Dusch, N., Eggeling, L., Eikmanns, B. J., Gaigalat, L., *et al.* (2003) The complete *Corynebacterium glutamicum* ATCC 13032 genome sequence and its impact on the production of L-aspartate-derived amino acids and vitamins. *J. Biotechnol.* **104**, 5–25.

Kawaji, H., and Hayashizaki, Y. (2008) Genome annotation. *Methods Mol. Biol.* **452**, 125–139.

Kjeldsen, K. R., and Nielsen, J. (2009) *In silico* genome-scale reconstruction and validation of the *Corynebacterium glutamicum* metabolic network. *Biotechnol. Bioeng.* **102**, 583–597.

Klein, S., and Heinzle, E. (2012) Isotope labeling experiments in metabolomics and fluxomics. *Wiley Interdiscip. Rev. Syst. Biol. Med.* **4**, 261–272.

Koch-Koerfges, A., Kabus, A., Ochrombel, I., Marin, K., and Bott, M. (2012) Physiology and global gene expression of a *Corynebacterium glutamicum* ΔF_1F_O-ATP synthase mutant devoid of oxidative phosphorylation. *Biochim. Biophys. Acta* **1817**, 370–380.

Koch-Koerfges, A., Pfelzer, N., Platzen, L., Oldiges, M., and Bott, M. (2013) Conversion of *Corynebacterium glutamicum* from an aerobic respiring to an aerobic fermenting bacterium by inactivation of the respiratory chain. *Biochim. Biophys. Acta* **1827**, 699–708.

Kulis-Horn, R. K., Persicke, M., and Kalinowski, J. (2014) Histidine biosynthesis, its regulation and biotechnological application in *Corynebacterium glutamicum*. *Microb. Biotechnol.* **7**, 5–25.

Latendresse, M., Malerich, J. P., Travers, M., and Karp, P. D. (2012) Accurate atom-mapping computation for biochemical reactions. *J. Chem. Inf. Model.* **52**, 12970–12982.

Lewis, N. E., Nagarajan, H., and Palsson, B. Ø. (2012) Constraining the metabolic genotype–phenotype relationship using a phylogeny of *in silico* methods. *Nat. Rev. Microbiol.* **10**, 291–305.

McCloskey, D., Palsson, B. Ø., and Feist, A. M. (2013) Basic and applied uses of genome-scale metabolic network reconstructions of *Escherichia coli*. *Mol. Syst. Biol.* **9**, 661.

Marx, A., de Graaf, A. A., Wiechert, W., and Eggeling, L. (1996) Determination of the fluxes in the central metabolism of *Corynebacterium glutamicum* by nuclear magnetic resonance spectroscopy combined with metabolite balancing. *Biotechnol. Bioeng.* **49**, 111–129.

Mo, M. L., Palsson, B. Ø., and Herrgård, M. J. (2009) Connecting extracellular metabolomic measurements to intracellular flux states in yeast. *BMC Syst. Biol.* **3**, 37.

Mu, F., Williams, R. F., Unkefer, C. J., Unkefer, P. J., Faeder, J. R., and Hlavacek, W. S. (2007) Carbon-fate maps for metabolic reactions. *Bioinformatics* **23**, 3193–3199.

Nantapong, N., Otofuji, A., Migita, C. T., Adachi, O., Toyama, H., and Matsushita, K. (2005) Electron transfer ability from NADH to menaquinone and from NADPH to oxygen of type II NADH dehydrogenase of *Corynebacterium glutamicum*. *Biosci. Biotechnol. Biochem.* **69**, 149–159.

Noack, S., Nöh, K., Moch, M., Oldiges, M., and Wiechert, W. (2011) Stationary versus non-stationary ^{13}C-MFA: a comparison using a consistent dataset. *J. Biotechnol.* **154**, 179–190.

Nöh, K., and Wiechert, W. (2011) The benefits of being transient: isotope-based metabolic flux analysis at the short time scale. *Appl. Microbiol. Biotechnol.* **91**, 1247–1265.

Nöh, K., Grönke, K., Luo, B., Takors, R., Oldiges, M., and Wiechert, W. (2007) Metabolic flux analysis at ultra short time scale: isotopically non-stationary ^{13}C labeling experiments. *J. Biotechnol.* **129**, 249–267.

Nolan, R. P., Fenley, A. P., and Lee, K. (2006) Identification of distributed metabolic objectives in the hypermetabolic liver by flux and energy balance analysis. *Metab. Eng.* **8**, 30–45.

van Ooyen, J., Noack, S., Bott, M., Reth, A., and Eggeling, L. (2012) Improved L-lysine production with *Corynebacterium glutamicum* and systemic insight into citrate synthase flux and activity. *Biotechnol. Bioeng.* **109**, 2070–2081.

Orth, J. D., Thiele, I., and Palsson, B. Ø. (2010) What is flux balance analysis? *Nat. Biotechnol.* **28**, 245–248.

Orth, J. D., Conrad, T. M., Na, J., Lerman, J. A., Nam, H., Feist, A. M., and Palsson, B. Ø. (2011) A comprehensive genome-scale reconstruction of *Escherichia coli* metabolism – 2011. *Mol. Syst. Biol.* **7**, 535.

Palsson, B. Ø. (2006) Systems Biology: Properties of Reconstructed Networks. Cambridge University Press, Cambridge, UK.

Petersen, S., de Graaf, A. A., Eggeling, L., Möllney, M., Wiechert, W., and Sahm, H. (2000) *In vivo* quantification of parallel and bidirectional fluxes in the anaplerosis of *Corynebacterium glutamicum*. *J. Biol. Chem.* **275**, 35932–35941.

Pitkänen, E., Akerlund, A., Rantanen, A., Jouhten, P., and Ukkonen, E. (2008) ReMatch: a web-based tool to construct, store and share stoichiometric metabolic models with carbon maps for metabolic flux analysis. *J. Integr. Bioinform.* **5**, doi: 10.2390/biecoll-jib-2008-102.

Price, N. D., Reed, J. L., and Palsson, B. Ø. (2004) Genome-scale models of microbial cells: evaluating the consequences of constraints. *Nat. Rev. Microbiol.* **2**, 886–897.

Ravikirthi, P., Suthers, P. F., and Maranas, C. D. (2011) Construction of an *E. coli* genome-scale atom mapping model for MFA calculations. *Biotechnol. Bioeng.* **108**, 1372–1382.

Reed, J. L. (2012) Shrinking the metabolic solution space using experimental datasets. *PLoS Comput. Biol.* **8**, e1002662.

Schellenberger, J., and Palsson, B. Ø. (2009) Use of randomized sampling for analysis of metabolic networks. *J. Biol. Chem.* **284**, 5457–5461.

Schellenberger, J., Que, R., Fleming, R. M. T., Thiele, I., Orth, J. D., Feist, A. M., Zielinski, D. C., Bordbar, A., Lewis, N. E., Rahmanian, S., *et al.* (2011) Quantitative prediction of cellular metabolism with constraint-based models: the COBRA Toolbox v2.0. *Nat. Protoc.* **6**, 1290–1307.

Schilling, C. H., Letscher, D., and Palsson, B. Ø. (2000) Theory for the systemic definition of metabolic pathways and their use in interpreting metabolic function from a pathway-oriented perspective. *J. Theor. Biol.* **203**, 229–248.

Schmidt, K., Marx, A., de Graaf, A. A., Wiechert, W., Sahm, H., Nielsen, J., and Villadsen, J. (1998) ^{13}C tracer experiments and metabolite balancing for metabolic flux analysis: comparing two approaches. *Biotechnol. Bioeng.* **58**, 254–257.

Schuetz, R., Kuepfer, L., and Sauer, U. (2007) Systematic evaluation of objective functions for predicting intracellular fluxes in *Escherichia coli*. *Mol. Syst. Biol.* **3**, 119.

Schuetz, R., Zamboni, N., Zampieri, M., Heinemann, M., and Sauer, U. (2012) Multidimensional optimality of microbial metabolism. *Science* **336**, 601–604.

Schuster, S., Pfeiffer, T., and Fell, D. A. (2008) Is maximization of molar yield in metabolic networks favoured by evolution? *J. Theor. Biol.* **252**, 497–504.

Shinfuku, Y., Sorpitiporn, N., Sono, M., Furusawa, C., Hirasawa, T., and Shimizu, H. (2009) Development and experimental verification of a genome-scale metabolic model for *Corynebacterium glutamicum*. *Microb. Cell Fact.* **8**, 43.

Sorgenfrei, O., Klopprogge, K., Heinzle, E., and Wittmann, C. (2004) In-depth profiling of lysine-producing *Corynebacterium glutamicum* by combined analysis of the transcriptome, metabolome, and fluxome. *J. Bacteriol.* **186**, 1769–1784.

Stepansky, A., and Leustek, T. (2006) Histidine biosynthesis in plants. *Amino Acids* **30**, 127–142.

Suthers, P. F., Burgard, A. P., Dasika, M. S., Nowroozi, F., Van Dien, S., Keasling, J. D., and Maranas, C. D. (2007) Metabolic flux elucidation for large-scale models using ^{13}C labeled isotopes. *Metab. Eng.* **9**, 387–405.

Thiele, I., and Palsson, B. Ø. (2010) A protocol for generating a high-quality genome-scale metabolic reconstruction. *Nat. Protoc.* **5**, 93–121.

Thiele, I., Swainston, N., Fleming, R. M. T., Hoppe, A., Sahoo, S., Aurich, M. K., Haraldsdottir, H., Mo, M. L., Rolfsson, O., Stobbe, M. D., *et al.* (2013) A community-driven global reconstruction of human metabolism. *Nat. Biotechnol.* **31**, 419–425.

Vallino, J. J., and Stephanopoulos, G. (1993) Metabolic flux distributions in *Corynebacterium glutamicum* during growth and lysine overproduction. *Biotechnol. Bioeng.* **41**, 633–646.

Weitzel, M., Nöh, K., Dalman, T., Niedenführ, S., Stute, B., and Wiechert, W. (2013) 13CFLUX2-high-performance software suite for ^{13}C-metabolic flux analysis. *Bioinformatics* **29**, 143–145.

Wendisch, V. F., Bott, M., Kalinowski, J., Oldiges, M., and Wiechert, W. (2006) Emerging *Corynebacterium glutamicum* systems biology. *J. Biotechnol.* **124**, 74–92.

Wiback, S. J., Famili, I., Greenberg, H. J., and Palsson, B. Ø. (2004) Monte Carlo sampling can be used to determine the size and shape of the steady-state flux space. *J. Theor. Biol.* **228**, 437–447.

Wiechert, W. (2001) ^{13}C metabolic flux analysis. *Metab. Eng.* **3**, 195–206.

Wiechert, W., and Noack, S. (2011) Mechanistic pathway modeling for industrial biotechnology: challenging but worthwhile. *Curr. Opin. Biotechnol.* **22**, 604–610.

Wiechert, W., and Nöh, K. (2013) Isotopically non-stationary metabolic flux analysis: complex yet highly informative. *Curr. Opin. Biotechnol.* **24**, 979–986.

Wittmann, C., and Heinzle, E. (2001) Application of MALDI-TOF MS to lysine-producing *Corynebacterium glutamicum*: a novel approach for metabolic flux analysis. *Eur. J. Biochem.* **268**, 2441–2455.

Wittmann, C., and Heinzle, E. (2002) Genealogy profiling through strain improvement by using metabolic network analysis: metabolic flux genealogy of several generations of lysine-producing corynebacteria. *Appl. Environ. Microbiol.* **68**, 5843–5859.

Woo, H. M., Noack, S., Seibold, G. M., Willbold, S., Eikmanns, B. J., and Bott, M. (2010) Link between phosphate starvation and glycogen metabolism in *Corynebacterium glutamicum*, revealed by metabolomics. *Appl. Environ. Microbiol.* **76**, 6910–6919.

Yasuda, K., Jojima, T., Suda, M., Okino, S., Inui, M., and Yukawa, H. (2007) Analyses of the acetate-producing pathways in *Corynebacterium glutamicum* under oxygen-deprived conditions. *Appl. Microbiol. Biotechnol.* **77**, 853–860.

Zamboni, N. (2011) ^{13}C metabolic flux analysis in complex systems. *Curr. Opin. Biotechnol.* **22**, 103–108.

Zhou, W., and Nakhleh, L. (2012) Quantifying and assessing the effect of chemical symmetry in metabolic pathways. *J. Chem. Inf. Model.* **52**, 2684–2696.

Metabolic Engineering of *Corynebacterium glutamicum* for Alternative Carbon Source Utilization

5

Lennart Leßmeier, Ahmed Zahoor, Steffen N. Lindner and Volker F. Wendisch

Abstract

Corynebacterium glutamicum is well known for the multi-million-ton-scale production of L-glutamate and L-lysine in biotechnological industry, which is primarily based on hexoses derived from starch and molasses. Only recently, metabolic engineering approaches were used to broaden the substrate spectrum of *C. glutamicum*, allowing access to alternative carbon sources such as glycerol, galactose, and pentose sugars present in lignocellulosic hydrolysates.

Unlike most microorganisms, *C. glutamicum* is able to co-utilize mixed carbon sources, making it a preferential host for utilization of complex carbon sources such as wastes from agricultural industry. Several industrial by-products have been successfully valorized by engineered *C. glutamicum* strains. Fermentative production of amino acids has been shown based on crude glycerol from biodiesel production, silage juice and rice straw hydrolysates.

In this review, we will present a glimpse into the state of the art of metabolic engineering of *C. glutamicum* towards utilization of alternative carbon sources, such as wastes and unprocessed sources.

Introduction

Industrial biotechnology is paving the way for a bio-based economy, which is proposed to complement and, in the long term, replace petro-based economy. The central parts of this technology are microorganisms, which carry out the production reactions such as fermentation, enzymatic reactions or whole-cell biocatalysis. Microbial processes have been used since many thousands years, but just the recombinant DNA technology and more recently the application of 'omics' technologies allowed scientists to rationally engineer these microbial factories for their purposes (Zahoor *et al.*, 2012).

Engineered microorganisms are widely used in large-scale biotechnological manufacturing of food and industrial or pharmaceutical products of immense economic importance. Most of these production processes, however, currently use carbon sources derived from starch and molasses, which are relatively expensive and have competing applications in food industry (Rumbold *et al.*, 2010).

A desirable advance of metabolic engineering aims towards broadening the spectrum of substrates usable by the microorganisms, developing production strains able to utilize or co-utilize wastes from agro- and food-industry as alternative carbon sources. Thus, environmental sustainability and economic efficiency of biotechnological processes could be improved. Increasingly, lignocellulosic hydrolysates are used as a feedstock for industrial fermentations. These biomass hydrolysates consist of complex mixtures of different fermentable sugars, but also contain inhibitors and salts that affect performance of the product-generating microorganisms (Matano *et al.*, 2013).

C. glutamicum is a rod-shaped, Gram-positive soil bacterium, isolated in Japan in the 1950s because of its ability to produce glutamate (Kinoshita *et al.*, 1957). This finding opened the way for thorough investigation and use of this organism in industrial biotechnology for more than 50 years. Today, this generally-regarded-as-save (GRAS) organism is used for annual production of 2,600,000 tons of L-glutamate (Ajinomoto, 2012). Another million-ton-scale product obtained with *C. glutamicum* is L-lysine (1,950,000 tons per year; Ajinomoto, 2013). After genetic engineering tools were developed, *C. glutamicum* genome was sequenced and published in 2003 (Ikeda and Nakagawa, 2003; Kalinowski *et al.*, 2003) and 'omics' technologies such as transcriptome studies were developed (Wendisch, 2003). These milestones marked the beginning of extensive research efforts to metabolically engineer *C. glutamicum*, initially for L-glutamate and L-lysine production and later on for the production of a variety of L-amino acids, e.g. L-proline (Jensen and Wendisch, 2013), and other products such as organic acids (e.g. glycolate; Zahoor *et al.*, 2014), alcohols, diamines or carotenoids (Heider *et al.*, 2012, 2014). Moreover, biotin prototrophic *C. glutamicum* strains have been developed and used for production (Ikeda *et al.*, 2013; Peters-Wendisch *et al.*, 2014).

Similarly, use of alternative carbon sources, that unlike the traditional sugars sucrose, glucose and fructose do not have competing applications in food industry, became possible by metabolic engineering (Zahoor *et al.*, 2012). Interestingly, *C. glutamicum* can well resist various growth inhibitors such as furans, phenols and acids that are formed during lignocellulose pre-treatment under non-growth conditions, thereby opening up the prospect of production processes from agro-wastes (Zahoor *et al.*, 2012).

The following pages are intended to give an impression of the current research towards utilization of underexploited carbon sources for industrial biotechnological processes involving *C. glutamicum*.

Natural carbon sources

C. glutamicum is able to utilize a broad spectrum of carbon sources for its growth and energy supply. Among these are sugars such as the monosaccharides glucose, fructose, mannose and ribose or disaccharides such as sucrose and maltose (Blombach and Seibold, 2010). Additionally, alcohols such as ethanol (Arndt *et al.*, 2008) or inositol (Krings *et al.*, 2006) and organic acids such as L-lactic acid (Stansen *et al.*, 2005), D-lactic acid (Kato *et al.*, 2010), acetic acid (Gerstmeir *et al.*, 2003), propionic acid (Claes *et al.*, 2002), pyruvic acid (Cocaign *et al.*, 1993) and gluconic acid (Lee *et al.*, 1998) can be utilized by *C. glutamicum*. Furthermore, it utilizes some amino acids such as L-glutamate and L-glutamine (Arndt and Eikmanns, 2008) and sugar alcohols such as arabitol and mannitol (Laslo *et al.*, 2012).

Interestingly, *C. glutamicum* wild type can grow on different aromatic compounds, which can be found e.g. as constituents of lignocellulosic hydrolysates and affect the performance of most product-generating microorganisms (Rumbold *et al.*, 2010). Among these are benzoate, phenol (Shen *et al.*, 2004), 3-hydrobenzoate, gentisate (Shen *et al.*, 2005), protocatechuate, vanillate, 4-hydroxybenzoate, 4-cresol (Shen and Liu, 2005; Qi *et al.*, 2007), resorcinol (Huang *et al.*, 2006), benzyl alcohol, 2,4-dihydroxybenzoate, 3,5-dihydroxytoluene (Shen and Liu, 2005), naphthalene (Lee *et al.*, 2010), vanillin and ferulic acid (Merkens *et al.*, 2005).

Currently, industrial production processes using *C. glutamicum* rely on sugars derived from starch (glucose) or molasses (sucrose, glucose and fructose). Use of molasses dominates in European, South American and Asian production plants, whereas starch hydrolysates are favoured in North America (Kimura, 2005).

In *C. glutamicum* these sugars are substrates of the main sugar uptake system, the phosphotransferase system (PTS), which phosphorylates its substrate by utilizing phosphoenolpyruvate (Parche *et al.*, 2001). However, PTS-independent uptake of glucose has been discovered and characterized in *C. glutamicum* (Lindner *et al.*, 2011a,b, 2013).

Co-utilization of carbon sources

A typical characteristic of *C. glutamicum* is the co-utilization of carbon sources, when supplemented as mixtures (Blombach and Seibold, 2010). For example, this has been studied during monophasic growth of *C. glutamicum* on glucose mixed with other sugars such as fructose (Dominguez *et al.*, 1997) or organic acids such as lactate, pyruvate, propionate or acetate (Wendisch *et al.*, 2000; Claes *et al.*, 2002; Gerstmeir *et al.*, 2003). Diauxic growth and sequential utilization of carbon sources occurs quite rarely in this organism, e.g. glucose is preferentially utilized in mixtures of glucose with glutamate (Kronemeyer *et al.*, 1995) or ethanol (Arndt and Eikmanns, 2007).

Co-utilization is advantageous in biotechnological applications based on complex carbon source mixtures, e.g. molasses of hydrolysates and is an argument for *C. glutamicum* as a host, since other hosts such as *Escherichia coli* (Kremling *et al.*, 2009) or *Bacillus subtilis* (Hueck and Hillen, 1995) show carbon catabolite repression, sequential utilization of substrates, and diauxic growth. Moreover, co-utilization of carbon sources also takes place, when additional carbon utilization pathways are introduced into *C. glutamicum*. This has been shown for, for example, growth on glycerol with glucose (Rittmann *et al.*, 2008) and also for different mixtures of the lignocellulosic compounds xylose, arabinose and cellobiose with glucose (Kawaguchi *et al.*, 2006; Sasaki *et al.*, 2009; Gopinath *et al.*, 2011). Blends of glucose and non-growth substrates such as serine (Netzer *et al.*, 2004) and, more recently, C1 compounds have been analysed to some extent (Witthoff *et al.*, 2012, 2013; Lessmeier *et al.*, 2013).

Optimization of endogenous substrate utilization pathways

Glucose

Glucose and other sugars are mainly utilized in *C. glutamicum* via the phosphotransferase system (PTS), which phosphorylates its substrates by utilizing phosphoenolpyruvate concomitant with their update into the cell (Parche *et al.*, 2001). Growth on glucose as carbon source could be accelerated by constitutive expression of the glucose-specific EII subunit gene *ptsG* (cg1537; Krause *et al.*, 2010).

Furthermore, *C. glutamicum* possesses a PTS-independent glucose utilization pathway, in which glucose is imported by IolT1/2 and then phosphorylated using ATP and/or polyphosphate-dependent glucokinases (Park *et al.*, 2000; Lindner *et al.*, 2010, 2011a,b). Increased lysine production was observed when glucose was only utilized via this system, probably because of a higher availability of the lysine precursor phosphoenolpyruvate (Ikeda *et al.*, 2011; Lindner *et al.*, 2011a). This alternative pathway might be advantageous for many production processes (Lindner *et al.*, 2011b).

Lactic acid

Lactic acid is the predominant by-product of *C. glutamicum* during growth under oxygen-deprived conditions. But, taking its high abundance in silages into account, it is also an interesting and potential carbon source. *C. glutamicum* possesses two lactate dehydrogenases for oxidization of lactic acids to pyruvate, one specific for L- and one for D-lactic acid (Stansen *et al.*, 2005; Kato *et al.*, 2010). The quinone-dependent L-lactate dehydrogenase gene (*lldD*; cg3227) is sufficiently expressed for fast growth with L-lactic acid, but the D-lactate dehydrogenase gene (*dld*; cg1027) needs to be overexpressed for efficient D-lactic acid utilization (Stansen *et al.*, 2005; Kato *et al.*, 2010; Neuner and Heinzle, 2011).

Lysine production from lactic acid could be obtained by overexpression of either malic enzyme or pyruvate carboxylase (pyc; cg0791) genes (Neuner and Heinzle, 2011). Recent studies showed that overexpression of the lactic acid forming NAD-dependent lactate dehydrogenase gene (*ldhA*; cg3219) rescued an *lldD* knockout strain (Sharkey *et al.*, 2011). This indicates that *ldhA* overexpression might prove beneficial to accelerate lactate catabolism.

Engineering of *Corynebacterium glutamicum* for alternative carbon sources

Several cheaply available and therefore interesting carbon sources cannot be utilized by *C. glutamicum*, either because the whole or part of the necessary genetic set-up is absent or because expression of the corresponding genes is not sufficient. Thus, construction of recombinant strains for utilization of new carbon sources is an important issue of metabolic engineering of this bacterium. An overview of genes used in these engineering approaches is provided in Table 5.1.

Sugars

Sugars such a glucose, fructose or sucrose are the preferred carbon sources in most fermentative

Table 5.1 Metabolic engineering approaches. The corresponding genes and their origin used in various metabolic engineering approaches to allow access to the indicated carbon sources are shown in rows

Substrate	Genes	Origin	Reference
Xylose	*xylA*	*Escherichia coli*	Kawaguchi *et al.* (2006), Sasaki *et al.*, 2009)
		Xanthomonas campestris	Meiswinkel *et al.* (2013a)
	xylB	*E. coli*	Kawaguchi *et al.* (2006), Sasaki *et al.* (2009)
		Corynebacterium glutamicum (cg0147)	Meiswinkel *et al.* (2013a)
	araE	*C. glutamicum* ATCC 31831	Sasaki *et al.* (2009)
Arabinose	*araB*, *araA*, *araD*	*E. coli*	Kawaguchi *et al.* (2008), Sasaki *et al.* (2009)
	araE	*C. glutamicum* ATCC 31831	Sasaki *et al.* (2009)
Lactose	*lacY*, *lacZ*	*E. coli*	Brabetz *et al.* (1991)
		Lactobacillus delbrueckii subsp. *bulgaricus* ATCC 11842	Barrett *et al.* (2004)
Galactose	*galE*, *galK*, *galM*, *galT*	*Lactococcus lactis* subsp. *cremoris* MG1363	Barrett *et al.* (2004)
Starch	*amyA*	*Streptomyces griseus*	Seibold *et al.* (2006)
		Streptococcus bovis	Tateno *et al.* (2007a)
	pgsA	*Bacillus subtilis*	Tateno *et al.* (2007a)
Cellulose	*celE*	*Clostridium cellulovorans*	Hyeon *et al.* (2011)
Cellobiose	*bglA* (cgR 2728)	*C. glutamicum R*	Sasaki *et al.* (2008)
	bglF (cgR 2729)	*C. glutamicum R*	Sasaki *et al.* (2008)
Glucosamine	*nagB* (cg2928)	*C. glutamicum*	Uhde *et al.* (2013)
	ptsG (cg1537)	*C. glutamicum*	Uhde *et al.* (2013)
Dicarboxylic acids	*dccA* (cg2870)	*C. glutamicum*	Youn *et al.* (2008, 2009)
	dccT (cg0277)	*C. glutamicum*	Youn *et al.* (2008, 2009)
Glycerol	*glpF*	*E. coli*	Rittmann *et al.* (2008)
	glpK, *glpD* (cg3198, cg1853)	*E. coli*	Rittmann *et al.* (2008)
		C. glutamicum	Meiswinkel *et al.* (2013b)

production processes. But the most abundant sources of monomeric sugars are complex polymers such as cellulose, hemicellulose or starch. These polymers cannot be directly utilized by most industrially relevant microorganisms such as *C. glutamicum* and costly pre-treatments are necessary, releasing accessible monomers. Among these monomers are, however, not only hexoses such as glucose and fructose, but also pentoses, which are not utilized by *C. glutamicum*. In order to reduce the costs of fermentations and broaden the substrate spectrum of *C. glutamicum*, e.g. towards pentoses, several metabolic engineering approaches have been carried out (Blombach and Seibold, 2010).

Xylose

Xylose, an aldopentose monosaccharide, is one of the most abundant sugars in lignocellulosic hydrolysates derived from agricultural or forestry wastes. Xylose is a main component of hemicelluloses and constitutes about 5–20% of the total dry mass of lignocellulosic biomass. Therefore, its consumption plays a key role in the efficient utilization of lignocellulosics (Aristidou and Penttila, 2000).

In bacteria, xylose is typically utilized via the pentose phosphate pathway after uptake, isomerization to xylulose and phosphorylation to xylulose-5-phosphate. *C. glutamicum* is not able to utilize xylose naturally, since its genome does not encode for the xylose isomerase XylA. But it possesses a gene (*xylB*; cg0147) encoding a functional xylulokinase enzyme responsible for the second reaction in the xylose utilization (Kawaguchi *et al.*, 2006). Heterologous expression of *xylA* from *E. coli* was sufficient to enable growth with xylose as sole carbon source, while growth of this recombinant *C. glutamicum* strain could be accelerated by additional expression of *E. coli xylB* (Kawaguchi *et al.*, 2006).

Further optimization experiments revealed that combining heterologous expression of *xylA* from *Xanthomonas campestris* with overexpression of the endogenous *xylB* gene allowed an even faster growth on xylose (Meiswinkel *et al.*, 2013a).

Uptake of xylose is slow at low concentrations and the xylose uptake system in *C. glutamicum* strains R and ATCC 13032 is not yet known. However, inspection of the genome reveals three candidate genes encoding hypothetical pentose sugar transporters. A high-affinity arabinose transporter (AraE), which probably was acquired via horizontal gene transfer, could be identified in the *C. glutamicum* strain ATCC 31831 (Kawaguchi *et al.*, 2009). When optimizing arabinose utilization, this transporter was also expressed in a recombinant xylose-utilizing *C. glutamicum* strain. This resulted in an increased xylose consumption rate and a 2.9-fold accelerated growth rate at low-xylose concentrations (Sasaki *et al.*, 2009). Xylose was co-utilized with glucose in all recombinant strains. However, some kind of repression or inhibition of the xylose transport was observed in presence of glucose (Sasaki *et al.*, 2009).

The biotechnological usability of xylose was demonstrated by anaerobic production of succinic and lactic acid (Kawaguchi *et al.*, 2006), as well as aerobic production of cadaverine (Buschke *et al.*, 2011), putrescine (Meiswinkel *et al.*, 2013a) and the amino acids L-glutamate and L-lysine (Gopinath *et al.*, 2011; Meiswinkel *et al.*, 2013a).

Arabinose

Like xylose, the aldopentose L-arabinose occurs in lignocellulosic hydrolysates. As component of hemicellulose, it constitutes a significant amount of the pentose sugars and about 1–5% of the dry mass in lignocellulosic biomasses (Aristidou and Penttila, 2000).

In bacteria, arabinose is taken up into the cell and typically converted via three enzymatic reactions before it enters the pentose phosphate pathway as xylulose-5-phosphate. In the first reaction, arabinose isomerase (AraA) converts arabinose to ribulose, which is then phosphorylated to ribulose-5-phosphate by ribulose kinase (AraB), which is finally converted to xylulose-5-phosphate by ribulose-5-phosphate 4-epimerase (AraD; Aristidou and Penttila, 2000).

C. glutamicum strains R and ATCC 13032 are not able to utilize arabinose naturally. But a cluster of arabinose utilization genes has been identified in strain ATCC 31831, which probably was obtained via horizontal gene transfer, and this strain can utilize arabinose as sole carbon source (Kawaguchi *et al.*, 2009).

Arabinose utilization in *C. glutamicum* strain R and ATCC 13032 was enabled via heterologous expression of the *araBAD* genes from *E. coli*. These strains were able to utilize arabinose as sole carbon source and also together with glucose (Kawaguchi *et al.*, 2008; Schneider *et al.*, 2011).

The ability of the recombinant *C. glutamicum* strains to grow on arabinose indicated the presence of a currently unknown endogenous arabinose transporter. As this strain was not able to grow with low arabinose concentrations, this transporter seemed to have a low affinity to arabinose. Additionally, the arabinose consumption was reduced in the presence of glucose (Kawaguchi *et al.*, 2008). For these reasons, additionally the high affinity arabinose transporter AraE from *C. glutamicum* ATCC 31831 (Kawaguchi *et al.*, 2009) was expressed in this strain (Sasaki *et al.*, 2009). The resulting strain was able to grow on low arabinose concentrations and the arabinose consumption rate under oxygen-deprived conditions was increased 2.8-fold (Sasaki *et al.*, 2009).

Arabinose-based production by recombinant *C. glutamicum* strains was established for organic acids (Sasaki *et al.*, 2009), the amino acids L-glutamate, L-lysine, L-ornithine and L-arginine (Schneider *et al.*, 2011) and the diamine putrescine (Meiswinkel *et al.*, 2013a).

Lactose and galactose

Cheese whey, a major by-product of dairy industries, can consist of up to 50 g/l lactose. With lactose providing the carbon source, it could be used as an abundant and renewable raw material for microbial fermentations, although often it is too dilute.

Whey is generated in high amounts and is an environmental problem since it is highly polluting. It is also necessary to find possibilities of adding value to this by-product. One of the major products obtained from whey is the production of whey protein concentrate (WPC). However, lactose-rich waste is still generated in the production of WPC. Although lactose is a very abundant by-product, most of the industrially relevant microorganisms lack the capacity for a lactose metabolism (Domingues *et al.*, 2010).

The *C. glutamicum* wild-type strain is not able to metabolize lactose or galactose. Growth on lactose has been enabled by heterologous expression of the *E. coli lac* genes (Brabetz *et al.*, 1991). Despite the completely different membrane composition, the lactose permease LacY from *E. coli* was shown to be active in the Gram-positive host (Brabetz *et al.*, 1993). Since *C. glutamicum* lacks enzymes for galactose catabolism, only glucose released by cleavage of lactose was metabolized in this approach but not galactose (Brabetz *et al.*, 1991).

Utilization of both glucose and galactose components of lactose has been allowed by additional expression of genes encoding for a galactose utilization pathway. In this approach, heterologous expression of genes encoding the lactose permease and β-galactosidase from *Lactobacillus delbrueckii* subsp. *bulgaricus* ATCC 11842 has been combined with expression of *Lactococcus lactis* subsp. *cremoris* MG1363 genes encoding for galactokinase (*galK*), UDP-glucose-1-P-uridylyltransferase (*galT*), UDP-galactose-4-epimerase (*galE*) and aldose-1-epimerase (*galM*). This combination of genes allowed *C. glutamicum* to utilize lactose completely. Whey-based L-lysine production was possible, although the growth in the whey-based medium was slower than that with glucose (Barrett *et al.*, 2004).

Starch

In industrial amino acid production, starch is hydrolysed before fermentation to gain access to monomeric glucose. Since cost of substrates have a high influence in the amino acid production costs, efficient use of the complex raw material is of high interest (Kimura, 2005). One possibility to reduce the costs in starch-based fermentations would be the direct use of soluble starch instead of hydrolysates.

In first engineering approaches towards direct utilization of starch, the α-amylase gene from *Streptomyces griseus* was expressed heterologous in *C. glutamicum* and the enzyme was excreted into the medium (Seibold *et al.*, 2006). This recombinant strain was able to grow with starch even faster than with glucose. However, growth on starch was only possible when small amounts of glucose were present. Additionally, some starch degradation products still remained in the medium (Seibold *et al.*, 2006).

In a different engineering approach, a surface displayed fusion protein was expressed, consisting of the α-amylase gene from *Streptococcus bovis* with *pgsA* from *B. subtilis* as an anchor protein. This strain was able to utilize starch as sole source of carbon (Tateno *et al.*, 2007a).

Starch-based production of L-glutamate (Yao *et al.*, 2009), L-lysine (Seibold *et al.*, 2006; Tateno *et al.*, 2007a,b), polyhydroxybutyrate (Song *et al.*, 2013) and cadaverine (Tateno *et al.*, 2009) has been reported and in the case of L-lysine, higher L-lysine production yields were reached with starch than with glucose (Tateno *et al.*, 2007a).

Cellulose and cellobiose

Cellulose, a major component of the cell wall of plants, is the most abundant renewable carbohydrate. Direct utilization of cellulosic materials by *C. glutamicum* wild type has not yet been reported. Thus, prior to fermentation, the cellulose polymers are usually hydrolysed to release monomeric glucose. An incomplete chemical depolymerization or enzymatic degradation of cellulose by endoglucanases and exoglucanases results mainly in cellobiose. These β-1,4 linked glucose dimers can be cleaved to glucose by β-glucosidase (Schwarz, 2001).

Since the production costs of these cellulolytic enzymes are very high, strategies reducing the amount of enzyme used such as heterologous gene expression in production hosts need to be developed.

Heterologous expression of cellulases in *C. glutamicum* has been reported previously (Paradis *et al.*, 1987; Adham *et al.*, 2001). And the efficiency of cellulose degradation could be recently increased in *C. glutamicum* ATCC 13032 by heterologous expression of a synthetic microcellulosome based on genes from *Clostridium cellulovorans*, including an endocellulase E gene (Hyeon *et al.*, 2011). However, cellulose degradation alone does not allow its utilization in *C. glutamicum*, because this organism is unable to utilize the disaccharide cellobiose. Therefore, β-glucan-based glutamate production could only be achieved when glucosidase was added (Tsuchidate *et al.*, 2011).

Utilization of the disaccharide cellobiose occurred in *C. glutamicum* R as a result of mutations in the phosphotransferase permease subunit gene *bglF* (cgR 2729), which is specific for the transport of methyl β-glucoside, arbutin and salicin. The mutations resulted in amino acid exchanges from valine at position 317 to methionine or alanine (Kotrba *et al.*, 2003). For the utilization of cellobiose also phospho-β-glucosidase is needed (*bglA*; cgR 2728). Both genes *bglF* and *bglA* are present in *C. glutamicum* R, but are absent from other strains. Mixtures of cellobiose, glucose and xylose could be used for the production of lactic and succinic acid under oxygen-deprivation conditions (Sasaki *et al.*, 2008, 2009).

Glucosamine and *N*-acetyl glucosamine

Amino sugars have great potential in biotechnological industry because they could serve both as carbon and nitrogen source for fermentation processes. *N*-acetyl-glucosamine (GlcNAc) is part of the bacterial cell wall and of chitin, the second most abundant natural polymer on earth. GlcNAc occurs in nature together with its deacetylated form, glucosamine (Stevenson, 1983; Appuhn *et al.*, 2004).

The *C. glutamicum* genome encodes the cytoplasmic *N*-acetylglucosamine-6-phosphate deacetylase NagA1 (cg2929), which is responsible for the first step of GlcNAc utilization. Since a transporter for GlcNAc is not encoded, *C. glutamicum* cannot use GlcNAc as substrate for growth (Uhde *et al.*, 2013).

The deacetylated amino sugar glucosamine, which is also very abundant in nature, can be utilized by *C. glutamicum* wild type, but not as efficiently as other carbon sources. Uptake of glucosamine is mediated via the glucose specific permease PtsG (cg1537) (Uhde *et al.*, 2013). The bottleneck of glucosamine utilization originates in low expression of the gene responsible for glucosamine deamination. Overproduction of the endogenous glucosamine-6-phosphate deaminase NagB gene (cg2928) allowed fast growth on glucosamine as sole source of carbon and nitrogen (Uhde *et al.*, 2013).

Various *C. glutamicum* production strains overexpressing endogenous *nagB* were shown to utilize glucosamine as substrate for efficient production of the amino acid L-lysine and of the diamine putrescine (Uhde *et al.*, 2013).

Dicarboxylic acids

The dicarboxylic acids succinic acid, fumaric acid and malic acid are intermediates of the tricarboxylic acid cycle and are utilized by many bacteria as carbon and/or energy source (Janausch *et al.*, 2002). Fast utilization of these substrates was observed only in spontaneous mutants of *C. glutamicum* wild type, which showed increased expression levels of genes encoding the proton- or sodium-dependent transporters DctA (cg2870) and DccT (cg0277), respectively. The increased expression was caused by mutations in the promoter regions of the corresponding genes *dctA* or *dccT* (Youn *et al.*, 2008, 2009).

Overexpression of these genes resulted in fast growth with succinic, fumaric and malic acid (Teramoto *et al.*, 2008; Youn *et al.*, 2008, 2009).

Glycerol

Glycerol is a versatile source of carbon and energy for growth of several microorganisms and is already used for the biotechnological production of more than 20 different chemicals in fermentative processes (Khanna *et al.*, 2012).

C. glutamicum cannot utilize glycerol naturally, while glycerol is a good carbon source for mycobacteria (Rittmann *et al.*, 2008). *C. glutamicum* could be engineered for glycerol utilization by heterologous expression of the *E. coli* genes for aerobic glycerol metabolism. Plasmid borne expression of genes for glycerol kinase (*glpK*) and glycerol-3-phosphate dehydrogenase (*glpD*) was shown to be sufficient for enabling growth on glycerol as sole carbon source. Additional expression of the gene for a transport facilitating aquaglyceroporin (*glpF*) was advantageous for the growth performance. By expression of these genes, a growth rate of 0.28/h and a biomass concentration of 4.25 g (DW) per litre could be obtained for growth on 110 mM glycerol. Furthermore, this strain was able to co-utilize glycerol with glucose (Rittmann *et al.*, 2008).

The *C. glutamicum* genome encodes endogenous homologues of *glpK* (cg3198) and *glpD* (cg1853) and ectopic expression of these genes also enabled growth on glycerol. However, the growth rate was significantly lower than expression of *glpFKD* from *E. coli* (Meiswinkel *et al.*, 2013b). *C. glutamicum* also encodes glycerol 3-phosphatase and deletion of its gene (cg1700) has been shown to abrogate glycerol production by *C. glutamicum* (Lindner *et al.*, 2012).

In comparison to glucose, similar L-glutamate and L-lysine production yields and rates were reached using glycerol as carbon source (Rittmann *et al.*, 2008). Recently, also efficient production of succinic acid (Litsanov *et al.*, 2013) from glycerol could be achieved. In addition, crude glycerol from biodiesel production could be utilized as carbon source for production of several amino acids and the diamine putrescine (Meiswinkel *et al.*, 2013b).

Complex carbon sources

Lignocellulosics

Lignocellulosic biomass from agricultural wastes is currently not as strongly utilized as possible. A proper recycling of remaining wastes is not always possible and their disposal can even be an ecological problem, e.g. when remains such as straw are burned, contributing to air pollution and causing health problems (Mussatto and Roberto, 2004).

Agro-wastes are available at large-scale level and represent viable candidates as alternative carbon sources. Therefore, utilization of lignocellulosic biomass for large-scale biotechnological production of value-added products such as amino acids or fine chemicals is promising.

Lignocellulose is composed mainly of cellulose (40–50%), hemicellulose (25–30%) and lignin (10–20%) (Wyman, 1999). These complex polymers cannot be directly utilized by industrially relevant microorganisms. Thus, lignocellulosic biomass is transformed into nutrients, mainly fermentable sugars, e.g. by acid or enzymatic hydrolysis (Rumbold *et al.*, 2010). Since these treatments are expensive, cheaper strategies such as heterologous expression of genes encoding for depolymerization enzymes are needed to be able to compete with established carbon sources such as molasses and starch hydrolysates.

The hemicellulose fraction is a heteropolymer of several hexose (glucose, galactose and mannose) and pentose (xylose and arabinose) subunits. Thus, lignocellulosic hydrolysates not only contain glucose as major component, but also a large fraction of xylose (5–20%) and arabinose (1–5%) (Aristidou and Penttila, 2000). Since *C. glutamicum* is not able to utilize these pentose sugars naturally, lignocellulose hydrolysates are not utilized completely.

Utilization of arabinose (Kawaguchi *et al.*, 2008), xylose (Kawaguchi *et al.*, 2006) and also cellobiose (Kotrba *et al.*, 2003), the dimeric component of cellulose, could be obtained by metabolic engineering. Co-utilization of these three substrates and glucose was also possible (Sasaki *et al.*, 2008).

Lignocellulose hydrolysates often present a challenge as they contain inhibitors and salts, which affect the performance of most product-generating microorganisms (Rumbold *et al.*, 2010). These inhibitors can be divided into three categories: furan derivatives such as 5-hydroxymethyl-2-furaldehyde, a big variety of phenolics and weak acids such as acetic, formic or laevulinic acid (Dunlop, 1948; Palmqvist and Hahn-Hagerdal, 2000; Almeida *et al.*, 2007). Therefore, a robust biotechnical system is very desirable, which can at least tolerate or even better utilize these inhibitors. For example, *C. glutamicum* can utilize acetic acid, which acts as an growth inhibitor to many microorganisms, as a carbon source (Gerstmeir *et al.*, 2003).

Phenolic compounds such as vanillin, vanillate, ferulic acid, benzoate or phenol are derived by hydrolysis of lignin and represent the main growth inhibitors in lignocellulosic hydrolysates (Klinke *et al.*, 2004; Parawira and Tekere, 2011). In *C. glutamicum*, acid production under oxygen deprived conditions is not affected by these substances, displaying a high tolerance to several phenolic inhibitors (Sakai *et al.*, 2007). In addition, *C. glutamicum* is even able to utilize some phenolic components of lignocellulosic hydrolysates such as benzoate, phenol (Shen and Liu, 2005), vanillin and ferulic acid (Merkens *et al.*, 2005), making it very interesting for bioprocesses based on agro-waste hydrolysates, such as rice straw.

Utilization of lignocellulosic biomass by *C. glutamicum* has recently been demonstrated by production of the amino acids glutamate and lysine, using rice straw or wheat bran hydrolysates as carbon source (Gopinath *et al.*, 2011, 2012; Meiswinkel *et al.*, 2013a).

Silage juice

Ensiling is a process of preserving wet plant material by applying anaerobic conditions, either in storage silos or wrapped in plastic. In this process, the green plant material is broken up mechanically and compression ensures anaerobic conditions. The anaerobic conditions inhibit remaining plant enzymes and aerobic microorganisms (McDonald *et al.*, 1991). The main compounds in sugar beet and corn silages are sucrose and starch, respectively, whereas grass silage does not have a major component (Neuner *et al.*, 2013). Owing to the conversion of most of the water soluble carbohydrates into lactic acid by lactic acid bacteria, the pH decreases, which preserves the silage from contamination by spoiling coliform bacteria and clostridia (McDonald *et al.*, 1991). Lactic acid itself is a valuable precursor for the production of biodegradable polymers (Sodergard and Stolt, 2002; Song *et al.*, 2012), but its extraction from silage is very expensive (Datta *et al.*, 1995).

Silage juice could be valorized without this cost-intensive recovery when used directly as a carbon source in fermentations. This has been demonstrated recently by L-lysine production by recombinant *C. glutamicum* based on grass and corn silages (Neuner *et al.*, 2013). *C. glutamicum* possesses lactate dehydrogenases encoded by *ldhA* (Sharkey *et al.*, 2011), *lldD* (Stansen *et al.*, 2005) and *dld* (Kato *et al.*, 2010). The recombinant strains used for valorizing grass juice overexpressed genes for D-lactate dehydrogenase (*dld*; Kato *et al.*, 2010), pyruvate carboxylase (*pyc*; Peters-Wendisch *et al.*, 1998), malic enzyme (*malE*; Gourdon *et al.*, 2000), fructose 1,6-bisphosphatase (*fbp*; Rittmann *et al.*, 2003) and glyceraldehyde 3-phosphate dehydrogenase (*gapX*) for efficient silage utilization (Neuner *et al.*, 2013).

Crude glycerol

Glycerol is a good source of carbon and energy for growth of several microorganisms and is suitable for the biotechnological production of a number of chemicals in fermentative processes. Previously, glycerol has been also an interesting product in biotechnology. But with arising biodiesel industry it became more and more a waste product. In the biodiesel production process, glycerol remains as a major by-product of the transesterification of oils and fats, with a stoichiometric yield of about 10% (w/w) with respect to biodiesel produced (Wendisch *et al.*, 2011). Crude glycerol formed in this process plays the main part of the annual production of about 1.78 million tons and the trend for biodiesel production and, thus, for glycerol production is growing (Khanna *et al.*, 2012). According the biorefinery concept, crude glycerol and other by-products of the biodiesel process, e.g. residual plant material, need to be valorized. Crude glycerol based production of amino acids and the diamine putrescine with *C. glutamicum* has been shown recently (Meiswinkel *et al.*, 2013b).

Chitin

Chitin, a polymer of *N*-acetyl glucosamine and glucosamine units, is the main component of cell walls of fungi, the radulas of molluscs, the beaks of cephalopods and the exoskeleton of arthropods and insects. The chitin fractions of arthropods contribute up to 12% to the organic nitrogen in soil (Parsons, 1981; Stevenson, 1982; Appuhn *et al.*, 2004). Although chitin is the second most abundant natural polysaccharide on earth, with a global annual turnover of 100 billion tons, it has still to find utilization in large quantities (Kurita, 2006). The glucosamine fraction of shellfish waste could represent an example of an alternative, renewable source for industrial fermentations

(Tharanathan and Kittur, 2003; Kurita, 2006) and is accessible by recombinant *C. glutamicum* strains (Uhde *et al.*, 2012). Ultimately, direct access to chitin is desired to avoid costs for chemical and/or enzymatic deacetylation and breakdown to monosaccharides.

Summary and outlook

Recently, the substrate spectrum for industrial microbial fermentations has been broadened, as shown here for *C. glutamicum*. In the future, 'omics' technologies might help improving production processes based on these alternative carbon sources. Initial successes in amino acid production based on alternative carbon sources by engineered *C. glutamicum* strains, such as L-lysine or putrescine production from rice straw hydrolysate or crude glycerol has been achieved. The current lab scale processes require future research for transfer to technical scale and process integration. The current focus on access to monomeric alternative carbon sources needs to be overcome to allow direct utilization of complex polymers such as starch, chitin or β-glucan.

References

Adham, S. A., Honrubia, P., Diaz, M., Fernandez-Abalos, J. M., Santamaria, R. I., and Gil, J. A. (2001) Expression of the genes coding for the xylanase Xys1 and the cellulase Cel1 from the straw-decomposing *Streptomyces halstedii* JM8 cloned into the amino-acid producer *Brevibacterium lactofermentum* ATCC13869. *Arch. Microbiol.* **177**, 91–97.

Ajinomoto Co., Inc. (2012) Food products business. Available at: http://www.ajinomoto.com/ir/pdf/Food-Oct2010.pdf (accessed 29 August 2012).

Ajinomoto Co., Inc. (2013) Feed-use amino acids business. Available at: http://www.ajinomoto.com/ir/pdf/Feed-useAA-Oct2011.pdf (accessed 29 August 2012).

Almeida, J. R. M., Modig, T., Petersson, A., Hahn-Hagerdal, B., Liden, G., and Gorwa-Grauslund, M. F. (2007) Increased tolerance and conversion of inhibitors in lignocellulosic hydrolysates by *Saccharomyces cerevisiae*. *J. Chem. Technol. Biotechnol.* **82**, 340–349.

Appuhn, A., Joergensen, R. G., Raubuch, M., Scheller, E., and Wilke, B. (2004) The automated determination of glucosamine, galactosamine, muramic acid, and mannosamine in soil and root hydrolysates by HPLC. *J. Plant Nutr. Soil Sci.* **167**, 17–21.

Aristidou, A., and Penttila, M. (2000) Metabolic engineering applications to renewable resource utilization. *Curr. Opin. Biotechnol.* **11**, 187–198.

Arndt, A., and Eikmanns, B. J. (2007) The alcohol dehydrogenase gene *adhA* in *Corynebacterium glutamicum* is subject to carbon catabolite repression. *J. Bacteriol.* **189**, 7408–7416.

Arndt, A., and Eikmanns, B. J. (2008) Regulation of carbon metabolism in *Corynebacterium glutamicum*. In: Burkovski, A. (ed.) Corynebacteria: Genomics and Molecular Biology. Caister Academic Press, Wymondham, UK, pp. 155–182.

Arndt, A., Auchter, M., Ishige, T., Wendisch, V. F., and Eikmanns, B. J. (2008) Ethanol catabolism in *Corynebacterium glutamicum*. *J. Mol. Microbiol. Biotechnol.* **15**, 222–233.

Barrett, E., Stanton, C., Zelder, O., Fitzgerald, G., and Ross, R. P. (2004) Heterologous expression of lactose- and galactose-utilizing pathways from lactic acid bacteria in *Corynebacterium glutamicum* for production of lysine in whey. *Appl. Environ. Microbiol.* **70**, 2861–2866.

Blombach, B., and Seibold, G. M. (2010) Carbohydrate metabolism in *Corynebacterium glutamicum* and applications for the metabolic engineering of L-lysine production strains. *Appl. Microbiol. Biotechnol.* **86**, 1313–1322.

Brabetz, W., Liebl, W., and Schleifer, K. H. (1991) Studies on the utilization of lactose by *Corynebacterium glutamicum*, bearing the lactose operon of *Escherichia coli*. *Arch. Microbiol.* **155**, 607–612.

Brabetz, W., Liebl, W., and Schleifer, K. H. (1993) Lactose permease of *Escherichia coli* catalyzes active beta-galactoside transport in a gram-positive bacterium. *J. Bacteriol.* **175**, 7488–7491.

Buschke, N., Schroder, H., and Wittmann, C. (2011) Metabolic engineering of *Corynebacterium glutamicum* for production of 1,5-diaminopentane from hemicellulose. *Biotechnol. J.* **6**, 306–317.

Claes, W. A., Pühler, A., and Kalinowski, J. (2002) Identification of two *prpDBC* gene clusters in *Corynebacterium glutamicum* and their involvement in propionate degradation via the 2-methylcitrate cycle. *J. Bacteriol.* **184**, 2728–2739.

Cocaign, M., Monnet, C., and Lindley, N. D. (1993) Batch kinetics of *Corynebacterium glutamicum* during growth on various carbon sources: use of substrate mixtures to localise metabolic bottlenecks. *Appl. Microbiol. Biotechnol.* **40**, 526–530.

Datta, R., Tsai, S. P., Bonsignore, P., Moon, S. H., and Frank, J. R. (1995) Technological and economic potential of poly (lactic acid) and lactic acid derivatives. *FEMS Microbiol. Rev.* **16**, 221–231.

Domingues, L., Guimaraes, P. M., and Oliveira, C. (2010) Metabolic engineering of *Saccharomyces cerevisiae* for lactose/whey fermentation. *Bioeng. Bugs* **1**, 164–171.

Dominguez, H., Cocaign-Bousquet, M., and Lindley, N. D. (1997) Simultaneous consumption of glucose and fructose from sugar mixtures during batch growth of *Corynebacterium glutamicum*. *Appl. Microbiol. Biotechnol.* **47**, 600–603.

Dunlop, A. P. (1948) Furfural formation and behavior. *Ind. Eng. Chem.* **40**, 204–209.

Gerstmeir, R., Wendisch, V. F., Schnicke, S., Ruan, H., Farwick, M., Reinscheid, D., and Eikmanns, B. J. (2003) Acetate metabolism and its regulation in

Corynebacterium glutamicum. *J. Biotechnol.* **104**, 99–122.

Gopinath, V., Meiswinkel, T. M., Wendisch, V. F., and Nampoothiri, K. M. (2011) Amino acid production from rice straw and wheat bran hydrolysates by recombinant pentose-utilizing *Corynebacterium glutamicum*. *Appl. Microbiol. Biotechnol.* **92**, 985–996.

Gopinath, V., Murali, A., Dhar, K. S., and Nampoothiri, K. M. (2012) *Corynebacterium glutamicum* as a potent biocatalyst for the bioconversion of pentose sugars to value-added products. *Appl. Microbiol. Biotechnol.* **93**, 95–106.

Gourdon, P., Baucher, M. F., Lindley, N. D., and Guyonvarch, A. (2000) Cloning of the malic enzyme gene from *Corynebacterium glutamicum* and role of the enzyme in lactate metabolism. *Appl. Environ. Microbiol.* **66**, 2981–2987.

Heider, S. A., Peters-Wendisch, P., and Wendisch, V. F. (2012) Carotenoid biosynthesis and overproduction in *Corynebacterium glutamicum*. *BMC Microbiol.* **12**, 198.

Heider, S. A., Peters-Wendisch, P., Netzer, R., Stafnes, M., Brautaset, T., and Wendisch, V. F. (2014) Production and glucosylation of C_{50} and C_{40} carotenoids by metabolically engineered *Corynebacterium glutamicum*. *Appl. Microbiol. Biotechnol.* **98**, 1223–1235.

Huang, Y., Zhao, K. X., Shen, X. H., Chaudhry, M. T., Jiang, C. Y., and Liu, S. J. (2006) Genetic characterization of the resorcinol catabolic pathway in *Corynebacterium glutamicum*. *Appl. Environ. Microbiol.* **72**, 7238–7245.

Hueck, C. J., and Hillen, W. (1995) Catabolite repression in *Bacillus subtilis*: a global regulatory mechanism for the Gram-positive bacteria? *Mol. Microbiol.* **15**, 395–401.

Hyeon, J. E., Jeon, W. J., Whang, S. Y., and Han, S. O. (2011) Production of minicellulosomes for the enhanced hydrolysis of cellulosic substrates by recombinant *Corynebacterium glutamicum*. *Enzyme Microb. Technol.* **48**, 371–377.

Ikeda, M., and Nakagawa, S. (2003) The *Corynebacterium glutamicum* genome: features and impacts on biotechnological processes. *Appl. Microbiol. Biotechnol.* **62**, 99–109.

Ikeda, M., Mizuno, Y., Awane, S., Hayashi, M., Mitsuhashi, S., and Takeno, S. (2011) Identification and application of a different glucose uptake system that functions as an alternative to the phosphotransferase system in *Corynebacterium glutamicum*. *Appl. Microbiol. Biotechnol.* **90**, 1443–1451.

Ikeda, M., Miyamoto, A., Mutoh, S., Kitano, Y., Tajima, M., Shirakura, D., Takasaki, M., Mitsuhashi, S., and Takeno, S. (2013) Development of biotin-prototrophic and -hyperauxotrophic *Corynebacterium glutamicum* strains. *Appl. Environ. Microbiol.* **79**, 4586–4594.

Janausch, I. G., Zientz, E., Tran, Q. H., Kröger, A., and Unden, G. (2002) C4-dicarboxylate carriers and sensors in bacteria. *Biochim. Biophys. Acta – Bioenerg.* **1553**, 39–56.

Jensen, J. V., and Wendisch, V. F. (2013) Ornithine cyclodeaminase-based proline production by *Corynebacterium glutamicum*. *Microb. Cell Fact.* **12**, 63.

Kalinowski, J., Bathe, B., Bartels, D., Bischoff, N., Bott, M., Burkovski, A., Dusch, N., Eggeling, L., Eikmanns, B. J., Gaigalat, L., *et al.* (2003) The complete *Corynebacterium glutamicum* ATCC 13032 genome sequence and its impact on the production of L-aspartate-derived amino acids and vitamins. *J. Biotechnol.* **104**, 5–25.

Kato, O., Youn, J. W., Stansen, K. C., Matsui, D., Oikawa, T., and Wendisch, V. F. (2010) Quinone-dependent D-lactate dehydrogenase Dld (Cg1027) is essential for growth of *Corynebacterium glutamicum* on D-lactate. *BMC Microbiol.* **10**, 321.

Kawaguchi, H., Vertes, A. A., Okino, S., Inui, M., and Yukawa, H. (2006) Engineering of a xylose metabolic pathway in *Corynebacterium glutamicum*. *Appl. Environ. Microbiol.* **72**, 3418–3428.

Kawaguchi, H., Sasaki, M., Vertes, A. A., Inui, M., and Yukawa, H. (2008) Engineering of an L-arabinose metabolic pathway in *Corynebacterium glutamicum*. *Appl. Microbiol. Biotechnol.* **77**, 1053–1062.

Kawaguchi, H., Sasaki, M., Vertes, A. A., Inui, M., and Yukawa, H. (2009) Identification and functional analysis of the gene cluster for L-arabinose utilization in *Corynebacterium glutamicum*. *Appl. Environ. Microbiol.* **75**, 3419–3429.

Khanna, S., Goyal, A., and Moholkar, V. S. (2012) Microbial conversion of glycerol: present status and future prospects. *Crit. Rev. Biotechnol.* **32**, 235–262.

Kimura, E. (2005) L-Glutamate production. In: Eggeling, L., and Bott, M. (eds.) Handbook of *Corynebacterium glutamicum*. CRC Press, Boca Raton, FL, USA, pp. 439–464.

Kinoshita, S., Udaka, S., and Shimono, M. (1957) Studies on the amino acid fermentation. Production of L-glutamic acid by various microorganisms. *J. Gen. Appl. Microbiol.* **3**, 193–205.

Klinke, H. B., Thomsen, A. B., and Ahring, B. K. (2004) Inhibition of ethanol-producing yeast and bacteria by degradation products produced during pre-treatment of biomass. *Appl. Microbiol. Biotechnol.* **66**, 10–26.

Kotrba, P., Inui, M., and Yukawa, H. (2003) A single V317A or V317M substitution in Enzyme II of a newly identified beta-glucoside phosphotransferase and utilization system of *Corynebacterium glutamicum* R extends its specificity towards cellobiose. *Microbiology* **149**, 1569–1580.

Krause, F. S., Henrich, A., Blombach, B., Kramer, R., Eikmanns, B. J., and Seibold, G. M. (2010) Increased glucose utilization in *Corynebacterium glutamicum* by use of maltose, and its application for the improvement of L-valine productivity. *Appl. Environ. Microbiol.* **76**, 370–374.

Kremling, A., Kremling, S., and Bettenbrock, K. (2009) Catabolite repression in *Escherichia coli*- a comparison of modelling approaches. *FEBS J.* **276**, 594–602.

Krings, E., Krumbach, K., Bathe, B., Kelle, R., Wendisch, V. F., Sahm, H., and Eggeling, L. (2006) Characterization of myo-inositol utilization by *Corynebacterium glutamicum*: the stimulon, identification of transporters, and influence on L-lysine formation. *J. Bacteriol.* **188**, 8054–8061.

Kronemeyer, W., Peekhaus, N., Kramer, R., Sahm, H., and Eggeling, L. (1995) Structure of the *gluABCD* cluster encoding the glutamate uptake system of *Corynebacterium glutamicum*. *J. Bacteriol.* **177**, 1152–1158.

Kurita, K. (2006) Chitin and chitosan: functional biopolymers from marine crustaceans. *Mar. Biotechnol. (NY)* **8**, 203–226.

Laslo, T., von Zaluskowski, P., Gabris, C., Lodd, E., Ruckert, C., Dangel, P., Kalinowski, J., Auchter, M., Seibold, G., and Eikmanns, B. J. (2012) Arabitol metabolism of *Corynebacterium glutamicum* and its regulation by AtlR. *J. Bacteriol.* **194**, 941–955.

Lee, H. W., Pan, J. G., and Lebeault, J. M. (1998) Enhanced L-lysine production in threonine-limited continuous culture of *Corynebacterium glutamicum* by using gluconate as a secondary carbon source with glucose. *Appl. Microbiol. Biotechnol.* **49**, 9.

Lee, S. Y., Le, T. H., Chang, S. T., Park, J. S., Kim, Y. H., and Min, J. (2010) Utilization of phenol and naphthalene affects synthesis of various amino acids in *Corynebacterium glutamicum*. *Curr. Microbiol.* **61**, 596–600.

Lessmeier, L., Hoefener, M., and Wendisch, V. F. (2013) Formaldehyde degradation in *Corynebacterium glutamicum* involves acetaldehyde dehydrogenase and mycothiol-dependent formaldehyde dehydrogenase. *Microbiology* **159**, 2651–2662.

Lindner, S. N., Knebel, S., Pallerla, S. R., Schoberth, S. M., and Wendisch, V. F. (2010) Cg2091 encodes a polyphosphate/ATP-dependent glucokinase of *Corynebacterium glutamicum*. *Appl. Microbiol. Biotechnol.* **87**, 703–713.

Lindner, S. N., Seibold, G. M., Henrich, A., Kramer, R., and Wendisch, V. F. (2011a) Phosphotransferase system-independent glucose utilization in *Corynebacterium glutamicum* by inositol permeases and glucokinases. *Appl. Environ. Microbiol.* **77**, 3571–3581.

Lindner, S. N., Seibold, G. M., Kramer, R., and Wendisch, V. F. (2011b) Impact of a new glucose utilization pathway in amino acid-producing *Corynebacterium glutamicum*. *Bioeng. Bugs* **2**, 291–295.

Lindner, S. N., Meiswinkel, T. M., Panhorst, M., Youn, J. W., Wiefel, L., and Wendisch, V. F. (2012) Glycerol-3-phosphatase of *Corynebacterium glutamicum*. *J. Biotechnol.* **159**, 216–224.

Lindner, S. N., Petrov, D. P., Hagmann, C. T., Henrich, A., Kramer, R., Eikmanns, B. J., Wendisch, V. F., and Seibold, G. M. (2013) Phosphotransferase system-(PTS-) mediated glucose uptake is repressed in phosphoglucoisomerase-deficient *Corynebacterium glutamicum* strains. *Appl. Environ. Microbiol.* **79**, 2588–2595.

Litsanov, B., Brocker, M., and Bott, M. (2013) Glycerol as a substrate for aerobic succinate production in minimal medium with *Corynebacterium glutamicum*. *Microb. Biotechnol.* **6**, 189–195.

McDonald, P., Henderson, A. R., and Heron, S. J. (1991) Microorganisms. In: McDonald, P., Henderson, A. R., and Heron, S. J. (eds.) The Biochemistry of Silage. Chalcombe Publications, Aberystwyth, Wales, pp. 81–152.

Matano, C., Meiswinkel, T. M., and Wendisch, V. F. (2013) Amino acid production from rice straw hydrolysates. In: Watson R. R., Preedy V. R., Zibaldi S. (eds.), Wheat and Rice in Disease Prevention and Health, Elsevier/Academic Press, Bielefeld University, Bielefeld, Germany, pp. 493–505.

Meiswinkel, T. M., Gopinath, V., Lindner, S. N., Nampoothiri, K. M., and Wendisch, V. F. (2013a) Accelerated pentose utilization by Corynebacterium glutamicum for accelerated production of lysine, glutamate, ornithine and putrescine. *Microb. Biotechnol.* **6**, 131–140.

Meiswinkel, T. M., Rittmann, D., Lindner, S. N., and Wendisch, V. F. (2013b) Crude glycerol-based production of amino acids and putrescine by *Corynebacterium glutamicum*. *Bioresour. Technol.* **145**, 254–258.

Merkens, H., Beckers, G., Wirtz, A., and Burkovski, A. (2005) Vanillate metabolism in *Corynebacterium glutamicum*. *Curr. Microbiol.* **51**, 59–65.

Mussatto, S. I., and Roberto, I. C. (2004) Optimal experimental condition for hemicellulosic hydrolyzate treatment with activated charcoal for xylitol production. *Biotechnol. Prog.* **20**, 134–139.

Netzer, R., Peters-Wendisch, P., Eggeling, L., and Sahm, H. (2004) Cometabolism of a nongrowth substrate: L-serine utilization by *Corynebacterium glutamicum*. *Appl. Environ. Microbiol.* **70**, 7148–7155.

Neuner, A., and Heinzle, E. (2011) Mixed glucose and lactate uptake by *Corynebacterium glutamicum* through metabolic engineering. *Biotechnol. J.* **6**, 318–329.

Neuner, A., Wagner, I., Sieker, T., Ulber, R., Schneider, K., Peifer, S., and Heinzle, E. (2013) Production of l-lysine on different silage juices using genetically engineered *Corynebacterium glutamicum*. *J. Biotechnol.* **163**, 217–224.

Palmqvist, E., and Hahn-Hagerdal, B. (2000) Fermentation of lignocellulosic hydrolysates. II: inhibitors and mechanisms of inhibition. *Bioresour. Technol.* **74**, 25–33.

Paradis, F. W., Warren, R. A., Kilburn, D. G., and Miller, R. C. Jr. (1987) The expression of *Cellulomonas fimi* cellulase genes in *Brevibacterium lactofermentum*. *Gene* **61**, 199–206.

Parawira, W., and Tekere, M. (2011) Biotechnological strategies to overcome inhibitors in lignocellulose hydrolysates for ethanol production: review. *Crit. Rev. Biotechnol.* **31**, 20–31.

Parche, S., Burkovski, A., Sprenger, G. A., Weil, B., Kramer, R., and Titgemeyer, F. (2001) *Corynebacterium glutamicum*: a dissection of the PTS. *J. Mol. Microbiol. Biotechnol.* **3**, 423–428.

Park, S. Y., Kim, H. K., Yoo, S. K., Oh, T. K., and Lee, J. K. (2000) Characterization of *glk*, a gene coding for glucose kinase of *Corynebacterium glutamicum*. *FEMS Microbiol. Lett.* **188**, 209–215.

Parsons, J. W. (1981) Chemistry and distribution of amino sugars in soils and soils organisms. In: Paul, E. A., and Ladd, J. N. (eds.) Soil Biochemistry. Marcel Dekker, New York, NY, USA, pp. 197–227.

Peters-Wendisch, P. G., Kreutzer, C., Kalinowski, J., Patek, M., Sahm, H., and Eikmanns, B. J. (1998) Pyruvate carboxylase from *Corynebacterium glutamicum*:

characterization, expression and inactivation of the *pyc* gene. *Microbiology* **144**, 915–927.

Peters-Wendisch, P., Gotker, S., Heider, S. A., Komati Reddy, G., Nguyen, A. Q., Stansen, K. C., and Wendisch, V. F. (2014) Engineering biotin prototrophic *Corynebacterium glutamicum* strains for amino acid, diamine and carotenoid production. *J. Biotechnol.* **192** (Pt B), 346–354.

Qi, S. W., Chaudhry, M. T., Zhang, Y., Meng, B., Huang, Y., Zhao, K. X., Poetsch, A., Jiang, C. Y., Liu, S., and Liu, S. J. (2007) Comparative proteomes of *Corynebacterium glutamicum* grown on aromatic compounds revealed novel proteins involved in aromatic degradation and a clear link between aromatic catabolism and gluconeogenesis via fructose-1,6-bisphosphatase. *Proteomics* **7**, 3775–3787.

Rittmann, D., Schaffer, S., Wendisch, V. F., and Sahm, H. (2003) Fructose-1,6-bisphosphatase from *Corynebacterium glutamicum*: expression and deletion of the *fbp* gene and biochemical characterization of the enzyme. *Arch. Microbiol.* **180**, 285–292.

Rittmann, D., Lindner, S. N., and Wendisch, V. F. (2008) Engineering of a glycerol utilization pathway for amino acid production by *Corynebacterium glutamicum*. *Appl. Environ. Microbiol.* **74**, 6216–6222.

Rumbold, K., van Buijsen, H. J., Gray, V. M., van Groenestijn, J. W., Overkamp, K. M., Slomp, R. S., van der Werf, M. J., and Punt, P. J. (2010) Microbial renewable feedstock utilization: a substrate-oriented approach. *Bioeng. Bugs* **1**, 359–366.

Sakai, S., Tsuchida, Y., Nakamoto, H., Okino, S., Ichihashi, O., Kawaguchi, H., Watanabe, T., Inui, M., and Yukawa, H. (2007) Effect of lignocellulose-derived inhibitors on growth of and ethanol production by growth-arrested *Corynebacterium glutamicum* R. *Appl. Environ. Microbiol.* **73**, 2349–2353.

Sasaki, M., Jojima, T., Inui, M., and Yukawa, H. (2008) Simultaneous utilization of D-cellobiose, D-glucose, and D-xylose by recombinant *Corynebacterium glutamicum* under oxygen-deprived conditions. *Appl. Microbiol. Biotechnol.* **81**, 691–699.

Sasaki, M., Jojima, T., Kawaguchi, H., Inui, M., and Yukawa, H. (2009) Engineering of pentose transport in *Corynebacterium glutamicum* to improve simultaneous utilization of mixed sugars. *Appl. Microbiol. Biotechnol.* **85**, 105–115.

Schneider, J., Niermann, K., and Wendisch, V. F. (2011) Production of the amino acids L-glutamate, L-lysine, L-ornithine and L-arginine from arabinose by recombinant *Corynebacterium glutamicum*. *J. Biotechnol.* **154**, 191–198.

Schwarz, W. H. (2001) The cellulosome and cellulose degradation by anaerobic bacteria. *Appl. Microbiol. Biotechnol.* **56**, 634–649.

Seibold, G., Auchter, M., Berens, S., Kalinowski, J., and Eikmanns, B. J. (2006) Utilization of soluble starch by a recombinant *Corynebacterium glutamicum* strain: growth and lysine production. *J. Biotechnol.* **124**, 381–391.

Sharkey, M. A., Maher, M. A., Guyonvarch, A., and Engel, P. C. (2011) Kinetic characterisation of recombinant *Corynebacterium glutamicum* NAD$^+$-dependent LDH over-expressed in *E. coli* and its rescue of an *lld*D- phenotype in *C. glutamicum*: the issue of reversibility re-examined. *Arch. Microbiol.* **193**, 731–740.

Shen, X., and Liu, S. (2005) Key enzymes of the protocatechuate branch of the beta-ketoadipate pathway for aromatic degradation in *Corynebacterium glutamicum*. *Sci. China C. Life Sci.* **48**, 241–249.

Shen, X. H., Liu, Z. P., and Liu, S. J. (2004) Functional identification of the gene locus NCgl2319 and characterization of catechol 1,2-dioxygenase in *Corynebacterium glutamicum*. *Biotechnol. Lett.* **26**, 575–580.

Shen, X. H., Jiang, C. Y., Huang, Y., Liu, Z. P., and Liu, S. J. (2005) Functional identification of novel genes involved in the glutathione-independent gentisate pathway in *Corynebacterium glutamicum*. *Appl. Environ. Microbiol.* **71**, 3442–3452.

Sodergard, A., and Stolt, M. (2002) Properties of lactic acid based polymers and their correlation with composition. *Prog. Polym. Sci.* **27**, 1123–1163.

Song, Y., Matsumoto, K., Yamada, M., Gohda, A., Brigham, C. J., Sinskey, A. J., and Taguchi, S. (2012) Engineered *Corynebacterium glutamicum* as an endotoxin-free platform strain for lactate-based polyester production. *Appl. Microbiol. Biotechnol.* **93**, 1917–1925.

Song, Y., Matsumoto, K., Tanaka, T., Kondo, A., and Taguchi, S. (2013) Single-step production of polyhydroxybutyrate from starch by using alpha-amylase cell-surface displaying system of *Corynebacterium glutamicum*. *J. Biosci. Bioeng.* **115**, 12–14.

Stansen, C., Uy, D., Delaunay, S., Eggeling, L., Goergen, J. L., and Wendisch, V. F. (2005) Characterization of a *Corynebacterium glutamicum* lactate utilization operon induced during temperature-triggered glutamate production. *Appl. Environ. Microbiol.* **71**, 5920–5928.

Stevenson, F. J. (1982) Organic forms of soil nitrogen. In: Stevenson, F. J. (ed.) Nitrogen in Agricultural Soils. American Society of Agronomy, Madison, WI, USA, pp. 101–104.

Stevenson, F. J. (1983) Isolation and identification of amino sugars in soil. *Soil Sci. Soc. Am. J.* **47**, 61–65.

Tateno, T., Fukuda, H., and Kondo, A. (2007a) Production of L-lysine from starch by *Corynebacterium glutamicum* displaying alpha-amylase on its cell surface. *Appl. Microbiol. Biotechnol.* **74**, 1213–1220.

Tateno, T., Fukuda, H., and Kondo, A. (2007b) Direct production of L-lysine from raw corn starch by *Corynebacterium glutamicum* secreting *Streptococcus bovis* alpha-amylase using *cspB* promoter and signal sequence. *Appl. Microbiol. Biotechnol.* **77**, 533–541.

Tateno, T., Okada, Y., Tsuchidate, T., Tanaka, T., Fukuda, H., and Kondo, A. (2009) Direct production of cadaverine from soluble starch using *Corynebacterium glutamicum* coexpressing alpha-amylase and lysine decarboxylase. *Appl. Microbiol. Biotechnol.* **82**, 115–121.

Teramoto, H., Shirai, T., Inui, M., and Yukawa, H. (2008) Identification of a gene encoding a transporter essential for utilization of C4 dicarboxylates in *Corynebacterium glutamicum*. *Appl. Environ. Microbiol.* **74**, 5290–5296.

Tharanathan, R. N., and Kittur, F. S. (2003) Chitin – the undisputed biomolecule of great potential. *Crit. Rev. Food Sci. Nutr.* **43**, 61–87.

Tsuchidate, T., Tateno, T., Okai, N., Tanaka, T., Ogino, C., and Kondo, A. (2011) Glutamate production from beta-glucan using endoglucanase-secreting Corynebacterium glutamicum. *Appl. Microbiol. Biotechnol.* **90**, 895–901.

Uhde, A., Youn, J. W., Maeda, T., Clermont, L., Matano, C., Kramer, R., Wendisch, V. F., Seibold, G. M., and Marin, K. (2013) Glucosamine as carbon source for amino acid-producing *Corynebacterium glutamicum*. *Appl. Microbiol. Biotechnol.* **97**, 1679–1687.

Wendisch, V. F. (2003) Genome-wide expression analysis in *Corynebacterium glutamicum* using DNA microarrays. *J. Biotechnol.* **104**, 273–285.

Wendisch, V. F., de Graaf, A. A., Sahm, H., and Eikmanns, B. J. (2000) Quantitative determination of metabolic fluxes during coutilization of two carbon sources: comparative analyses with *Corynebacterium glutamicum* during growth on acetate and/or glucose. *J. Bacteriol.* **182**, 3088–3096.

Wendisch, V. F., Lindner, S. N., and Meiswinkel, T. M. (2011) Use of glycerol in biotechnological applications. In: Montero, D. G. (ed.) Biodiesel – Quality, Emissions and By-Products. InTech., Rijeka, Croatia, pp. 305–340.

Witthoff, S., Eggeling, L., Bott, M., and Polen, T. (2012) *Corynebacterium glutamicum* harbours a molybdenum cofactor-dependent formate dehydrogenase which alleviates growth inhibition in the presence of formate. *Microbiology* **158**, 2428–2439.

Witthoff, S., Muhlroth, A., Marienhagen, J., and Bott, M. (2013) C1 Metabolism in *Corynebacterium glutamicum*: an endogenous pathway for oxidation of methanol to carbon dioxide. *Appl. Environ. Microbiol.* **79**, 6974–6983.

Wyman, C. E. (1999) Production of low cost sugars from biomass: progress, opportunities, and challenges. In: Overend, R. P., and Chornet, E. (eds.) Biomass: A Growth Opportunity in Green Energy and Value Added Products. Pergamon Press, Oxford, UK, pp. 867–872.

Yao, W., Chu, C., Deng, X., Zhang, Y., Liu, M., Zheng, P., and Sun, Z. (2009) Display of alpha-amylase on the surface of *Corynebacterium glutamicum* cells by using NCgl1221 as the anchoring protein, and production of glutamate from starch. *Arch. Microbiol.* **191**, 751–759.

Youn, J. W., Jolkver, E., Krämer, R., Marin, K., and Wendisch, V. F. (2008) Identification and characterization of the dicarboxylate uptake system DccT in *Corynebacterium glutamicum*. *J. Bacteriol.* **190**, 6458–6466.

Youn, J. W., Jolkver, E., Krämer, R., Marin, K., and Wendisch, V. F. (2009) Characterization of the dicarboxylate transporter DctA in *Corynebacterium glutamicum*. *J. Bacteriol.* **191**, 5480–5488.

Zahoor, A., Lindner, S. N., and Wendisch, V. F. (2012) Metabolic engineering of *Corynebacterium glutamicum* aimed at alternative carbon sources and new products. *Comput. Struct. Biotechnol. J.* **3**, e201210004.

Zahoor, A., Otten, A., and Wendisch, V. F. (2014) Metabolic engineering of *Corynebacterium glutamicum* for glycolate production. *J. Biotechnol.* **192** (Pt B), 366–375.

Manipulation of Nitrogen Metabolism and Alternative Nitrogen Sources for *Corynebacterium glutamicum*

6

Nadine Rehm, Julia Bürger and Andreas Burkovski

Abstract

Nitrogen is an essential component of nearly all macromolecules in a bacterial cell, e.g. proteins, nucleic acids and cell wall components and their corresponding building blocks. To satisfy the nitrogen demand and to ensure a sufficient nitrogen supply even in situations of nitrogen limitation, microorganisms have evolved sophisticated uptake and assimilation mechanisms for different nitrogen sources. This review focuses on nitrogen metabolism and its control in the biotechnology workhorse *Corynebacterium glutamicum*, which is used for the industrial production of almost four million tons of L-amino acids each year. Ammonium assimilation and connected control mechanisms on activity and transcription level are summarized and the influence of mutations on amino acid pools and production of L-glutamate, L-glutamine and L-lysine is discussed.

Ammonium assimilation in *Corynebacterium glutamicum*

The isolation of *Corynebacterium glutamicum* in 1957 (Kinoshita *et al.*, 1957; Udaka, 2008) laid the foundation of a highly successful fermentation industry for the bulk production of amino acids and other chemicals (see also Chapter 1). Nowadays, about 2.9 million tons of L-glutamate and 1.9 million tons of L-lysine are produced annually using different *C. glutamicum* strains in addition to smaller amounts of L-arginine, L-histidine, L-tryptophan and L-valine and other chemicals (Ajinomoto, 2013a,b; Becker and Wittmann, 2012). L-Glutamine-excreting strains were described as well; however, information about production of this amino acid is scarce (Kusumoto, 2001; Li *et al.*, 2007). The role of *C. glutamicum* as a biotechnology workhorse was further supported and strengthened in the last years by the sequencing and publication of the *C. glutamicum* genome (Ikeda and Nakagawa, 2003; Kalinowski *et al.*, 2003). This landmark of *C. glutamicum* research was the basis of the development of multiple -omics techniques (Wittmann and Heinzle, 2008; Kalinowski *et al.*, 2008; Wendisch, 2008; Poetsch *et al.*, 2011; see also Chapter 2), which allow a global view on different aspects of physiology. Since amino acid synthesis is directly linked to nitrogen metabolism, uptake and assimilation of nitrogen sources in *C. glutamicum* are within focus of research for more than 30 years (for reviews, see Burkovski, 2005, 2007; Hänßler and Burkovski, 2008; Amon *et al.*, 2010; Rehm and Burkovski, 2011).

Ammonium is a preferred nitrogen source of *C. glutamicum* and a standard component of defined growth media. It is almost exclusively assimilated via glutamate dehydrogenase and glutamine synthetase. Other enzymes such as alanine or diaminopimelate dehydrogenase do not significantly contribute to ammonium assimilation as deduced from *in vivo* flux measurements of ammonium assimilation (Tesch *et al.*, 1999).

When ammonium is present in concentrations above 5 mM, it is primarily assimilated via glutamate dehydrogenase. With K_m values of 3.08 mM for ammonium, 5.72 mM for 2-oxoglutarate and 100 mM to L-glutamate, glutamate dehydrogenase shows not low affinities to its substrates but especially to its product (Shiio and Ozaki, 1970). Despite its high activity in the wild type

(Börmann *et al.*, 1992; Hänßler *et al.*, 2009), the analysis of a *gdh* mutant showed that the enzyme is not essential for growth and glutamate production (Börmann-El Kholy *et al.*, 1993; Tesch *et al.*, 1998). In case of glutamate dehydrogenase absence ammonium assimilation is taken over by the glutamine synthetase. Even without ammonium limitation, about 28% of the ammonium metabolized is assimilated via this enzyme as shown by *in vivo* flux measurements (Tesch *et al.*, 1999). In case of ammonium limitation, i.e. when concentrations decrease to 5 mM and below, ammonium is almost exclusively assimilated via glutamine synthetase and L-glutamate demand is satisfied by conversion of α-ketoglutarate and glutamine to L-glutamate by glutamate synthase (Tochikura *et al.*, 1984; Tesch *et al.*, 1998; Beckers *et al.*, 2001).

In contrast to the observations made for glutamate dehydrogenase, glutamine synthetase activity is essential for growth. Although two different glutamine synthetase-encoding genes are located on the *C. glutamicum* genome, *glnA* and *glnA2*, only *glnA* seems to encode an active glutamine synthetase since deletion of this gene results in a glutamine-auxotrophic phenotype, while deletion of the *glnA2* gene had no effect (Jakoby *et al.*, 1997; Nolden *et al.*, 2001a).

As described above, glutamate dehydrogenase and glutamine synthetase are the main ammonium-assimilating enzymes in *C. glutamicum* (Tesch *et al.*, 1999). Nevertheless, at least partially also the *ddh* gene product diaminopimelate dehydrogenase contributes to ammonium assimilation. ^{13}C- and ^{1}H-NMR spectroscopy experiments revealed a pronounced flux partitioning in the split pathway of L-lysine biosynthesis. About one third of L-lysine is synthesized by diaminopimelate dehydrogenase and two-thirds via the succinylase pathway at the beginning of fermentations, while later L-lysine is almost exclusively synthesized via the succinylase pathway (Sonntag *et al.*, 1993). The observed time-dependent activity of diaminopimelate dehydrogenase is caused by its low affinity to ammonium with a K_m of 36 mM (Misono and Soda, 1980) and the decreased ammonium concentration in later stages of cultivation (Sahm *et al.*, 2000). In *C. glutamicum* strains with an increased flow of precursors to D,L-diaminopimelate, the inactivation of diaminopimelate dehydrogenase resulted in a reduced L-lysine formation (Schrumpf *et al.*, 1991).

Regulation of nitrogen metabolism

For fast adaptation to changing ammonium supply and glutamine demand, glutamine synthetase activity is regulated at post-translational level by successive adenylylation/deadenylylation of glutamine synthetase subunits by adenylyltransferase GlnE (Jakoby *et al.*, 1999). With tyrosyl residue 405 of glutamine synthetase a typical adenylylation site was identified (Jakoby *et al.*, 1997). When this amino acid is changed to phenylalanine (Jakoby *et al.*, 1999) or the *glnE* gene is deleted (Nolden *et al.*, 2001a; Rehm *et al.*, 2010a) activity regulation of glutamine synthetase via post-translational modification is abolished.

For long-term adaptation, ammonium assimilation is regulated at the transcriptional level by AmtR, the master regulator of nitrogen metabolism in *C. glutamicum* and other corynebacteria (Jakoby *et al.*, 2000; Beckers *et al.*, 2005; Walter *et al.*, 2007; Muhl *et al.*, 2009; Hasselt *et al.*, 2011). The *C. glutamicum* AmtR regulon comprises at least 35 genes involved in nitrogen metabolism (Beckers *et al.*, 2005; Buchinger *et al.*, 2009); encoding transporters of ammonium (*amtA*, *amtB*) and alternative nitrogen sources (*crnT*, *gluABCD*, *urtABCDE*), enzymes for ammonium assimilation (*dapD*, *gdh*, *glnA*, *gltBD*), utilization of creatinine (*codA*) and urea (*ureABCEFGD*) as well as nitrogen signal transduction proteins (*glnK*, *glnD*).

AmtR is a member of the TetR family of transcriptional regulators (for review, see Ramos *et al.*, 2005). AmtR is released from its *cis* elements on the DNA upon interaction with the P_{II}-type signal transduction protein GlnK (Beckers *et al.*, 2005). As a prerequisite for this interaction, the trimeric GlnK complex has to be adenylylated at tyrosyl residue 51 located in the T-loop of the protein by GlnD (Strösser *et al.*, 2004). Based on RNA hybridization analyses in combination with various growth conditions, ammonium and α-ketoglutarate are favoured as molecules, which flag the nitrogen status of *C. glutamicum* (Nolden *et al.*, 2001b; Müller *et al.*, 2006; Rehm *et al.*, 2010b); however, direct experimental evidence for binding of α-ketoglutarate to GlnK or a validated receptor for ammonium binding are unknown.

Manipulation of nitrogen metabolism for amino acid production

Besides manipulation of carbon supply (Chapter 5), carbon flux (Chapter 3) and amino acid synthesis pathways, nitrogen metabolism is a prime target for the improvement of amino acid production (Rehm and Burkovski, 2011; Yasueda, 2014).

Overexpression, deletion and heterologous expression of glutamate dehydrogenase

Owing to its key function in glutamate synthesis, overexpression of glutamate dehydrogenase was an obvious strategy to enhance glutamate production. In fact, a corresponding *C. glutamicum* strain showed higher glutamate dehydrogenase activity and an elevated L-glutamate pool; however, no influence on secretion of L-glutamate induced by biotin limitation was observed (Börmann-El Kholy *et al.*, 1993). Vice versa, deletion of *gdh* did affect L-glutamate production severely (Börmann-El Kholy *et al.*, 1993; Tesch *et al.*, 1998; Müller *et al.*, 2006). Müller and co-workers (Müller *et al.*, 2006) showed that glutamate dehydrogenase deficiency leads to a partial deregulation of nitrogen starvation response in *C. glutamicum*. As a result, ammonium assimilation by glutamate dehydrogenase is taken over by glutamine synthetase and L-glutamate is subsequently produced from L-glutamine – and α-ketoglutarate – by glutamate synthase, complementing the lack of glutamate dehydrogenase activity. This idea is supported by metabolic flux analyses, which revealed that α-ketoglutarate flux is not altered significantly in response to changes in glutamate dehydrogenase activity (Shimizu *et al.*, 2003).

Glutamate dehydrogenase manipulation was also carried out to improve the synthesis of other amino acids. For example, increased glutamate dehydrogenase activity was described in patents to be advantageous for L-glutamine production (Nakamura *et al.*, 2002, 2004). The physiological background for this claim is unclear and needs further investigation.

More obvious, an impact of glutamate dehydrogenase on L-lysine biosynthesis was proposed. Ammonium assimilation by glutamate dehydrogenase requires NADPH and indeed, glutamate dehydrogenase consumes about 50% of the *C. glutamicum* NADPH pool. Since L-lysine biosynthesis relies on sufficient NADPH as well, the *gdh* gene of *C. glutamicum* was deleted and the NADH-dependent glutamate dehydrogenase enzyme from *Peptostreptococcus asaccharolyticus* was expressed instead to improve L-lysine production. In contrast to the expectations, L-lysine biosynthesis and intracellular L-lysine pool were not increased, since the decreased NADPH demand due to NADH-dependent glutamate dehydrogenase was counterbalanced by a decreased NADPH synthesis in the corresponding strain (Marx *et al.*, 1999).

Overexpression of glutamine synthetases

A number of patent applications were filed in respect to L-glutamine production with *C. glutamicum*. Claims include increased glutamine synthetase activity due to overexpression (Nakamura *et al.*, 2002, 2004; Liu *et al.*, 2008), use of recombinant GlnA and GlnA2 variants (Nakamura *et al.*, 2002; Hayashi *et al.*, 2008; Adachi *et al.*, 2012) as well as deregulated GSI activity due to lack of adenylylation and decreased GlnK activity (Nakamura *et al.*, 2002, 2004; Rehm *et al.*, 2010a; Reimer *et al.*, 2014). Additionally, patent applications regarding L-lysine and L-arginine production were filed, which claimed the use of enhanced glutamine synthetase activity due to reduced or eliminated adenylylation and disruption of the *amtR* gene (Matsuzaki *et al.*, 2004). The physiological background of these patent claims is sometimes obvious such as the increase of GS activity for glutamine production, sometimes it is unclear or contradictory to work carried out in wild-type strains as outlined before especially in the case of GlnK (Rehm and Burkovski, 2011). In *C. glutamicum* wild type, lack of GlnK activity leads to in a constantly repressed AmtR regulon. As a result *glnA* transcription is repressed and glutamine synthetase activity is low. Furthermore, it was shown that GlnE works independently from GlnK, and consequently a *glnK* deletion in the wild type has no effect on adenylylation of glutamine synthetase (Strösser *et al.*, 2004).

Influence of glutamate synthase on L-glutamate biosynthesis

Inactivation of the *gltBD* operon – and as a consequence a complete loss of glutamate synthase activity – has no detrimental effect on growth when cells are cultured in ammonium-containing standard minimal medium. Only a minor increase in the doubling time of the corresponding mutant strains was observed in the presence of limiting amounts of ammonium or urea (Beckers *et al.*, 2001). In combination with a *gdh* deletion, loss of glutamate synthase activity has a more pronounced negative effect of on growth and the intracellular glutamate pool (Rehm *et al.*, 2010b). However, the synthesis of L-glutamate is still possible in a *gdh*/*gltBD* double mutant since a corresponding strain is not glutamate-auxotrophic and a considerable L-glutamate concentration was observed within the cells (Rehm *et al.*, 2010b). It was assumed that L-glutamate is synthesized by transamination of α-ketoglutarate using another amino acid as amino donor (Rehm *et al.*, 2010b; Rehm and Burkovski, 2011). However, this requires an alternative ammonium assimilation reaction. A possible pathway in this respect might be the reversible conversion of fumarate and ammonia to L-aspartate by aspartate ammonia lyase (aspartase). Subsequently, L-aspartate might serve as amino donor for the transamination of 2-oxoglutarate to glutamate, presumably by the transaminase AspT (Marienhagen *et al.*, 2005). Probably, there are other possibilities; however, transaminase reactions have only been partially characterized today (Marienhagen *et al.*, 2005; Marienhagen and Eggeling, 2008) and more work has to be carried out.

Changing ammonium assimilation and amino acid production by manipulation of α-ketoglutarate supply

In *C. glutamicum* L-glutamate is predominantly synthesized under consumption of α-ketoglutarate by glutamate dehydrogenase or glutamate synthase depending on the nitrogen supply. The 2-oxoglutarate dehydrogenase complex (ODHC), which catalyses the oxidative decarboxylation of α-ketoglutarate to succinyl-CoA as part of the tricarboxylic acid cycle is a direct competitor for α-ketoglutarate. Since glutamate synthase is expressed only in response to nitrogen starvation (Beckers *et al.*, 2001, 2005), under standard L-glutamate production conditions only glutamate dehydrogenase and the 2-oxoglutarate dehydrogenase complex are competing for α-ketoglutarate. With a K_m of 0.08 mM (Shiio and Ujigawa-Takeda, 1980) ODHC has a much higher affinity for α-ketoglutarate than glutamate dehydrogenase, which has a K_m of about 5.72 mM (Shiio and Ozaki, 1970). Vice versa, with between 2 and 30 mU/mg protein (Shimizu *et al.*, 2003; Uy *et al.*, 2003, 2005; Asakura *et al.*, 2007) activity of ODHC is much lower than the glutamate dehydrogenase activity of about 2 U/mg protein (Börmann *et al.*, 1992; Hänßler *et al.*, 2009). Consequently, increasing glutamate dehydrogenase activities have no significant influence on L-glutamate production, while attenuation of ODHC showed a strong impact (Shimizu *et al.*, 2003). In fact, it was shown that L-glutamate excretion induced by biotin limitation, treatment with antibiotics (penicillin, ethambutol) or detergents (Tween) is accompanied by reduced ODHC activity (Kawahara *et al.*, 1997; Radmacher *et al.*, 2005; Shirai *et al.*, 2005; Kim *et al.*, 2009).

As known so far, activity of ODHC is regulated by a small 15.4 kDa protein named OdhI, which directly interacts with the OdhA subunit of ODHC, thereby inhibiting ODHC activity (Niebisch *et al.*, 2006; Krawczyk *et al.*, 2010). This inhibitory protein–protein interaction can be blocked by phosphorylation of OdhI by PknG, one of the four eukaryotic-like serine/threonine protein kinases that have been identified in *C. glutamicum*, and as a consequence ODHC activity remains high (Niebisch *et al.*, 2006; Schultz *et al.*, 2009). L-Glutamine was proposed to be an effector of PknG-mediated OdhI phosphorylation (Bott, 2007). Dephosphorylation of OdhI is catalysed by Ppp, the only serine/threonine phosphatase present in *C. glutamicum* (Schultz *et al.*, 2007, 2009).

A correlation of the phosphorylation status of OdhI and glutamate production was observed by Boulahya and co-workers (Boulahya *et al.*, 2010). Depending on glutamate excretion conditions the level of dephosphorylated OdhI increases. However, since a strict relation between L-glutamate production rate and OdhI phosphorylation was not observed, the phosphorylation status seems to only one parameter that determines strain performance (Boulahya *et al.*, 2010). Nevertheless, as expected, an *odhI* deletion strain exhibited increased ODHC activity and impaired L-glutamate formation (Schultz *et al.*, 2007). Vice versa, an *odhI* overexpression mutant showed reduced ODHC activity and L-glutamate production without further induction of L-glutamate excretion (Kim *et al.*, 2010; Zhang *et al.*, 2012). Obviously, increased expression of OdhI overcomes activity regulation by PknG. In addition, deletion of *pknG* results in an accumulation of unphosphorylated OdhI, in inhibition of ODHC activity and enhanced L-glutamate production (Schultz *et al.*, 2007).

In case of a complete loss of ODHC activity due to *odhA* deletion or reduced protein levels due to antisense RNA expression, a positive effect on L-glutamate excretion was observed (Asakura *et al.*, 2007; Kim *et al.*, 2009). Interestingly, Nakamura and co-workers (Nakamura *et al.*, 2007) reported that an *odhA* disruption alone is not responsible for L-glutamate excretion even under non-inducing conditions, but also involves mutations of the NCgl1221 gene encoding a putative mechanosensitive channel homologue in L-glutamate-excreting *odhA* mutants. This observation further strengthens the idea that low ODHC activity is essential but not the only prerequisite for efficient L-glutamate production.

Influence of ammonium and glutamate transport systems on amino acid production

In the presence of high ammonium concentrations as found in fermentation media, diffusion of ammonia (NH_3), which is in equilibrium with ammonium (NH_4^+), across the cytoplasmic membrane is sufficient for growth and amino acid production. During nitrogen shortage, diffusion becomes limiting and ammonium transporters AmtA and AmtB are expressed to ensure nitrogen supply (Siewe *et al.*, 1996; Meier-Wagner *et al.*, 2001; Walter *et al.*, 2008). Interestingly, Sindelar and Wendisch (2007) found indications for a positive effect of these permeases on amino acid production, when transcriptome patterns of wild type and different L-lysine producer strains were compared. Overexpression of *amtA* (together with *ocd* and *soxA*) in strain MH20–22B resulted in a significantly enhanced L-lysine production (Sindelar and Wendisch, 2007) supporting the idea of a positive effect of additional ammonium supply for L-lysine synthesis.

A screening of *C. glutamicum* transposon mutants for increased L-glutamate excretion led to the identification of the secondary L-glutamate uptake system GltS. Deletion of this transporter led to slightly increased L-glutamate production, while overexpression had a negative effect (Trötschel *et al.*, 2003), most likely due to a re-uptake of excreted L-glutamate by GltS.

In contrast to the negative influence of GltS, the mechanosensitive channel encoded by NCgl1221 seems to have a positive effect on glutamate production under certain circumstances. Specific mutations in NCgl1221 lead to glutamate overproduction without *odhA* inactivation (Nakamura *et al.*, 2007). Overexpression of NCgl1221 increases L-glutamate secretion, while production is abolished upon NCgl1221 inactivation in *C. glutamicum*. As reported recently, the *N*-terminal domain of the NCgl1221 mechanosensitive channel is sufficient for L-glutamate secretion (Yamashita *et al.*, 2013) and has also a positive effect on glutamate production when fused to *Escherichia coli* mechanosensitive channel MscS (Becker *et al.*, 2013). In summary, NCgl1221 is a putative glutamate exporter of *C. glutamicum*.

Manipulation of nitrogen regulation: influences on metabolite pools

Nitrogen metabolism is regulated at post-transcriptional level for fast adaptation and at transcriptional level for long-term adaptation to changes in nitrogen supply. The only known enzyme involved in nitrogen assimilation which is post-translationally modified is glutamine synthetase, which can be reversibly inactivated by adenylylation. The responsible adenylyltransferase GlnE was target of several studies. Mutation of the *glnE* gene resulted in a loss of activity regulation of glutamine synthetase (Nolden *et al.*, 2001a; Rehm *et al.*, 2010a), induction of nitrogen limitation response at transcript level and global metabolite alterations (Rehm *et al.*, 2010a). Pools of glycolysis, pentose phosphate pathway and tricarboxylic acid cycle metabolites were significantly decreased in the Δ*glnE* mutant, including α-ketoglutarate. L-Glutamate and the aspartate-derived metabolites β-alanine, homoserine, L-methionine, L-threonine and L-isoleucine showed a significant decrease in the Δ*glnE* mutant, while L-lysine, L-leucine as well as L-glutamine concentrations were elevated in the mutant compared with the wild type.

An *amtR* deletion strain, lacking the master regulator of nitrogen control AmtR, showed a complete derepression of the AmtR regulon. Metabolites of the preparatory part of glycolysis and pentose phosphate pathway metabolites showed increased concentrations in the mutant compared with the wild type. In contrast, concentrations of metabolites in the energy-delivering part of glycolysis as well as metabolites of the tricarboxylic acid cycle were decreased in the mutant compared with the wild type. L-Glutamate and the L-aspartate-derived metabolites β-alanine, homoserine, L-methionine and L-isoleucine showed higher concentrations in the wild type, while L-lysine and L-glutamine concentrations were strongly increased in the *amtR* mutant. This positive effect of weakening of AmtR repression for L-amino acid production was filed in frame of a patent (Jeßberger *et al.*, 2014).

In summary, metabolite patterns changed were in majority not directly connected to nitrogen metabolism. Owing to a *glnE* deletion more metabolites were altered compared with the effect of an *amtR* deletion, showing that the effect of loss of transcriptional control can be compensated by activity control mechanisms.

Assimilation of alternative nitrogen sources

Ammonium is a preferred nitrogen source for *C. glutamicum*, but urea, creatinine and different amino acids can also be utilized (for review, see Burkovski, 2005; Hänßler and Burkovski, 2008). In order to improve *C. glutamicum* as a platform organism for different biotechnological processes in respect to flexibility of feedstock as well as hypoxic stress resistance, the utilization of nitrate as a nitrogen and/or energy source might be favourable. In contrast to many other organisms, *C. glutamicum* is unable to use nitrate or nitrite as nitrogen source (Nishimura *et al.*, 2007). Nitrate utilization systems found in bacteria include membrane-associated proteins for dissimilatory nitrate reduction (NAR), periplasmic nitrate reduction proteins also involved in dissimilatory processes (NAP) as well as assimilatory nitrate reductases located in the cytoplasm of the cell (NAS) (Richardson *et al.*, 2001).

For *C. glutamicum* strains R and ATCC 13032 NAR activity was reported recently, while nitrite reductase activity, which is essential for ammonium generation, was not observed (Nishimura *et al.*, 2007; Takeno *et al.*, 2007). In contrast, NAS was found in different bacteria such as *Synechococcus* sp. PCC7942, *Bacillus subtilis* and *Mycobacterium smegmatis* (Khan *et al.*, 2008; Amon *et al.*, 2009). In *M. smegmatis* also putative genes encoding a NarK uptake system as well as nitrate and nitrite reductase were annotated (Khan *et al.*, 2008; Amon *et al.*, 2009).

Heterologous expression of nitrite reductase genes of *M. smegmatis* in *C. glutamicum* led to a recombinant strain, which was able to metabolize nitrite as sole nitrogen source. Assimilation of nitrate by coupling of dissimilatory nitrate reductase from *C. glutamicum* and nitrite reductase from *M. smegmatis* was not achieved (Rehm, unpublished). Since nitrite is highly toxic, the assimilation of this compound by *C. glutamicum* might be favourable for bioremediation processes.

Nitrogen-rich wastewaters are readily available; however, they are often contaminated and not useful for production. Fore biotechnological production processes, side products from food industry are much more interesting. An example in this respect might be whey, which is remaining after removal of milk fat and casein from milk, e.g. during the production of cheese. Although of nutritional value, the liquid is one of the major disposal problems of dairy industry (Panesar and Kennedy, 2012). Since whey is rich in protein and other organic content including lactose, attempts were made already ten years ago to use whey as carbon and energy source for recombinant lysine-producing *C. glutamicum*; however, without much success since growth as slow and strain instabilities were observed (Barrett *et al.*, 2004). Nevertheless, growth of *C. glutamicum* was possible showing the potential of this feedstock, which should provide an excellent nitrogen source.

Chitin is one of the most abundant biopolymers on earth. It can be extracted from crab and shrimp shells and by acid hydrolysis of chitin, the amino sugar glucosamine and its acetylated derivative *N*-acetylglucosamine can be obtained (Liu *et al.*, 2013). Recombinant *C. glutamicum* strains are able to use these compounds not only as carbon but also as nitrogen sources (Uhde *et al.*, 2013; Matano *et al.*, 2014). Proof of principle of *N*-acetylglucosamine-based production of L-lysine and lycopene by recombinant *C. glutamicum* was already published (Matano *et al.*, 2014); however, if these amino sugars will be applied in production processes as novel feedstocks depends primarily on their market prize. Although chitin is a readily available renewable resource, application of these amino sugars in food, cosmetics and pharmaceutical industry (Liu *et al.*, 2013) results in high competition in respect to these resources.

Concluding remarks

C. glutamicum was successfully established as biotechnological platform organism in the last years (Becker and Wittmann, 2012; Chapter 1). One reason for this success might be its ability to assimilate vast amounts of ammonium by highly active assimilation pathways. As a consequence, high internal pools of L-glutamate and L-glutamine are established already in the wild type (Nolden *et al.*, 2001b; Rehm *et al.*, 2010b), which might serve as a nitrogen buffer and be beneficial at least for L-glutamate and L-glutamine production. In fact, a high robustness of ammonium assimilation and glutamate pool control was observed (for review, see Rehm and Burkovski, 2011), which might be one reason for the versatility and robustness of *C. glutamicum* in biotechnological processes.

References

Adachi, Y., Hayashi, M., Senoo, A., Yonetani, Y., and Hashimoto, S.-i. (2012) Method for production of L-glutamine. Patent No US 8,187,843 B2.

Ajinomoto Co., Inc. (2013a) Feed-use amino acids business. Available at: http://www.ajinomoto.com/en/ir/pdf/Feed-useAA-Oct2013.pdf.

Ajinomoto Co., Inc. (2013b) First quarter-FY2013 market and other information. Available at: http://www.ajinomoto.com/en/ir/pdf/FY13_Data_E.pdf.

Amon, J.,Titgemeyer, F., and Burkovski, A. (2009) A genomic view on nitrogen metabolism and nitrogen control in mycobacteria. *J. Mol. Microbiol. Biotechnol.* **17**, 20–29.

Amon, J., Titgemeyer, F., and Burkovski, A. (2010) Common patterns – unique features: nitrogen metabolism and regulation in Gram-positive bacteria. *FEMS Microbiol. Rev.* **34**, 588–605.

Asakura, Y., Kimura, E., Usuda, Y., Kawahara, Y., Matsui, K., Osumi, T., and Nakamutsa, T. (2007) Altered metabolic flux due to deletion of *odhA* causes L-glutamate overproduction in *Corynebacterium glutamicum*. *Appl. Environ. Microbiol.* **73**, 1308–1319.

Barrett, E., Stanton, C., Zelder, O., Fitzgerald, G., and Ross, R. P. (2004) Heterologous expression of lactose- and galactose-utilizing pathways from lactic acid bacteria in *Corynebacterium glutamicum* for production of lysine in whey. *Appl. Environ. Microbiol.* **70**, 2861–2866.

Becker, J., and Wittmann, C. (2012) Bio-based production of chemicals, materials and fuels – *Corynebacterium glutamicum* as versatile cell factory. *Curr. Opin. Biotechnol.* **23**, 631–640.

Becker, M., Börngen, K., Nomura, T., Battle, A. R., Marin, K., Martinac, B., and Krämer, R. (2013) Glutamate efflux mediated by *Corynebacterium glutamicum* MscCG, *Escherichia coli* MscS, and their derivatives. *Biochim. Biophys. Acta* **1828**, 1230–1240.

Beckers, G., Nolden, L., and Burkovski, A. (2001) Glutamate synthase of *Corynebacterium glutamicum* is not essential for glutamate synthesis and is regulated by the nitrogen status. *Microbiology* **147**, 2961–2970.

Beckers, G., Strösser, J., Hildebrandt, U., Kalinowski, J., Farwick, M., Krämer, R., and Burkovski, A. (2005) Regulation of AmtR-controlled gene expression in *Corynebacterium glutamicum*: mechanism and characterization of the AmtR regulon. *Mol. Microbiol.* **58**, 580–595.

Bendt, A. K., Beckers, G., Silberbach, M., Wittmann, A., and Burkovski, A. (2004) Utilization of creatinine as an alternative nitrogen source in *Corynebacterium glutamicum*. *Arch. Microbiol.* **181**, 443–450.

Börmann, E. R., Eikmanns, B. J., and Sahm, H. (1992) Molecular analysis of the *Corynebacterium glutamicum gdh* gene encoding glutamate dehydrogenase. *Mol. Microbiol.* **6**, 317–326.

Börmann-El Kholy, E. R., Eikmanns, B. J., Gutmann, M., and Sahm, H. (1993) Glutamate dehydrogenase is not essential for glutamate formation in *Corynebacterium glutamicum*. *Appl. Environ. Microbiol.* **59**, 2329–2331.

Bott, M. (2007) Offering surprises: TCA cycle regulation in *Corynebacterium glutamicum*. *Trends Microbiol.* **15**, 417–425.

Boulahya, K.-A., Guedon, E., Delaunay, S., Schultz, C., Boudrant, J., Bott, M., and Goergen, J.-L. (2010) OdhI dephosphorylation kinetics during different glutamate production processes involving *Corynebacterium glutamicum*. *Appl. Microbiol. Biotechnol.* **87**, 1867–1874.

Buchinger, S., Strösser, J., Rehm, N., Hänßler, E., Hans, S., Bathe, B., Schomburg, D., Krämer, R., and Burkovski, A. (2009) A combination of transcriptome and metabolome analyses reveals new targets of the *Corynebacterium glutamicum* nitrogen regulator AmtR. *J. Biotechnol.* **140**, 68–74.

Burkovski, A. (2005) Nitrogen metabolism and its regulation. In: Bott, M., and Eggeling, L. (eds.) Handbook of *Corynebacterium glutamicum*. CRC Press LLC, Boca Raton, FL, USA, pp. 333–349.

Burkovski, A. (2007) Nitrogen control in *Corynebacterium glutamicum*: proteins, mechanisms, signals. *J. Microbiol. Biotechnol.* **17**, 187–194.

Hänßler, E., and Burkovski, A. (2008) Molecular mechanisms of nitrogen control in Corynebacteria. In: Burkovski, A. (ed.) Corynebacteria: Genomics and Molecular Biology. Caister Academic Press, Norfolk, UK, pp. 183–201.

Hänßler, E., Müller, T., Palumbo, K., Patek, M., Brocker, M., Krämer, R., and Burkovski, A. (2009) A game with many players: regulation of *gdh* transcription in *Corynebacterium glutamicum*. *J. Biotechnol.* **142**, 114–122.

Hasselt, K., Rankl, S., Worsch, S., and Burkovski, A. (2011) Adaptation of AmtR-controlled gene expression by modulation of AmtR binding activity in *Corynebacterium glutamicum*. *J. Biotechnol.* **154**, 156–162.

Hayashi, M., Maeda, M., and Yonetani, Y. (2008) Method for production of L-glutamine. Patent No EP1978094A1.

Ikeda, M., and Nakagawa, S. (2003) The *Corynebacterium glutamicum* genome: features and impacts on biotechnological processes. *Appl. Microbiol. Biotechnol.* **62**, 99–109.

Jakoby, M., Tesch, M., Sahm, H., Krämer, R., and Burkovski, A. (1997) Isolation of the *Corynebacterium glutamicum glnA* gene encoding glutamine synthetase I. *FEMS Microbiol. Lett.* **154**, 81–88.

Jakoby, M., Krämer, R., and Burkovski, A. (1999) Nitrogen regulation in *Corynebacterium glutamicum*: isolation of genes involved and biochemical characterization of corresponding proteins. *FEMS Microbiol. Lett.* **173**, 303–310.

Jakoby, M., Nolden, L., Meier-Wagner, J., Krämer, R., and Burkovski, A. (2000) AmtR, a global repressor in the nitrogen regulation system of *Corynebacterium glutamicum*. *Mol. Microbiol.* **37**, 964–977.

Jeßberger, N., Burkovski, A., Bathe, B., and Reth, A. (2014) Process for preparing L-amino acids. Patent No US 8,697,850 B2.

Kalinowski, J., Bathe, B., Bischoff, N., Bott, M., Burkovski, A., Dusch, N., Eggeling, L., Eikmanns, B. J., Gaigalat, L., Goesmann, A., *et al.* (2003) The complete *Corynebacterium glutamicum* ATCC 13032 genome sequence and its impact on the production of L-aspartate-derived amino acids and vitamins. *J. Biotechnol.* **104**, 5–25.

Kalinowski, J., Wolters, D., and Poetsch, A. (2008) Proteomics of *Corynebacterium glutamicum* and other corynebacteria. In: Burkovski, A. (ed.) Corynebacteria: Genomics and Molecular Biology. Caister Academic Press, Norfolk, UK, pp 55–78.

Kawahara, Y., Takahashi-Fuke, K., Shimizu, E., Nakamatsu, T., and Nakamori, S. (1997) Relationship between the glutamate production and the activity of 2-oxoglutarate dehydrogenase in *Brevibacterium lactofermentum*. *Biosci. Biotech. Biochem.* **61**, 1109–1112.

Khan, A., Akhtar, S., Ahmad, J. N., and Sarkar, D. (2008) Presence of a functional nitrate assimilation pathway in *Mycobacterium smegmatis*. *Microbial Path.* **44**, 71–77.

Kim, J., Hirasawa, T., Sato, Y., Nagahisa, K., Furusawa, C., and Shimizu, H. (2009) Effect of *odhA* overexpression and *odhA* antisense RNA expression on Tween-40-triggered glutamate production by *Corynebacterium glutamicum*. *Appl. Microbiol. Biotechnol.* **81**, 1097–1106.

Kim, J., Fukuda, H., Hirasawa, T., Nagahisa, K., Nagai, K., Wachi, M., and Shimizu, H. (2010) Requirement of *de novo* synthesis of the OdhI protein in penicillin-induced glutamate production in *Corynebacterium glutamicum*. *Appl. Microbiol. Biotechnol.* **86**, 911–910.

Kinoshita, S., Udaka, S., and Shimono, M. (1957) Studies on the amino acid fermentation part.1. Production of L-glutamic acid by various microorganisms. J. *Gen. Appl. Microbiol.* **3**, 193–205.

Krawczyk, S., Raasch, K., Schultz, C., Hoffelder, M., Eggeling, L., and Bott, M. (2010) The FHA domain of OdhI interacts with the carboxyterminal 2-oxoglutarate dehydrogenase domain of OdhA in *Corynebacterium glutamicum*. *FEBS Lett.* **584**, 1463–1468.

Kusumoto, I. (2001) Industrial production of L-glutamine. *J. Nutr.* **131**, 2552S–2555S.

Li, J., Ma, C., Ma, Y., Li, Y., Zhou, W., and Xu, P. (2007) Medium optimization by combination of response surface methodology and desirability function: an application in glutamine production. *Appl. Microbiol. Biotechnol.* **74**, 563–571.

Liu, L., Liu, Y., Shin, H.-d., Shen, R., Li, J., Du, G., and Chen, J. (2013) Microbial production of glucosamine and *N*-acetylglucosamine: advances and perspectives. *Appl. Microbiol. Biotechnol.* **97**, 6149–6158.

Liu, Q., Zhang, J., Wei, X.-X., Ouyang, S.-P., Wu, Q., and Chen, G.-Q. (2008) Microbial production of L-glutamate and L-glutamine by recombinant *Corynebacterium glutamicum* harboring *Vitreoscilla* hemoglobin gene *vgb*. *Appl. Microbiol. Biotechnol.* **77**, 1297–1304.

Marx, A., Eikmanns, B. J., Sahm, H., de Graaf, A. A., and Eggeling, L. (1999) Response of the central metabolism in *Corynebacterium glutamicum* to the use of an NADH-dependent glutamate dehydrogenase. *Metab. Eng.* **1**, 35–48.

Marienhagen, J., and Eggeling, L. (2008) Metabolic function of *Corynebacterium glutamicum* aminotransferases AlaT and AvtA and impact on L-valine production. *Appl. Environ. Microbiol.* **74**, 7457–7462.

Marienhagen, J., Kennerknecht, N., Sahm, H., and Eggeling, L. (2005) Functional analysis of all aminotransferase proteins inferred from the genome sequence of *Corynebacterium glutamicum*. *J. Bacteriol.* **187**, 7639–7646.

Matsuzaki, Y., Nakamura, J., and Hashiguchi, K. (2004) Method for producing L-arginine or L-lysine by fermentation. Patent No EP1460128A1.

Meier-Wagner, J., Nolden, L., Jakoby, M., Siewe, R., Krämer, R., and Burkovski, A. (2001) Multiplicity of ammonium uptake systems in *Corynebacterium glutamicum*: role of Amt and AmtB. *Microbiology* **147**, 135–143.

Misono, H., and Soda, K. (1980) Properties of *meso*-α,ε-diaminopimelate D-dehydrogenase from *Bacillus sphaericus*. *J. Biol. Chem.* **255**, 10599–10605.

Muhl, D., Jeßberger, N., Hasselt, K., Jardin, C., Sticht, H., and Burkovski, A. (2009) DNA binding by *Corynebacterium glutamicum* TetR-type transcription regulator AmtR. *BMC Mol. Biol.* **10**, 73.

Müller, T., Strösser, J., Buchinger, S., Nolden, L., Wirtz, A., Krämer, R., and Burkovski, A. (2006) Mutation-induced metabolite pool alterations in *Corynebacterium glutamicum*: towards the identification of nitrogen control signals. *J. Biotechnol.* **126**, 440–453.

Nakamura, J., Izui, H., Moriguchi, K., Kawashima, H., Nakamatsu, T., and Kurahashi, O. (2002) Method for producing L-glutamine by fermentation and L-glutamine producing bacterium. Patent No EP1229121A2.

Nakamura, J., Izui, H., Moriguchi, K., Kawashima, H., Nakamatsu, T., and Kurahashi, O. (2004) Method for producing L-glutamine by fermentation and L-glutamine producing bacterium. Patent No EP1424398A2.

Nakamura, J., Hirano, S., Ito, H., and Wachi, M. (2007) Mutations of the *Corynebacterium glutamicum* NCgl1221 gene, encoding a mechanosensitive channel homolog, induce L-glutamic acid production. *Appl. Environ. Microbiol.* **73**, 4491–4498.

Niebisch, A., Kabus, A., Schultz, C., Weil, B., and Bott, M. (2006) Corynebacterial protein kinase G controls 2-oxoglutarate dehydrogenase activity via the phosphorylation status of the OdhI protein. *J. Biol. Chem.* **281**, 12300–12307.

Nishimura, T., Vertes, A. A., Shinoda, Y., Inui, M., and Yukawa, H. (2007) Anaerobic growth of *Corynebacterium glutamicum* using nitrate as a terminal electron acceptor. *Appl. Microbiol. Biotechnol.* **75**, 889–897.

Nolden, L., Farwick, M., Krämer, R., and Burkovski, A. (2001a) Glutamine synthetases in *Corynebacterium glutamicum*: transcriptional control and regulation of activity. *FEMS Microbiol. Lett.* **201**, 91–98.

Nolden, L., Ngouoto-Nkili, C.-E., Bendt, A. K., Krämer, R., and Burkovski, A. (2001b) Sensing nitrogen limitation in *Corynebacterium glutamicum*: the role of *glnK* and *glnD*. *Mol. Microbiol.* **42**, 1281–1295.

Panesar, P. S., and Kennedy, J. F. (2012) Biotechnological approaches for the value addition of whey. *Crit. Rev. Biotechnol.* **32**, 327–348.

Poetsch, A., Haußmann, U., and Burkovski, A. (2011) Proteomics of Corynebacteria: from biotechnology workhorses to pathogens. *Proteomics* **11**, 3244–3255.

Radmacher, E., Stansen, K. C., Besra, G. S., Alderwick, L. J., Maughan, W. N., Hollweg, G., Sahm, H., Wendisch, V. F., and Eggeling, L. (2005) Ethambutal, a cell wall inhibitor of *Mycobacterium tuberculosis*, elicits L-glutamate efflux of *Corynebacterium glutamicum*. *Microbiology* **151**, 1359–1368.

Ramos, J. L., Martinéz-Bueno, M., Molina-Henares, A. J., Terán, W., Watanabe, K., Zhang, X., Gallegos, M. T., Brennan, R., and Tobes, R. (2005) The TetR family of transcriptional repressors. *Microbiol. Mol. Biol. Rev.* **69**, 326–356.

Rehm, N., and Burkovski, A. (2011) Engineering of nitrogen metabolism and its regulation in *Corynebacterium glutamicum*: influence on amino acid pools and production. *Appl. Microbiol. Biotechnol.* **89**, 239–248.

Rehm, N., Buchinger, S., Strösser, J., Walter, B., Hans, S., Bathe, B., Schomburg, D., Krämer, R., and Burkovski, A. (2010a) Impact of adenylyltransferase GlnE on nitrogen starvation response in *Corynebacterium glutamicum*. *J. Biotechnol.* **145**, 244–252.

Rehm, N., Georgi, T., Hiery, E., Degner, U., Schmiedl, A., Burkovski, A., and Bott, M. (2010b) L-Glutamine as nitrogen source for *Corynebacterium glutamicum*: derepression of the AmtR regulon and implications for nitrogen sensing. *Microbiology* **156**, 3180–3193.

Reimer, L. C., Spura, J., Schmidt-Hohagen, K., and Schomburg, D. (2014) High-throughput screening of a *Corynebacterium glutamicum* mutant library on genomic and metabolic level. *PLoS One* **9**, e86799.

Richardson, D. J., Berks, B. C., Russell, D. A., Spiro, S., and Taylor, C. J. (2001) Functional, biochemical and genetic diversity of prokaryotic nitrate reductases. *Cell. Mol. Life Sci.* **58**, 165–178.

Sahm, H., Eggeling, L., and de Graaf, A. A. (2000) Pathway analysis and metabolic engineering in *Corynebacterium glutamicum*. *Biol. Chem.* **381**, 899–910.

Schrumpf, B., Schwarzer, A., Kalinowski, J., Pühler, A., Eggeling,L., and Sahm, H. (1991) A functionally split pathway for lysine synthesis in *Corynebacterium glutamicum*. *J. Bacteriol.* **173**, 4510–4516.

Schultz, C., Niebisch, A., Gebel, L., and Bott, M. (2007) Glutamate production by *Corynebacterium glutamicum*: dependence on the oxoglutarate dehydrogenase inhibitor protein OdhI and protein kinase PknG. *Appl. Microbiol. Biotechnol.* **76**, 691–700.

Schultz, C., Niebisch, A., Schwaiger, A., Viets, U., Metzger, S., Bramkamp, M., and Bott, M. (2009) Genetic and biochemical analysis of the serine/threonine protein kinases PknA, PknB, PknG and PknL of *Corynebacterium glutamicum*: evidence for non-essentiality and for phosphorylation of OdhI and FtsZ by multiple kinases. *Mol. Microbiol.* **74**, 724–741.

Shiio, I., and Ozaki, H. (1970) Regulation of nicotinamide adenine dinucleotide phosphate-specific glutamate dehydrogenase from *Brevibacterium flavum*, a glutamate-producing bacterium. *J. Biochem.* **68**, 633–647.

Shiio, I., and Ujigawa-Takeda, K. (1980) Presence and regulation of α-ketoglutarate dehydrogenase complex in a glutamate-producing bacterium, *Brevibacterium flavum*. *Agric. Biol. Chem.* **44**, 1897–1904.

Shimizu, H., Tanaka, H., Nakato, A., Nagahisa, K., Kimura, E., and Shioya, S. (2003) Effects of the changes in enzyme activities on metabolic flux redistribution around the 2-oxoglutarate branch in glutamate production by *Corynebacterium glutamicum*. *Bioprocess Biosyst. Eng.* **25**, 291–298.

Shirai, T., Nakato, A., Izutani, N., Nagahisa, K., Shioya, S., Kimura, E., Kawarabayasi, Y., Yamagishi, A., Gojobori, T., and Shimizu, H. (2005) Comparative study of flux redistribution of metabolic pathway in glutamate production by two coryneform bacteria. *Metab. Eng.* **7**, 59–69.

Siewe, R. M., Weil, B., Burkovski, A., Eikmanns, B. J., Eikmanns, M., and Krämer, R. (1996) Functional and genetic characterization of the (methyl)ammonium uptake carrier of *Corynebacterium glutamicum*. *J. Biol. Chem.* **271**, 5398–5403.

Sindelar, G., and Wendisch, V. F. (2007) Improving lysine production by *Corynebacterium glutamicum* through DNA microarray-based identification of novel target genes. *Appl. Microbiol. Biotechnol.* **76**, 677–689.

Sonntag, K., Eggeling, L., de Graaf, A. A., and Sahm, H. (1993) Flux partitioning in the split pathway of lysine synthesis in *Corynebacterium glutamicum*. Quantification by ^{13}C- and ^{1}H-NMR spectroscopy. *Eur. J. Biochem.* **213**, 1325–1331.

Strösser, J., Lüdke, A., Schaffer, S., Krämer, R., and Burkovski, A. (2004) Regulation of GlnK activity: modification, membrane sequestration, and proteolysis as regulatory principles in the network of nitrogen control in *Corynebacterium glutamicum*. *Mol. Microbiol.* **54**, 132–147.

Takeno, S., Ohnishi, J., Komatsu, T., Masaki, T., Sen, K., and Ikeda, M. (2007) Anaerobic growth and potential for amino acid production by nitrate respiration in *Corynebacterium glutamicum*. *Appl. Microbiol. Biotechnol.* **75**, 1173–1182.

Tesch, M., Eikmanns, B. J., de Graaf, A. A., and Sahm, H. (1998) Ammonia assimilation in *Corynebacterium glutamicum* and a glutamate dehydrogenase-deficient mutant. *Biotechnol. Lett.* **20**, 953–957.

Tesch, M., de Graaf, A. A., and Sahm, H. (1999) *In vivo* fluxes in the ammonium-assimilatory pathways in *Corynebacterium glutamicum* studied by ^{15}N nuclear magnetic resonance. *Appl. Environ. Microbiol.* **65**, 1099–1109.

Tochikura, T., Sung, H.-C., Tachiki, T., and Kumagai, H. (1984) Occurence of glutamate synthase in *Brevibacterium flavum*. *Agric. Biol. Chem.* **48**, 2149–2150.

Trötschel, C., Kandirali, S., Diaz-Achirica, P., Meinhardt, A., Morbach, S., Krämer, R., and Burkovski, A. (2003) GltS, the sodium-coupled L-glutamate uptake system of *Corynebacterium glutamicum*: identification of the corresponding gene and impact on L-glutamate production. *Appl. Microbiol. Biotechnol.* **60**, 738–742.

Udaka, S. (2008) The discovery of *Corynebacterium glutamicum* and birth of amino acid fermentation industry in Japan. In: Burkovski, A. (ed.) Corynebacteria – Genomics and Molecular Biology. Caister Academic Press, Norfolk, UK, pp. 1–6.

Uhde, A., Youn, J.-W., Maeda, T., Clermont, L., Matano, C., Krämer, R., Wendisch, V. F., Seibold, G. M., and Marin, K. (2013) Glucosamine as carbon source for amino acid-producing *Corynebacterium glutamicum*. *Appl. Microbiol. Biotechnol.* **97**, 1679–1687.

Uy, D., Delauny, S., Germain, P., Engasser, J.-E., and Goergen, J.-L. (2003) Instability of glutamate production by *Corynebacterium glutamicum* 2262 in continuous culture using the temperature-triggered process. *J. Biotechnol.* **104**, 173–184.

Uy, D., Delauny, S., Goergen, J.-L., and Engasser, J.-E. (2005) Dynamics of glutamate synthesis and excretion fluxes in batch and continuous cultures of temperature-triggered *Corynebacterium glutamicum*. *Bioprocess Biosyst. Eng.* **27**, 153–162.

Walter, B., Hänßler, E., Kalinowski, J., and Burkovski, A. (2007) Nitrogen metabolism and nitrogen control in corynebacteria: variations of a common theme. *J. Mol. Microbiol. Biotechnol.* **12**, 131–138.

Walter, B., Küspert, M., Ansorge, A., Krämer, R., and Burkovski, A. (2008) Dissection of ammonium uptake systems in *Corynebacterium glutamicum*: mechanism and energetics of AmtA and AmtB. *J. Bacteriol.* **190**, 2611–2614.

Wendisch, V. F. (2008) DNA microarray-based transcriptome analysis of *Corynebacterium glutamicum*. In: Burkovski, A. (ed.) Corynebacteria: Genomics and Molecular Biology. Caister Academic Press, Norfolk, UK, pp. 33–54.

Wittmann, C., and Heinzle, E. (2008) Metabolic network analysis and design in *Corynebacterium glutamicum*. In: Burkovski, A. (ed.) Corynebacteria: Genomics and Molecular Biology. Caister Academic Press, Norfolk, UK, pp. 79–112.

Yamashita, C., Hashimoto, K.-i., Kumagai, K., Maeda, T., Takada, A., Yabe, I., Kawasaki, H., and Wachi, M. (2013) L-Glutamate secretion by the N-terminal domain of the *Corynebacterium glutamicum* NCgl1221 mechanosensitive channel. *Biosci. Biotechnol. Biochem.* **77**, 1008–1013.

Yasueda, H. (2014) Overproduction of L-glutamate in *Corynebacterium glutamicum*. In: Anazawa, H., and Shimizu, S. (eds.) Microbial Production: From Genome Design to Cell Engineering. Springer, Tokyo, Japan, pp. 165–176.

Zhang, Y., Shang, X., Lai, S., Zhang, G., Liang, Y., and Wen, T. (2012) Development and application of an arabinose-inducible expression system by facilitating inducer uptake in *Corynebacterium glutamicum*. *Appl. Environ. Microbiol.* **78**, 5831–5838.

Transport, Degradation and Assimilation of Aromatic Compounds and their Regulation in *Corynebacterium glutamicum*

7

Xi-Hui Shen, Tang Li, Ying Xu, Ning-Yi Zhou and Shuang-Jiang Liu

Abstract

The closing of the last gap of the genome sequence of *Corynebacterium glutamicum* opened a new door to insight in physiology and genetics. For example, long-overlooked pathways for aromatic degradation and assimilation by *C. glutamicum* were discovered by the access to genome data, and broad knowledge on the physiology and genetics of aromatic degradation and assimilation by *C. glutamicum* has been accumulated in the last 10 years. This chapter summarizes the current understanding of transport and metabolism of aromatic compounds by *C. glutamicum* as well as their regulation.

Introduction

Aromatic compounds widely occur in nature and are structurally originated from one or more benzene-rings. They are extensively used in agriculture, medicine and chemical industries. Although many commercially used aromatic compounds are chemically synthesized, lignin, that makes up 25–35% of woody biomass, is the most abundant natural aromatic compound (Tuck *et al.*, 2012). Many aromatic compounds are toxic or carcinogenic and are serious pollutants when improperly discharged into environments. For example, phenol is one of the major organic pollutants in coking wastewater. As known so far, many microorganisms have the ability to use aromatic compounds as sole carbon and energy sources for growth, and this ability is exploited for bioremediation of polluted environments.

As described in other chapters, *Corynebacterium glutamicum* plays a key role in today's fermentation industry for many products such as amino acids and vitamins. Hexose has been the major carbon source used for *C. glutamicum* fermentation. Efforts have been made to generate strains that use pentose, disaccharides or sugar derivatives (Blombach and Seibold, 2010; Meiswinkel *et al.*, 2013). However, the ability to use aromatic compounds by *C. glutamicum* had long been overlooked until its genome was sequenced. Genome annotation predicted that *C. glutamicum* might be able to degrade and assimilate various aromatic compounds. This chapter summarizes the current knowledge of transport, degradation and assimilation of aromatic compounds by *C. glutamicum*, and the corresponding regulation.

What do the *Corynebacterium glutamicum* genomes predict for degradation and assimilation of aromatic compounds?

Corynebacterium glutamicum genomes

The complete genome of *C. glutamicum* strain ATCC13032 was published independently by Ikeda and Nakagawa (2003) and Kalinowski *et al.* (2003). Originally isolated from soil (Udaka, 1960), it might be assumed that *C. glutamicum* is adapted to diverse organic substances including various aromatic compounds and *C. glutamicum* is able to use aromatic compounds as carbon source for growth. Microbial degradation of aromatic compounds includes various peripheral pathways, aromatic ring-cleavage reactions and central carbon metabolism (Díaz, 2004; Cao *et al.*, 2009; Fuchs *et al.*, 2011) (see Fig. 7.1). An early study based on the *C. glutamicum* genome sequence revealed that a gene annotated to encode a subunit of protocatechuate 3,4-dioxygenase encodes a catechol 1,2-dioxygenase, which is involved in phenol degradation (Shen *et al.*, 2004). This observation invoked further interest to get deeper insight into the potential of aromatic degradation by *C. glutamicum*. Further genome data-mining revealed that *C. glutamicum* carries additional genes involved in the degradation of 4-cresol, benzoate, 4-hydroxybenzoate, vanillate, vanillin, resorcinol, 3,5-dihydroxytolunene and 2,4-dihydroxybenzoate (Shen *et al.*, 2005a). At the time of writing this chapter, we revisited the *C. glutamicum* genome and new annotations have been made for genes involved in aromatic degradation and assimilation (Table 7.1).

Many of the genes listed in Table 7.1 have been functionally identified, and there gene products are involved in multiple peripheral pathways and key ring-cleavage reactions for aromatic compounds metabolism (see references listed in Table 7.1). However, there are still genes, e.g. encoding a putative intradiol ring-cleavage dioxygenase (*ncgl0234/cg0291*), that need to be functionally characterized. Interestingly and consistently to the fact that *C. glutamicum* is not able to degrade L-Phe, L-Tyr and L-Trp, none of the aromatic amino acid degradation pathways has been found in the *C. glutamicum* genomes (Ikeda and Nakagawa, 2003; Kalinowski *et al.*, 2003; Zhao *et al.*, 2011).

So far there are five genomes of *C. glutamicum* publicly available (Table 7.1), with sizes ranging from 3.23 to 3.36 Mb. The numbers of aromatic degradation pathways per Mb across the *C. glutamicum* was calculated to be 1.55 to 1.79. For other well-known aromatic compound degraders such as *Pseudomonas putida* KT2440 (Jiménez *et al.*, 2002), *Rhodococcus jostii* RHA1 (McLeod *et al.*, 2006), *Burkholderia xenovorans* LB400 (Chain *et al.*, 2006) and *Cupriavidus necator* JMP134 (Lykidis *et al.*, 2010), these numbers are 0.97, 0.82, 1.13 and 1.69, respectively. The high enrichment of genes encoding aromatic compound degradation pathways indicates that *C. glutamicum* possesses robust metabolic capabilities, making it a potential chassis bacterium for engineering of efficient strains for biodegradation and bioremediation of aromatic compounds.

Other genomes of the genus *Corynebacterium*

Currently, the genus *Corynebacterium* consists of more than 100 species, from 24 of them the genomes were sequenced (Shin *et al.*, 2011). Data-mining of those genomes indicated that although *C. glutamicum* strains are well equipped with metabolic potentials for aromatic compounds, this capability is highly variable among different *Corynebacterium* species (Table 7.2). *Corynebacterium* species originated from human and animal sources carry less genes for aromatic degradation pathways than those species from environmental samples. In particular, aromatic degradation pathways are absent from *C. diphtheriae, C. accolens, C. amycolatum, C. jeikeium, C. kroppenstedtii, C. lipophiloflavum, C. matruchotii, C. pseudogenitalium, C. resistens, C. tuberculostearicum* and *C. ulcerans*. The only exception is *C. ammoniagenes*, a species isolated from human intestinal tract, which carries 11 oxygenase markers in its genome (Perez-Pantoja *et al.*, 2012). The distribution of aromatic catabolic pathways among different *Corynebacterium* species suggests that this metabolic capability is highly influenced by their ecological niches.

The most broadly distributed aromatic ring-cleavage reactions are pathways for protocatechuate degradation by protocatechuate 3,4-dioxygenase (P34D) and catechol degradation by catechol 1,2-dioxygenase (C12D) as key enzymes, which were observed in almost all genomes from soil-dwelling *Corynebacterium* species and also in the intestinal *C. ammoniagenes*. A recently identified phenylacetic acid (PAA) degradation pathway (Chen *et al.*, 2012) in *C. glutamicum* strain AS 1.542 is also encoded in the genomes of *C. efficiens, C. ammoniagenes, C. casei, C. nuruki* and *C. striatum*. A gentisate degradation pathway with gentisate 1,2-dioxygenase (G12D) as key enzyme was found in *C.*

Table 7.1 New annotations for genes involved in aromatic compound degradation with *Corynebacterium glutamicum* genomes. Five genomes have been explored for *C. glutamicum* strains ATCC 13032 (Ikeda and Nakagawa, 2003; Kalinowski *et al.*, 2003), R (Yukawa *et al.*, 2007), ATCC 14067 (Lv *et al.*, 2012), K051 (Binder *et al.*, 2012), and S9114 (Lv *et al.*, 2011), respectively. Those genes that have been functionally characterized are shown in bold and references are listed

Corynebacterium glutamicum strains					Gene names	Products	References
ATCC 13032	R	ATCC 14067	K051	S9114			
ncgl2299/cg2615	***cgR_2264***	**KIQ_00590**	**WA5_2299**	**CgS9114_08891**	***vanR***	**PadR-like family transcriptional regulator**	Merkens *et al.* (2005)
ncgl2300/cg2616	***cgR_2265***	**KIQ_00585**	**WA5_2300**	**CgS9114_08886**	***vanA***	**Vanillate-*O*-demethylase subunit A**	Merkens *et al.* (2005)
ncgl2301/cg2617	***cgR_2266***	**KIQ_00580**	**WA5_2301**	**CgS9114_08881**	***vanB***	**Vanillate-*O*-demethylase subunit B**	Merkens *et al.* (2005)
ncgl2302/cg2618	***cgR_2267***	**KIQ_00575**	**WA5_2302**	**CgS9114_08876**	***vanK***	**Vanillate transporter (MFS superfamily)**	Chaudhry *et al.* (2007)
ncgl2306/cg2622	*cgR_2271*	KIQ_00555	WA5_2306	CgS9114_08856	*pcaJ*	β-Ketoadipate succinyl-CoA transferase subunit B	
ncgl2307/cg2623	*cgR_2272*	KIQ_00550	WA5_2307	CgS9114_08851	*pcaI*	β-Ketoadipate succinyl-CoA transferase subunit A	
ncgl2308/cg2624	*cgR_2273*	KIQ_00545	WA5_2308	CgS9114_08846	*pcaR*	Transcriptional activator (IclR family)	
ncgl2309/cg2625	*cgR_2274*	KIQ_00540	WA5_2309	CgS9114_08841	*pcaF*	β-Ketoadipyl CoA thiolase	
ncgl2310/cg2626	*cgR_2275*	KIQ_00535	WA5_2310	CgS9114_08836	*pcaD*	β-Ketoadipate enolactone hydrolase	
ncgl2311/cg2627	***cgR_2276***	**KIQ_00530**	**WA5_2311**	**CgS9114_08831**	***pcaO***	**LuxR-family transcriptional regulator**	Zhao *et al.* (2010)
ncgl2312/cg2628	*cgR_2277*	KIQ_00525	WA5_2312	CgS9114_08826	*pcaC*	γ-Carboxymucono-lactone decarboxylase	
ncgl2313/cg2629	*cgR_2278*	KIQ_00520	WA5_2313	CgS9114_08821	*pcaB*	β-Carboxy-cis,cis-muconate cycloisomerase	
ncgl2314/cg2630	***cgR_2279***	**KIQ_00515**	**WA5_2314**	**CgS9114_08816**	***pcaG***	**Protocatechuate 3,4-dioxygenase α subunit**	Shen and Liu (2005)
ncgl2315/cg2631	***cgR_2280***	**KIQ_00510**	**WA5_2315**	**CgS9114_08811**	***pcaH***	**Protocatechuate 3,4-dioxygenase β subunit**	Shen and Liu (2005)
ncgl2317/cg2634	*cgR_2282*	KIQ_00500	WA5_2317	CgS9114_08801	*catC*	Muconolactone isomerase	
ncgl2318/cg2635	*cgR_2283*	KIQ_00495	WA5_2318	CgS9114_08796	*catB*	Cis,cis-muconate cycloisomerase	
ncgl2319/cg2636	***cgR_2284***	**KIQ_00490**	**WA5_2319**	**CgS9114_08791**	***catA***	**Catechol 1,2-dioxygenase**	Shen *et al.* (2004)
ncgl2320/cg2637	*cgR_2285*	KIQ_00485	WA5_2320	CgS9114_08786	*benA*	Benzoate dioxygenase large subunit	
ncgl2321/cg2638	*cgR_2286*	KIQ_00480	WA5_2321	CgS9114_08781	*benB*	Benzoate dioxygenase small subunit	
ncgl2322/cg2639	*cgR_2287*	KIQ_00475	WA5_2322	CgS9114_08776	*benC*	Benzoate dioxygenase reductase subunit	
ncgl2323/cg2640	*cgR_2288*	KIQ_00470	WA5_2323	CgS9114_08771	*benD*	2-Hydro-1,2-dihydroxy-benzoate dehydrogenase	
ncgl2324/cg2641	*cgR_2289*	KIQ_00465	WA5_2324	CgS9114_08766	*benR*	LuxR-family transcriptional regulator	
ncgl2325/cg2642	***cgR_2290***	**KIQ_00460**	**WA5_2325**	**CgS9114_08761**	***benK***	**Benzoate transport protein**	Haußmann *et al.* (2009)
ncgl2326/cg2643	***cgR_2291***	**KIQ_00455**	**WA5_2326**	**CgS9114_08756**	***benE***	**Benzoate membrane transport protein**	Haußmann *et al.* (2009)

Table 7.1 (Continued)

Corynebacterium glutamicum strains					Gene names	Products	References
ATCC 13032	R	ATCC 14067	K051	S9114			
	cgR_2295			CgS9114_08736		Rieske non-haem iron oxygenase (RO) family	
	cgR_2296			CgS9114_08731		Aromatic hydrocarbon dioxygenase small subunit	
	cgR_2297			CgS9114_08726		Short-chain dehydrogenase/reductase	
	cgR_2298			CgS9114_08721		Ferredoxin	
	cgR_2299			CgS9114_08716		NarL family two-component response regulator	
	cgR_2301			CgS9114_08706		Major facilitator superfamily permease	
	cgR_2302			CgS9114_08696		Monooxygenase	
	cgR_2303			CgS9114_08691		Hypothetical protein	
	cgR_2304			CgS9114_08686		TRAP dicarboxylate transporter subunit Dct	
	cgR_2305			CgS9114_08681		TRAP dicarboxylate transporter subunit DctQ	
	cgR_2306			CgS9114_08676		TRAP dicarboxylate transporter subunit DctM	
ncgl2918/cg3349	***cgR_2908***	**KIQ_06322**	**WA5_2918**	**CgS9114_03900**	***genM***	**Maleylpyruvate isomerase**	Feng *et al.* (2006)
ncgl2919/cg3350	***cgR_2909***	**KIQ_06327**	**WA5_2919**	**CgS9114_03905**	***genF***	**Fumarylpyruvate hydrolase**	Shen *et al.* (2005b)
ncgl2920/cg3351	***cgR_2910***	**KIQ_06332**	**WA5_2920**	**CgS9114_03910**	***genD***	**Gentisate 1,2-dioxygenase**	Shen *et al.* (2005b)
ncgl2921/cg3352	***cgR_2911***	**KIQ_06337**	**WA5_2921**	**CgS9114_03915**	***genR***	**Transcriptional regulator (IclR family)**	Shen *et al.* (2005b)
ncgl2922/cg3353	***cgR_2912***	**KIQ_06342**	**WA5_2922**	**CgS9114_03920**	***genK***	**Gentisate transporter**	Xu *et al.* (2012)
ncgl2923/cg3354	***cgR_2913***	**KIQ_06347**	**WA5_2923**	**CgS9114_03925**	***genH***	**3-Hydroxybenzoate hydroxylase**	Shen *et al.* (2005b)
ncgl1110/cg1308	***cgR_1239***		**WA5_1110**	**CgS9114_03403**	***rolR***	**TetR family transcriptional regulator**	Li *et al.* (2012)
ncgl1111/cg1309	***cgR_1240***		**WA5_1111**	**CgS9114_03398**	***rolH***	**Resorcinol hydroxylase**	Huang *et al.* (2006)
ncgl1112/cg1310	***cgR_1241***		**WA5_1112**	**CgS9114_03393**	***rolM***	**Maleylacetate reductase**	Huang *et al.* (2006)
ncgl1113/cg1311	***cgR_1242***		**WA5_1113**	**CgS9114_03388**	***rolD***	**Hydroxyquinol 1,2-dioxygenase**	Huang *et al.* (2006)
ncgl2950/cg3384	*cgR_2940*	KIQ_06477	WA5_2950	CgS9114_04060		TetR family transcriptional regulator	
ncgl2951/cg3385	***cgR_2941***	**KIQ_06482**	**WA5_2951**	**CgS9114_04065**		**Hydroxyquinol 1,2-dioxygenase**	Huang *et al.* (2006)
ncgl2952/cg3386	***cgR_2942***	**KIQ_06487**	**WA5_2952**	**CgS9114_04070**		**Maleylacetate reductase**	Huang *et al.* (2006)
ncgl2954/cg3388	*cgR_2944*	KIQ_06497	WA5_2954	CgS9114_04080		IclR family transcriptional regulator	
ncgl0405/cg0500	***cgR_0491***	**KIQ_03790**	**WA5_0405**	**CgS9114_05292**	***qsuR***	**LysR-type transcriptional regulator**	Teramoto *et al.* (2009)

ncgl0406/cg0501	***cgR_0492***	**KIQ_03795**	**WA5_0406**	**CgS9114_05297**	***qsuA***	**Shikimate transporter**	Teramoto *et al*. (2009)
ncgl0407/cg0502	***cgR_0493***	**KIQ_03800**	**WA5_0407**	**CgS9114_05302**	***qsuB***	**Dehydroshikimate dehydratases**	Teramoto *et al*. (2009)
ncgl0408/cg0503	***cgR_0494***	**KIQ_03805**	**WA5_0408**	**CgS9114_05307**	***qsuC***	**Type II dehydroquinate dehydratase**	Teramoto *et al*. (2009)
ncgl0409/cg0504	***cgR_0495***	**KIQ_03810**	**WA5_0409**	**CgS9114_05312**	***qsuD***	**Quinate/shikimate dehydrogenase**	Teramoto *et al*. (2009)
	cgR_0016			CgS9114_04430	*maoA*	Amine oxidase	
	cgR_0018			CgS9114_04440	*padA*	Phenylacetaldehyde dehydrogenase	
	cgR_0643			CgS9114_01918	*paaT*	Shikimate transporter	
	cgR_0644			CgS9114_01923	*paaK*	Phenylacetate-CoA ligase	
	cgR_0645			**CgS9114_01928**	***paaR***	**TetR family transcriptional regulator**	Chen *et al*. (2012)
	cgR_0646			CgS9114_01933	*paaI*	Thioesterase, most active with ring-hydroxylated phenylacetyl-coenzyme A thioesters	
	cgR_0647			CgS9114_01938	*paaA*	Ring 1,2-phenylacetyl-CoA epoxidase subunit A	
	cgR_0648			CgS9114_01943	*paaB*	Ring 1,2-phenylacetyl-CoA epoxidase subunit B	
	cgR_0649			CgS9114_01948	*paaC*	Ring 1,2-phenylacetyl-CoA epoxidase subunit C	
	cgR_0650			CgS9114_01953	*paaD*	Ring 1,2-phenylacetyl-CoA epoxidase subunit D	
	cgR_0651			CgS9114_01958	*paaE*	Ring 1,2-phenylacetyl-CoA epoxidase NAD(P)H oxidoreductase component	
	cgR_0652			CgS9114_01963	*paaG*	Ring 1,2-epoxyphenylacetyl-CoA isomerase	
	cgR_0653			CgS9114_01968	*paaJ*	3-Oxoadipyl-CoA/3-oxo-5,6-dehydrosuberyl-CoA thiolase	
	cgR_0654			CgS9114_01973	*paaF*	2,3-Dehydroadipyl-CoA hydratase	
	cgR_0655			CgS9114_01978	*paaH*	3-Hydroxyadipyl-CoA dehydrogenase	
	cgR_0656			CgS9114_01983		Thioesterase superfamily protein	
	cgR_0657			CgS9114_01988	*paaZ*	Oxepin-CoA hydrolase/3-oxo-5,6-dehydrosuberyl-CoA semialdehyde dehydrogenase	
ncgl1031/cg1225	***cgR_1159***	**KIQ_09355**	**WA5_1031**	**CgS9114_14242**	***pcaK***	**4-Hydroxybenzoate transporter**	Chaudhry *et al*. (2007)
ncgl1032/cg1226	***cgR_1160***	**KIQ_09350**	**WA5_1032**	**CgS9114_14247**	***pobA***	**4-Hydroxybenzoate hydroxylase**	Huang *et al*. (2008)
ncgl0388/cg0480	*cgR_0472*	KIQ_01913	WA5_0388	CgS9114_05192	*fcs*	Feruloyl-CoA synthetase	
ncgl0446/cg0548	*cgR_0566*	KIQ_04035	WA5_0446	CgS9114_01543	*ech*	Enoyl-CoA hydratase/aldolase	
ncgl2578/cg2953	*cgR_2572*	KIQ_12390	WA5_2578	CgS9114_00910	*vdh*	Vanillin dehydrogenase	

Table 7.1 (Continued)

Corynebacterium glutamicum strains					Gene names	Products	References
ATCC 13032	R	ATCC 14067	K051	S9114			
ncgl0049/cg0067	*cgR_0068*	KIQ_05878	WA5_0049	CgS9114_04685		Aldehyde dehydrogenase	
ncgl0463/cg0567	*cgR_0583*	KIQ_04120	WA5_0463	CgS9114_01628		Aldehyde dehydrogenase	
ncgl0523/cg0637	*cgR_0659*	KIQ_04420	WA5_0523	CgS9114_01998		Aldehyde dehydrogenase	
ncgl0525/cg0639	*cgR_0661*	KIQ_04430	WA5_0525	CgS9114_02008		Ferredoxin reductase	
ncgl0527/cg0641	*cgR_0663*	KIQ_04440	WA5_0527	CgS9114_02018		Short-chain dehydrogenase	
ncgl0531/cg0646	*cgR_0667*	KIQ_04460	WA5_0531	CgS9114_02038		IclR family transcriptional regulator	
ncgl2587/cg2965	*cgR_2583*	KIQ_12315	WA5_2587	CgS9114_00855		AraC family transcriptional regulator	
ncgl2588/cg2966	*cgR_2584*	KIQ_12310	WA5_2588	CgS9114_00850		Phenol 2-monooxygenase	
ncgl0050/cg0070			WA5_0050			Phenol 2-monooxygenase	
ncgl0234/cg0291	*cgR_0311*	KIQ_13651	WA5_0234	CgS9114_05662		Intradiol ring-cleavage dioxygenase/chlorocatechol 1,2-dioxygenase	

Table 7.2 Surveys on genes potentially involved in aromatic compound degradation from *Corynebacterium* genomes

Strains	Sources	Genome (in Mb)	P34D	H23D	C12D	C23D	G12D	H12D	Icd	Paa	Dhp	BenDO	Pmo	VanOD	Mhb6H	Res4H	Phb3H	Rhd	Sal1H	Mhp
C. glutamicum strains																				
ATCC 13032	Soil	3.28	+		+		+	++	+			+	++	+	+	+	+			
R	Soil	3.36	+		+		+	++	+	+		+	+	+	+	+	+	+		
ATCC 14067	Soil	3.23	+		+		+	+	+			+	+	+	+		+			
K051	Soil	3.28	+		+		+	++	+			+	++	+	+	+	+			
S9114	Soil	3.29	+		+		+	++	+	+		+	+	+	+	+	+	+		
C. efficiens	Soil	3.22	+		+		+			+		+	+	+	+		+		+	
C. casei	Cheese	3.11	+	+					+	+			+							
C. nuruki	Alcohol fermentation starter	3.11	+		+					+		+					+	+		
C. variabile	Cheese	3.43							+					+						
C. ammoniagenes	Human	2.76	+	+	+	+				+	+	+	+				+	+		+
C. bovis	Cattle	2.52		+									+							
C. aurimucosum	Human	2.82		+									+							
C. diphtheriae	Human	2.54									+									
C. genitalium	Human	2.35																	+	
C. glucuronolyticum	Human	2.85		+									+							
C. striatum	Human	2.72								+			+							
C. urealyticum	Human	2.37																	+	

+ and ++ denote the presence of one and two copies, respectively, of the catabolic gene marker sequence in the genome.
P34D, protocatechuate 3,4-dioxygenase; C12D, catechol 1,2-dioxygenase; H12D, hydroxyquinol 1,2-dioxygenase; C23D, catechol 2,3-dioxygenase; H23D, homoprotocatechuate 2,3-dioxygenase; G12D, gentisate 1,2-dioxygenase; Paa, phenylacetyl-CoA oxygenase/reductase; Dhp, 2,3-dihydroxyphenylpropionate 1,2-dioxygenase; Mhp, 3-(3-hydroxyphenyl)propionate hydroxylase; Pmo, phenol monooxygenase; Phb3H, 4-hydroxybenzoate 3-hydroxylase; Mhb6H, 3-hydroxybenzoate 6-hydroxylase; Sal1H, salicylate 1-hydroxylase; Res4H, resorcinol 4-hydroxylase; BenDO, benzoate 1,2-dioxygenase; VanOD, vanillate 3-*O* demethylase; Icd, putative intradiol ring-cleavage dioxygenase; Rhd, putative Rieske (2Fe–2S) domain-containing ring-hydroxylating dioxygenase.
Aromatic degradation pathways are absent from *C. diphtheriae*, *C. accolens*, *C. amycolatum*, *C. jeikeium*, *C. kroppenstedtii*, *C. lipophiloflavum*, *C. matruchotii*, *C. pseudogenitalium*, *C. resistens*, *C. tuberculostearicum*, and *C. ulcerans*.

glutamicum and *C. efficiens*, and the hydroxyquinol degradation pathway was exclusively found in *C. glutamicum*. Additional aromatic compound degradation mechanisms such homoprotocatechuate 2,3-dioxygenase, catechol 2,3-dioxygenase and 2,3-dihydroxyphenylpropionate 1,2-dioxygenase pathways were observed in the genomes of other *Corynebacterium* species than *C. glutamicum*.

Corynebacterium glutamicum grows on various aromatic compounds

C. glutamicum grows on the following aromatic compounds as sole carbon sources: benzoate and phenol (Shen *et al.*, 2004), gentisate and 3-hydroxybenzoate (Shen *et al.*, 2005b), 4-hydroxybenzoate and protocatechuate and 4-cresol (Shen and Liu, 2005; Qi *et al.*, 2007), resorcinol, benzyl alcohol, 2,4-dihydroxybenzoate and 3,5-dihydroxytoluene (Shen *et al.*, 2005a), naphthalene (Lee *et al.*, 2010b), vanillin, vanillate and ferulic acid (Merkens *et al.*, 2005), caffeate, cinnamate and 4-coumarate (unpublished data). These structurally diverse aromatic compounds are metabolized through various peripheral pathways into a number of key intermediates and are subsequently channelled into central metabolic pathways (Fig. 7.1), which are similar to aromatic compound degradation as described for other bacteria (Díaz, 2004; Cao *et al.*, 2009; Fuchs *et al.*, 2011).

Physiological adaptation of *Corynebacterium glutamicum* growing on aromatic compounds compared with carbohydrates

During the past decades, proteomics has been emerged as a powerful tool to investigate various physiological processes, including microbial aromatic compound metabolism. Early studies were focused on identification of proteins that were directly involved in degradative pathways (Pessione *et al.*, 2003; Kim *et al.*, 2004b, 2006; Zhao *et al.*, 2004; Tomás-Gallardo *et al.*, 2006). In addition to such proteins involved in aromatic degradation, the proteomic analysis of *C. glutamicum* cells grown on various aromatic compounds revealed that central carbon metabolism and aromatic degradation were linked to each other (Qi *et al.*, 2007). The abundance of fructose-1,6-bisphosphatase (Fbp), catalysing the conversion of fructose-1,6-bisphosphate to fructose-6-phosphate as an essential enzyme for gluconeogenesis, consistently increased compared with glucose-dependent growth by a factor of 2.0–2.7 when *C. glutamicum* was grown on various aromatic compounds. Besides Fbp, the gluconeogenesis enzyme phosphoglycerate kinase (Pgk) was also increased when using phenol, 4-cresol, resorcinol and gentisate as sole carbon source (Qi *et al.*, 2007). The abundances of enzymes involved in TCA cycles such as citrate synthase and aconitase A were also changed when aromatic compounds served as carbon sources. These results indicated that the central carbon metabolism changed when cells grow on benzoate, 4-cresol, phenol or resorcinol (Qi *et al.*, 2007), in comparison to the cells grown on glucose.

Besides the above-described physiological changes, other physiological adaptation to growth on aromatic compounds were observed by comparative proteomics of *C. glutamicum*. Two benzoate transporters-encoding genes, *benK* and *benE*, were expressed when benzoate was used as sole carbon and energy source (Haußmann *et al.*, 2009). Up-regulation of proteins related to energy metabolism, such as subunits of succinate dehydrogenase, the cytochrome bc_1-aa_3 super-complex and ATP synthase, suggested that cells up-regulated TCA cycle and oxidative phosphorylation to meet increased energy requirements during growth on benzoate, which might be resulting from the lack of reduction equivalents and ATP from glycolysis (Haußmann *et al.*, 2009). Similar physiological adaptations were also observed for cells grown with protocatechuate as sole carbon and energy source (Haußmann and Poetsch, 2012). Down-regulation of the enzymes involved in amino acid biosynthesis and up-regulation of amino acid and putative sugar transporters seemed to be general starvation responses in cells grown with aromatic compounds (Haußmann *et al.*, 2009; Haußmann and Poetsch, 2012). Another aspect is that the proteins under regulation of McbR, the master

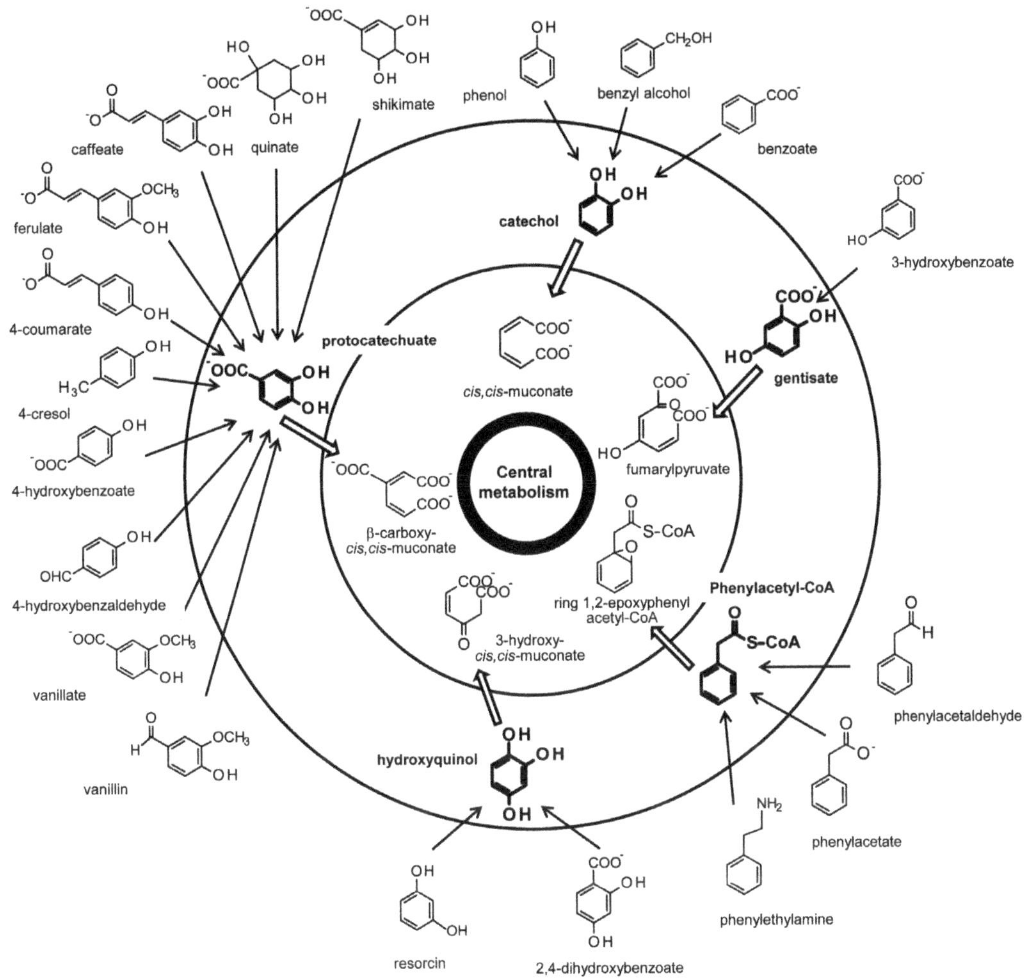

Figure 7.1 Illustration of microbial degradation of aromatic compounds and piling up of various aromatic compounds assimilated by *Corynebacterium glutamicum*. Circles from inside to outside indicate the central pathways, key aromatic ring cleavage and peripheral pathways of aromatic compound degradation.

regulator of sulfur metabolism in *C. glutamicum* (Rey *et al.*, 2005), were down-regulated in cells grown with benzoate (Haußmann *et al.*, 2009) or protocatechuate (Haußmann and Poetsch, 2012), suggesting a change of sulfur metabolism. Additionally, proteins related to cell wall biosynthesis showed differential expression during growth on protocatechuate. For example, MurA (UDP-*N*-acetylglucosamine 1-carboxyvinyltransferase; (Brown *et al.*, 1995)) was down-regulated while Wzz, a protein determining the length on polysaccharide chains (Franco *et al.*, 1998), was up-regulated compared with glucose-grown cells (Haußmann and Poetsch, 2012). Owing to their hydrophobicity, aromatic compounds can interact with membranes and induce stress responses in cells (Sikkema *et al.*, 1994); e.g. phenol and catechol induced a heat shock-like response and an oxidative stress response, respectively, in *Bacillus subtilis* (Tam le *et al.*, 2006). Additionally, in cells of *C. glutamicum* grown with protocatechuate, some stress-associated proteins such as UspA3 (Kvint *et al.*, 2003), DnaJ2 (Caplan *et al.*, 1993) and ClpC (Krüger *et al.*, 1994) were up-regulated (Haußmann and Poetsch, 2012). Taken together, *C. glutamicum* showed global physiological adaptation for aromatic compound metabolism, including altered energy metabolism, starvation response, oxidative stress response and other responses.

Uptake and transport of aromatic compounds in *Corynebacterium glutamicum*

The uptake of aromatic compounds into the cell is an essential step for their assimilation. Although some aromatic compounds can enter cells via passive diffusion, active transport increases the efficiency and rate of substrate acquisition especially when these substrates are present at low (micromolar) concentrations, resulting in higher survival chances of these microbes in natural environments. According to the classification of transporters by Ren *et al.* (Ren *et al.*, 2004), the majority of the currently known and functionally identified aromatic compound transport systems belong to the aromatic acid/ H^+ symporter (AAHS) family (2.A.1.15) within the major facilitator superfamily (MFS) (2.A.1) in the TCDB database (http://www.tcdb.org/). The MFS includes 77 families (updated on 24 January 2013), which are functionally diverse and involved in the transport of a variety of solutes and metabolites (Pao *et al.*, 1998). The AAHS family contains 12 members, six of which were functionally identified according to uptake assays with ^{14}C-labelled substrates. Examples of the AAHS family transporters are PcaK (4-hydroxybenzoate and protocatechuate transporter) of *P. putida* PRS2000 (Nichols and Harwood, 1997), BenK (benzoate transporter) of *Acinetobacter* sp. strain ADP1 (Collier *et al.*, 1997), TfdK (2,4-dichlorophenoxyacetate transporter) of *Ralstonia eutropha* (Leveau *et al.*, 1998) and MhbT (3-hydroxybenzoate transporter) of *Klebsiella pneumoniae* M5a1 (Xu *et al.*, 2012a).

The robust ability of *C. glutamicum* to grow on a variety of aromatic compounds partially relies on its multiple transporters for aromatic compounds (Chaudhry *et al.*, 2007; Xu *et al.*, 2012b). Genome data mining and experimental results confirmed that the *C. glutamicum* genome is encoding four AAHS family transporters: the benzoate transporter BenK (NCgl2325) (Chaudhry *et al.*, 2007; Wang, 2011), the gentisate transporter GenK (NCgl2922) (Shen *et al.*, 2005b; Xu *et al.*, 2006, 2012b), the protocatechuate/4-hydroxybenzoate transporter PcaK (NCgl1031) and the vanillate transporter VanK (NCgl2302) (Chaudhry *et al.*, 2007). They all have hydrophobic 12 α-helix transmembrane (TM) regions, a characteristic of transporters for the AAHS family. A second, novel type of benzoate transporters, BenE (NCgl2326), was identified (Chaudhry *et al.*, 2007; Haußmann *et al.*, 2009), which belongs to the benzoate:H^+ symporter family (2.A.46) (Marshak-Rothstein *et al.*, 1981).

Benzoate transporters: BenK and BenE

As mentioned above, *C. glutamicum* possesses two benzoate transporters, BenK (NCgl2325) and BenE (NCgl2326), which are adjacently located in the same operon according to operon prediction in CoryneRegNet (http://www.coryneregnet.de/). BenK and BenE belong to the AAHS family and the benzoate: H^+ symporter (Marshak-Rothstein *et al.*, 1981) family (2.A.46), respectively. The proteins show low sequence identity (15%) to each other, while BenK from *C. glutamicum* shares 29% identity with the BenK from *Acinetobacter* sp. strain ADP1 (Collier *et al.*, 1997). Membrane topology analysis using TMHMM (Krogh *et al.*, 2001) predicted that BenK exhibits 12 transmembrane α-helices. Under confocal microscopy, the fusion of BenK and green fluorescence protein (BenK-GFP) were located at the membranes in both *E. coli* and *C. glutamicum* cells. Further observation demonstrated that BenK was an inner membrane binding protein (Wang *et al.*, 2011). BenE (NCgl2326) was predicted to have 11 α-helices using TMHMM (Krogh *et al.*, 2001). BenE homologues were also identified from genomes of *Agrobacterium tumefaciens*, *Acinetobacter* sp. strain ADP1 (AAC46440), *R. jostii* RHA1 (YP_705462), *Moorella thermoacetica* ATCC 39073 (ABC20073) and *P. putida* (AAF63453), but the function of these homologues has not been experimentally characterized.

Single disruption of *benK* (*ncgl2325*) or *benE* (*ncgl2326*) in *C. glutamicum* did not result in any phenotypic variation for growth on benzoate compared with the parental strain RES167. However, the double-knockout mutant strain

RES167Δ*ncgl*(*2325–2326*) lost the ability to grow on benzoate, without any effect on growth on 3-hydroxybenzoate, 4-hydroxybenzoate, protocatechuate, gentisate, vanillate and resorcinol. When either one of these two transporter genes was introduced into the double-knockout strain, the ability to grow on benzoate was restored (Chaudhry *et al.*, 2007). Uptake assays of benzoate with resting cells indicated that there was no benzoate transport for the double-knockout mutant strain RES167Δ*ncgl*(2325–2326). However, the single-knockout mutant strains RES167Δ*ncgl2325* and RES167Δ*ncgl2326* still retained the ability to take up benzoate, and the uptake rates for benzoate in strain RES167Δ*ncgl2326* were higher than in strain RES167Δ*ncgl2325* (Chaudhry *et al.*, 2007).

The kinetics of benzoate transport by BenK were measured with *C. glutamicum* RES167Δ*ncgl*(2325–2326) expressing BenK with ^{14}C-labelled benzoate. The V_{max} of benzoate transport was 0.19 ± 0.01 nmol/min/mg of dry weight and the K_m was 1.11 ± 0.24 μM (Wang *et al.*, 2011). While mono- or dihydroxylated benzoates (salicylate, 3-hydroxybenzoate, 4-hydroxybenzoate, gentisate and protocatechuate) strongly inhibited benzoate transport (retaining only 50% or less of the original activity), mono-nitro-substituted benzoates (2-nitrobenzoate, 3-nitrobenzoate and 4-nitrobenzoate) showed less inhibition of benzoate transport than the hydroxylated benzoates, while 3,5-dinitrobenzoate and catechol had virtually no impact on transport activity. Four conserved charged amino acids of BenK located in hydrophobic TM regions (Asp-35, Arg-119, Glu-139 and Arg-386) and three conserved residues located at cytoplasmic loops (Gly-80, Asp-84, and Asp-312) are critical for benzoate transport activity (Wang *et al.*, 2011).

Gentisate transporter: GenK

Today, GenK (NCgl2922) from *C. glutamicum* is the only functionally characterized gentisate transporter. GenK exhibits moderate identities (24–34%) to a group of membrane proteins belonging to the AAHS family in other bacteria. Topology analyses suggested that GenK has 12 transmembrane-spanning α-helices. Furthermore, GenK contains the GXXX(D/E)(R/K)XG(R/K)(R/K) motif in two cytoplasmic hydrophilic loops (Xu *et al.*, 2012b), which is another feature of AAHS members. The fusion protein GenK-GFP was visualized in the periphery of *E. coli* cells by confocal microscopy, confirming its cytoplasmic membrane localization (Xu *et al.*, 2006). The *genK* mutant strain RES167Δ*genK* lost the ability to grow on gentisate as sole carbon source but retained the ability to grow with 3-hydroxybenzoate and other substituted benzoates. The ability of mutant RES167Δ*genK* to grow on gentisate was restored by providing the *genK* gene *in trans* (Shen *et al.*, 2005b). *Ralstonia* sp. strain U2 was characterized to degrade naphthalene via the gentisate pathway (Zhou *et al.*, 2001). However, strain U2 was unable to utilize gentisate as growth substrate due to the absence of a gentisate transporter. Heterologous expression of the *genK* from *C. glutamicum* converted strain U2 to a gentisate utilizer (Xu *et al.*, 2006). Uptake assays indicated that the ability to transport ^{14}C-labelled gentisate in *C. glutamicum* is associated with the intact *genK* gene. When *genK* was disrupted, cells virtually lost the ability to transport gentisate, but the transport ability was restored with the complementation of *genK* (Xu *et al.*, 2012b). The kinetics experiments revealed that *C. glutamicum* transported ^{14}C-labelled gentisate with a V_{max} of 3.06 ± 0.16 nmol/min/mg of dry weight, and a K_m of 10.71 ± 0.11 μM (Xu *et al.*, 2012b). Transport of ^{14}C-labelled 3-hydroxybenzoate and benzoate by GenK was not detected. Amino acid residues that were critical for gentisate transport activity were determined to be five conserved charged amino acids (Arg-103, Trp-309, Asp-312, Arg-313 and Ile-317) located at cytoplasmic loops and the three conserved residues (Asp-54, Asp-57 and Arg-386) located within hydrophobic TM regions. However, when Asp-54 and Asp-57 in the DGXD motif of TM1 were individually substituted with Glu residues, the mutated GenKs (D54E and D57E) retained approximately 50% activity in comparison with wild-type GenK. Furthermore, the two recombinant GenK proteins (with D54E and D57E exchange, respectively) conferred RES167Δ*genK* the ability to grow on gentisate, but with lower growth rates than the parent strain RES167 (Xu *et al.*, 2012b).

4-Hydroxybenzoate and protocatechuate transporter: PcaK

The PcaK encoded by *ncgl1031* of *C. glutamicum* had 26% identity to a 4-hydroxybenzoate and protocatechuate transporter of *P. putida* PRS2000 (Nichols and Harwood, 1997). Genetic disruption of the *ncgl1031* in *C. glutamicum* resulted in the loss of the ability to grow on 4-hydroxybenzoate and protocatechuate, yet did not impair bacterial growth on benzoate and other substituted benzoates as sole carbon source. The ability to grow on 4-hydroxybenzoate and protocatechuate was restored by complementation with *ncgl1031*. This indicated that PcaK (NCgl1031) is essential for the growth of *C. glutamicum* on 4-hydroxybenzoate and protocatechuate (Chaudhry *et al.*, 2007).

Vanillate transporter: VanK

The VanK encoded by *ncgl2302* shares 45% and 42% amino acid identity with the vanillate transporters from *P. putida* PRS2000 (AAL27558) and *Acinetobacter* sp. ADP1 (AAC27108), respectively. Since the deletion of *ncgl2302* in *C. glutamicum* led to slow, biphasic growth on protocatechuate as sole carbon source, the *ncgl2302* had been initially suggested to encode a protocatechuate transporter (Merkens *et al.*, 2005). Further studies showed that RES167_*ncgl2302* lost the ability to grow on vanillate, but not on benzoate, 3-hydroxybenzoate, 4-hydroxybenzoate, protocatechuate, gentisate or resorcinol. The ability to grow on vanillate was restored with the complementation of *ncgl2302*. Uptake assay confirmed that RES167 was able to transport vanillate, but mutant RES167Δ*ncgl2302* was not (Chaudhry *et al.*, 2007). Combining the above results, *ncgl2302* was proposed to encode a vanillate transporter in *C. glutamicum* (Chaudhry *et al.*, 2007). In contrast to VanK in *Acinetobacter* sp. strain ADP1 (Caplan *et al.*, 1993), VanK from *C. glutamicum* was not involved in transport of protocatechuate or 4-hydroxybenzoate.

Aromatic amino acid transporters: AroP$_{Cg}$ and PheP$_{Cg}$

During the study of aromatic compound transportation, a new L-Phe transporter was identified in *C. glutamicum* (Zhao *et al.*, 2011), which is encoded by *ncgl1108* (*pheP*$_{Cg}$) and located upstream of the regulator-encoding gene (*ncgl1110*) for resorcinol degradation (Huang *et al.*, 2006). Cellular localization of PheP$_{Cg}$–GFP fusion proteins with confocal microscopy indicated that PheP$_{Cg}$ is localized at cellular membrane. Furthermore, ^{14}C-labelled aromatic amino acid uptake assays revealed that *pheP*$_{Cg}$ was involved in the uptake of L-Phe but not in resorcinol uptake or degradation. The uptake of L-Phe was inhibited by L-Tyr. The disruption of *pheP*$_{Cg}$ resulted in decreased growth on L-Phe (as nitrogen source) but not on L-Tyr or L-Trp. Uptake assays with unlabelled and ^{14}C-labelled L-Phe and L-Tyr indicated that strain RES167Δ*pheP*$_{Cg}$ showed significant reduction in L-Phe uptake than strain RES167. Complementation of *pheP*$_{Cg}$ in strain RES167Δ*pheP*$_{Cg}$ or RES167Δ(*pheP*$_{Cg}$–*aroP*$_{Cg}$) restored their ability to uptake L-Phe and to grow on L-Phe as nitrogen source. The K_m and V_{max} values of L-Phe uptake by PheP$_{Cg}$ were 10.4 ± 1.5 μM and 1.2 ± 0.1 nmol/min/mg of dry weight, respectively, which were higher than those of the double-knockout strain RES167Δ(*pheP*$_{Cg}$–*aroP*$_{Cg}$) (4.0 ± 0.4 μM and 0.6 ± 0.1 nmol/min/mg of dry weight) (Zhao *et al.*, 2011).

The AroP$_{Cg}$ from *C. glutamicum* catalysed the uptake of aromatic amino acids (L-Tyr, L-Phe and L-Trp). In comparison with L-Phe and L-Trp, the most pronounced transport effect was observed with L-Tyr, presumably because that membrane of *C. glutamicum* has a very low passive permeability for L-Tyr. The addition of 200 μM L-Trp or L-Phe reduced the uptake of 20 μM ^{14}C-labelled L-Tyr to less than 14%. The K_m of L-Tyr uptake was determined to be 3.0 ± 0.3 μM with *C. glutamicum*/pJCaroP cells (Wehrmann *et al.*, 1995).

Aromatic compounds degraded via protocatechuate branch of the β-ketoadipate pathway

The protocatechuate branch of the β-ketoadipate pathway plays an important role in aromatic compound degradation in natural environments. For example, the aromatic compounds derived from lignin are further degraded via the protocatechuate branch (Harwood and Parales, 1996). Genes encoding the enzymes catalysing reactions of the protocatechuate branch, namely *pca* genes, were identified in *R. opacus* 1CP (Eulberg *et al.*, 1998), *R. jostii* RHA1 (Patrauchan *et al.*, 2005), *Streptomyces* sp. 2065 (Iwagami *et al.*, 2000) and *Streptomyces coelicolor* A3 (Davis and Sello, 2010). Similar to those aromatic degraders, the *pca* genes in *C. glutamicum* are organized in a continuous gene cluster of three operons, i.e. *pcaHGBC*, *pcaIJ* and *pcaRFDO*. Protocatechuate 3,4-dioxygenase, the key enzyme for this branch, is encoded by *ncgl2314* (*pcaG*) and *ncgl2315* (*pcaH*). Deletion of the *pcaGH* genes in the *C. glutamicum* resulted in a mutant strain that lost protocatechuate 3,4-dioxygenase activity and was unable to utilize 4-hydroxybenzoate, protocatechuate and 4-cresol as the sole carbon source for growth (Shen and Liu, 2005). Two genes *pcaD* and *pcaC* encode β-ketoadipate enol-lactone hydrolase and γ-carboxymuconolactone decarboxylase, respectively, in *C. glutamicum*. Interestingly, the function of *pcaD* and *pcaC* are replaced in *R. opacus* 1CP (Eulberg *et al.*, 1998), *R. jostii* RHA1 (Patrauchan *et al.*, 2005) and *Streptomyces* sp. 2065 (Iwagami *et al.*, 2000) by a single gene, *pcaL*, that encodes a bifunctional enzyme exhibiting both β-ketoadipate enol-lactone hydrolase and γ-carboxymuconolactone decarboxylase activities. The physiological function of other *pca* genes involved in the β-ketoadipate pathway was deduced from sequence identity searches (Shen and Liu, 2005), and this deduction is supported by the observation with proteomic tools, in which their expression was specifically induced during growth on protocatechuate (Haußmann and Poetsch, 2012).

4-Hydroxybenzoate and vanillate degradation

Protocatechuate is converted from 4-hydroxybenzoate by a single step of hydroxylation. The 4-hydroxybenzoate hydroxylase is encoded by the *pobA* (*ncgl1032*) that is neighboured to the 4-hydroxybenzoate and protocatechuate transporter gene, *pcaK* (*ncgl1031*), and is distantly located to other *pca* genes (Shen *et al.*, 2005a). This 4-hydroxybenzoate 3-hydroxylase is unique in its preference for NADPH instead of NADH as substrate, although its sequence is similar to other 4-hydroxybenzoate 3-hydroxylases that prefer NADH as a substrate (Huang *et al.*, 2008).

A cluster of four genes (*ncgl2299* to *ncgl2302*) associated with vanillate metabolism including the *vanA*, *vanB* and *vanK* genes and a divergently transcribed *vanR* (*ncgl2299*) gene was functionally identified (Merkens *et al.*, 2005). The amino acids of VanAB in *C. glutamicum* showed significant similarity to other bacterial vanillate demethylases (Kalinowski *et al.*, 2003). The *vanA* deletion mutant was unable to grow on vanillate, whereas growth on protocatechuate was not impaired (Merkens *et al.*, 2005). Vanillate significantly induced the expression of the *vanABK* genes (Merkens *et al.*, 2005).

4-Cresol degradation

C. glutamicum also grows on 4-cresol. Proteomic data indicated that this compound is assimilated via the protocatechuate branch of the β-ketoadipate pathway (Qi *et al.*, 2007). A *cre* gene cluster (*ncgl0521-ncgl0531*, named *creABCDEFGHIJR*), was identified to be involved in 4-cresol degradation and six of the genes were functionally characterized (Li *et al.*, 2014). The genes *creD* (*ncgl0524*), *creE* (*ncgl0525*) and *creJ* (*ncgl0530*) are encoding proteins involved in the oxidation of 4-cresol into 4-hydroxybenzyl alcohol. It was interesting to find that CreJ, which is a cytochrome P450 and was deduced to compose a class I P450 system with CreE and CreF (NCgl0526), was involved in initial catabolism of 4-cresol in *C. glutamicum*. 4-hydroxybenzyl alcohol is further catabolized into 4-hydroxybenzaldehyde by CreG (NCgl0527) with NAD^+ as electron acceptor. Two other genes, *creH* (*ncgl0528*) and *creI* (*ncgl0529*), seem to encode proteins involved in catabolism of 4-hydroxybenzaldehyde into 4-hydroxybenzoate (Li *et al.*, 2014). As so many genes being involved, the degradation of 4-cresol might be a more complicated process relatively to that of other aromatic compounds in *C. glutamicum*. The biochemical properties of most enzymes involving in 4-cresol degradation remain to be experimentally identified in the future.

Phenylpropenoid compound degradation

Phenylpropenoid compounds (i.e. ferulate, coumarate, caffeate, etc.) from the natural turnover of lignin constitute a common carbon sources for soil microorganisms (Jiménez *et al.*, 2002). So far, the known enzymes involved in the degradation of 4-coumarate, caffeate and ferulate are feruloyl-CoA synthetase (Fcs), enoyl-CoA hydratase/aldolase (Ech) and vanillin dehydrogenase (Vdh). Their catalytic intermediates 4-hydroxybenzoate, protocatechuate and vanillate, respectively are further degraded through the protocatechuate branch of the β-ketoadipate pathway (Venturi *et al.*, 1998; Mitra *et al.*, 1999). Homologues of Fcs-, Ech- and Vdh-encoding genes were mapped on the *C. glutamicum* genome, and they are scattered over different loci of the genome; however, their functionalities still need to be verified.

Hydroaromatic compound degradation

C. glutamicum uses the hydroaromatic compounds quinate and shikimate as carbon source for growth (Teramoto *et al.*, 2009). A gene cluster (*qsuABCD*) involved in the quinate/shikimate utilization via protocatechuate was identified within the genome sequence of *C. glutamicum*. Reverse transcriptase PCR analyses revealed that the expression of *qsuABCD* genes was markedly induced during growth with either quinate or shikimate. Moreover, the quinate/shikimate dehydrogenase gene (*qsuD*) deletion mutant lost the capability to grow on quinate and shikimate, while growth on glucose was not impaired. Further investigation indicated that the *qsuABCD* gene cluster is involved in quinate metabolism in *C. glutamicum*. Interestingly, the three proteins QsuD (quinate/shikimate dehydrogenase), QsuC (dehydroquinate dehydratase) and QsuB (dehydroshikimate dehydratase) in *C. glutamicum* were not homologous to their counterparts such as QuiA, QuiB and QuiC in *Acinetobacter* (Elsemore and Ornston, 1994). Instead, they can be classified into the families of fungal enzymes QutB, QutE and QutC (Hawkins *et al.*, 1993), although the degrees of amino acid sequence identities are relatively low (approximately 25%). These findings provided insights into the complex evolutionary history of the quinate/shikimate pathway in *C. glutamicum* (Teramoto *et al.*, 2009).

Aromatic compounds degraded via the catechol branch of the β-ketoadipate pathway

The catechol branch of the β-ketoadipate pathway, for which the involved genes are named *cat* genes, converts catechol into β-ketoadipate. The putative *catABC* operon is located within a 30 kb β-ketoadipate pathway metabolic island and shares over 96% sequence identity to the corresponding genes in *R. opacus* 1CP. The gene order (*catABC*) is conserved in the Gram-positive *R. opacus* and *C. glutamicum* but not in the Gram-negative strains such as *P. putida* (*catBCA*). The three enzymes catechol 1,2-dioxygenase, *cis,cis*-muconate lactonizing enzyme and muconolactone isomerase transform catechol to β-ketoadipate enol-lactone. The gene, *ncgl2319*, encoding the catechol 1,2-dioxygenase was confirmed to be involved in the degradation of benzoate, phenol and benzyl alcohols (Shen *et al.*, 2004).

Benzoate degradation

Contiguous to the *C. glutamicum cat* operon, the *benABCDRKE* genes are located which encode the proteins responsible for the transformation of benzoate into catechol. The *benABCDRKE* cluster is transcribed in opposite direction to the catechol 1,2-dioxygenase gene (*ncgl2319*). This genetic organization is different from the *benABCDK* cluster in the genome of *P. putida* KT2440 (Jiménez *et al.*, 2002). Two steps are needed to convert benzoate to catechol: The first step is catalysed by the benzoate dioxygenase complex encoded putatively by *ncgl2320*, *ncgl2321* and *ncgl2322*, which is thought to catalyse the oxidation of benzoate to benzoate diol. Following this oxidation, benzoate diol is decarboxylated by 2-hydro-1,2-dihydroxybenzoate dehydrogenase (putatively encoded by *ncgl2323*) to form catechol. The benzoate degrading enzymes BenA (NCgl2320) and BenD (NCgl2323) as well as the putative LuxR-type transcriptional regulator of the *ben* gene cluster, BenR (NCgl2324), are induced during growth on benzoate (Qi *et al.*, 2007; Haußmann *et al.*, 2009).

Phenol degradation

Phenol is one of the most prevalent pollutants in wastewater and toxic even at low concentrations.

It is usually degraded by hydroxylation of the aromatic ring and subsequently the ring is cleaved via *meta* or *ortho* cleavage. Our previous studies showed that the degradation of phenol in *C. glutamicum* occurs through the *ortho* cleavage route (Shen *et al.*, 2005a). Phenol is converted to catechol by two putative phenol monooxygenases encoded by *ncgl2588* and *ncgl0050* (Table 7.1).

Degradation of 3-hydroxybenzoate and gentisate: the mycothiol-dependent gentisate pathway

Gentisate is one of major metabolic intermediates for the degradation of many aromatic compounds. The degradation pathway for gentisate was initially reported by Lack (1959). *C. glutamicum* utilizes gentisate as a sole carbon and energy source for growth. A gene cluster, *ncgl2918-ncgl2923* (*genMFDRKH*), was identified to encode enzymes involved in the degradation of 3-hydroxybenzoate (Shen *et al.*, 2005b). Genes encoding gentisate 1,2-dioxygenase (*genD*) and fumarylpyruvate hydrolase (*genF*) were identified by cloning and expression in *E. coli*. GenM (NCgl2918), initially annotated as a hypothetical protein, was subsequently shown to be essential for gentisate assimilation (Shen *et al.*, 2005b) and finally identified as a mycothiol-dependent maleylpyruvate isomerase (Feng *et al.*, 2006). This enzyme is different from the previously known glutathione-dependent maleylpyruvate isomerase from Gram-negative *Ralstonia* sp. strain U2 (Zhou *et al.*, 2001). Additional genes *genR* (*ncgl2921*), *genK* (*ncgl2922*) and *genH* (*ncgl2923*) of the gentisate degradation locus are located upstream of *genD* (*ncgl2920*). GenR is a positive regulator and it up-regulated gentisate 1,2-dioxygenase, maleylpyruvate isomerase and fumarylpyruvate hydrolase activities (Shen *et al.*, 2005b; Chao and Zhou, 2013). As stated in previous sections, GenK is a gentisate transporter protein (Xu *et al.*, 2006, 2012b). GenH is a 3-hydroxybenzoate 6-monooxygenase converting 3-hydroxybenzoate to gentisate (Yang *et al.*, 2010).

C. glutamicum was reported to be able to grow on naphthalene. When cultured with 4.2 mM naphthalene in *C. glutamicum*, aspartate and glutamate production increased 1.5- and 1.3-fold, respectively, compared with the control without naphthalene (Lee *et al.*, 2010b).

Degradation of resorcinol and 2,4-dihydroxybenzoate via hydroxyquinol 1,2-dioxygenase cleavage

C. glutamicum grows on resorcinol (1,3-dihydroxybenzene), 2,4-dihydroxybenzoate and 3,5-dihydroxytoluene, assimilating these compounds through the hydroxyquinol pathway (Shen *et al.*, 2005a). This pathway is existing in many aerobic bacteria and responsible for the degradation of a large variety of single-ring aromatic compounds, such as resorcinol (Chapman and Ribbons, 1976a), 4-nitrocatechol (Wei *et al.*, 2010), 4-nitrophenol (Zhang *et al.*, 2012), 4-fluorophenol (Ferreira *et al.*, 2008), 4-chlorophenol (Nordin *et al.*, 2005), 2,6-dihydroxybenzoate (Yoshida *et al.*, 2007) and 4-aminophenol (Takenaka *et al.*, 2003). The chloro-, fluoro-, amino- and nitroaromatic compounds are first hydroxylated by different monooxygenases and then converted to hydroxyquinol (Takenaka *et al.*, 2003; Kitagawa *et al.*, 2004; Nordin *et al.*, 2005; Ferreira *et al.*, 2008; Shen *et al.*, 2010; Wei *et al.*, 2010; Zhang *et al.*, 2012). The intradiol dioxygenase hydroxyquinol 1,2-dioxygenase catalyses the *ortho* cleavage of hydroxyquinol to form 3-hydroxy-*cis,cis*-muconate, which spontaneously isomerize into maleylacetate (Pieper *et al.*, 2002). Maleylacetate is subsequently degraded via the β-ketoadipate pathway (Harwood and Parales, 1996).

Resorcinol and its derivatives are produced in nature as secondary plant products and are also important industrial chemicals used in photography, tanning, manufacturing of rubber products, adhesives and cosmetics preparation (Levin, 1971; Lynch *et al.*, 2002; Dayan *et al.*, 2007). Moreover, resorcinol is also found in effluents from production of coal tar products and waters from aquifers rich in coal and shale deposits (Lynch *et al.*, 2002). Resorcinol is converted to pyrogallol and further degraded by a pyrogallol 1,2-dioxygenase in *Azotobacter vinelandii* (Groseclose and Ribbons, 1981). In *P. putida* and *Rhizobium* species, resorcinol is converted to hydroxyquinol and further degraded by a hydroxyquinol 1,2-dioxygenase (Chapman and Ribbons, 1976b; Yoshida *et al.*, 2007). In *C. glutamicum*, two genetic clusters, *rolRHMD* (*ncgl1110–ncgl1113*) and *ncgl2950–ncgl2953*, are involved in resorcinol degradation (Huang *et al.*, 2006). Genetic and biochemical analysis indicated that the *rolH* (*ncgl1111*) gene encodes a resorcinol hydroxylase for the conversion of resorcinol to hydroxyquinol. Hydroxyquinol is ring-cleaved by the hydroxyquinol 1,2-dioxygenases encoded by *rolD* (*ncgl1113*) or *ncgl2951*. Both *rolM* (*ncgl1112*) and *ncgl2952* encoded a functional maleylacetate reductase that catalyses the conversion of maleylacetate to β-ketoadipate (Huang *et al.*, 2006). Gene disruption and complementation indicated that the operon *rolHMD* rather than the operon *ncgl2950–ncgl2953* were required for resorcinol metabolism, indicating that the *ncgl2951* and *ncgl2952* genes were redundant (Huang *et al.*, 2006).

Metabolism of phenylacetate via phenylacetyl-CoA pathway

Unlike the above mentioned aerobic metabolism of aromatic compounds, which usually starts with hydroxylation of the aromatic ring and subsequent ring-cleavage by dioxygenases, the aerobic catabolism of phenylacetate is initiated by CoA thioesterification of the acetyl branch. The CoA thioesterification is a conventional strategy for anaerobic degradation of aromatic compounds (Mohamed *et al.*, 2002; Luengo *et al.*, 2007). Investigations showed that phenylacetate is activated by an aerobically induced phenylacetate-CoA ligase (Martínez-Blanco *et al.*, 1990; El-Said Mohamed, 2000; Erb *et al.*, 2008). Subsequently, phenylacetyl-CoA undergoes epoxidation, isomerization to a seven-membered C-O-heterocycle (oxepin-CoA) and then non-oxygenolytic ring-cleavage to generate β-ketoadipyl-CoA (Teufel *et al.*, 2010, 2011). β-Ketoadipyl-CoA is further degraded to succinyl-CoA and acetyl-CoA and channelled into the tricarboxylic acid cycle (Nogales *et al.*, 2007). Besides phenylacetic acid, other structurally related compounds such as styrene, phenylacetaldehyde, 2-phenylethylamine, ethylbenzene, phenylacetaldoxime, tropic acid, other esters of phenylacetic acid and *n*-phenylalkanoic acid (where *n* is an even number of carbon atoms) are also converted into phenylacetyl-CoA via peripheral catabolic pathways (Luengo *et al.*, 2007). These different peripheral catabolic pathways are integrated finally into the core phenylacetyl-CoA pathway to compose a complex catabolic unit, called phenylacetyl-CoA catabolon (Luengo *et al.*, 2001, 2007).

The phenylacetic acid (*paa*) pathway has been identified in the genomes of *C. glutamicum* strain R, *C. efficiens*, and serine-auxotrophic strain AS 1.542, but not in strain ATCC 13032 (Chen *et al.*, 2012). The *paa* cluster comprises 15 genes and has a unique genetic organization compared with those in other bacteria such as *E. coli* (Ferrández *et al.*, 1998). A TetR-type regulator-encoding gene *paaR*, instead of the GntR-type regulator-encoding gene *paaX* in *E. coli*, was identified in the *paa* cluster of *C. glutamicum*. In contrast to three transcription units organized by 14 *paa* genes in *E. coli*, the *paa* gene cluster of strain AS 1.542 is organized in six transcription units. Although their genetic organizations are different, the gene products of *paa* cluster in *C. glutamicum* have significant amino acid sequence identity (32–66%) to the corresponding ones in *E. coli* (Chen *et al.*, 2012). When the *paa* cluster of strain AS 1.542 was introduced into strain ATCC 13032, the recombinant strain gained the ability to grow on phenylacetate as sole carbon source (Chen *et al.*, 2012).

Regulation of aromatic metabolism in *Corynebacterium glutamicum*

The regulation of aromatic metabolism generally occurs at transcriptional level in bacteria (Gerischer, 2002). A sum of 158 potential regulators that represent 5.3% of the totally 3002 predicted proteins were identified in the *C. glutamicum* ATCC 13032 genome by bioinformatics tools (Brinkrolf *et al.*, 2007). These potential regulators can be assigned into 36 regulatory protein families (Schroder and Tauch, 2010). Up to date, regulators belonging to IclR, LuxR, LysR, PadR and TetR families were experimentally shown to be involved in aromatic metabolism (Table 7.3). The regulator-encoding genes involved in aromatic

Table 7.3 The transcriptional regulators involved in aromatic compound metabolism in *Corynebacterium glutamicum*

Protein family	Regulator	Coding gene[a]	Metabolic pathway involved	Regulation type	Operator sequence (5′ to 3′)	Number of target genes	Target genes	References
IclR	PcaR	*cg2624/ ncgl2308*	Protocatechuate pathway	Repression	GTTCGC-N_3-GCGAAC	11	*pcaIJ*, *pcaFDO*, *pcaHGBC*[b], *pobA-pcaK*[b]	Brinkrolf *et al*. (2006)
	GenR	*cg3352/ ncgl2921*	Gentisate pathway	Activation/ repression	ATTCC-$N^{7(5)}$-GGAAT	6	*genDFM*, *genKH*, *genR*	Brinkrolf *et al*. (2006), Shen *et al*. (2005b), Chao and Zhou (2013)
	–	*cg3388/ ncgl2954*	Putative hydroxyquinol pathway	Unknown	–	3	*ncgl2951–2953*[b]	Brinkrolf *et al*. (2006), Huang *et al*. (2006)
	–	*cg0646/ ncgl0531*	Putative 4-cresol pathway	Unknown	–	–	–	Qi *et al*. (2007)
TetR	RolR	*cg1308/ ncgl1110*	Hydroxyquinol pathway	Repression	AAAATCTGAACCCTTGTTCATTTATGAAT	4	*rolR*, *rolHMD*	Brinkrolf *et al*. (2006), Huang *et al*. (2006), Li *et al*. (2011, 2012)
	PaaR	*paaR*[c]	Phenylacetate pathway	Repression	ACTnACCGnnCGnnCGGTnAGT	15	*paaABCDEGJFH*, *paaTK*, *paaR*, *paaI*, *paaY*, *paaZ*	Chen *et al*. (2012)
LuxR	PcaO	*cg2627/ ncgl2311*	Protocatechuate pathway	Activation	AACCCCTGACCTTCGGGGTT	2	*pcaHG*	Zhao *et al*. (2010)
	BenR	*cg2641/ ncgl2324*	Catechol pathway	Activation	–	–	–	Brinkrolf *et al*. (2006), Haußmann *et al*. (2009)
LysR	QsuR	*cg0500/ ncgl0405*	Quinate/shikimate pathway	Activation	–	4	*qsuABCD*	Teramoto *et al*. (2009)
PadR	VanR	*cg2615/ ncgl2299*	Vanillate pathway	Repression	TAAACT-N_7-AGTTTA	3	*vanABK*	Brinkrolf *et al*. (2006)

[a]The coding gene ID shows the gene ID and alternative gene ID of *Corynebacterium glutamicum* ATCC 13032 in CoryneRegNet 6.0.
[b]The regulation for target genes has not been verified by experimental data.
[c]The *paaR* gene exists in the genome of *Corynebacterium glutamicum* strain AS1.542.

metabolism are usually clustered with aromatic metabolism genes and the coding proteins only control the expression of nearby genes (Fig. 7.2). Consequently, they are specific regulators (Brinkrolf *et al.*, 2006). Besides specific regulators, global regulators are also involved in aromatic metabolism in *C. glutamicum* (Kohl and Tauch, 2009). The global regulators together with specific regulators provide a precise and efficient regulation on aromatic metabolism in *C. glutamicum*.

The IclR-type regulators: PcaR and GenR

The IclR-type regulator was firstly identified in *E. coli* to regulate the glyoxylate bypass route (Sunnarborg *et al.*, 1990; Molina-Henares *et al.*, 2006). In *C. glutamicum*, two IclR proteins namely PcaR and GenR were confirmed to be involved in regulation of the protocatechuate (Brinkrolf *et al.*, 2006) and 3-hydroxybenzoate/gentisate pathway (Chao and Zhou, 2013), respectively.

PcaR negatively regulated the transcription of genes involved in protocatechuate pathway. PcaR bound to a 15-bp operator sequence, which were found in the *pcaI–pcaR* intergenic region as well as the upstreams of *pcaH* and *pobA* (Brinkrolf *et al.*, 2006). Disruption of *pcaR* elevated the transcription of *pcaIJ* and *pcaFDO*, but not the *pcaHGBC* and *pobA-pcaK* (Brinkrolf *et al.*, 2006). It was found later that LuxR type regulator PcaO controlled the transcription of *pcaHG* (Zhao *et al.*, 2010). Thus, PcaR and PcaO together co-regulated the genes involved in the protocatechuate branch of the β-ketoadipate pathway.

GenR positively regulated the metabolism of gentisate and other aromatic compounds that were metabolized via the gentisate pathway in *C. glutamicum* (Shen *et al.*, 2005b; Brinkrolf *et al.*, 2006). The *genR* was located upstream of the *genDFM* genes that encoded gentisate 1,2-dioxygenase, fumarylpyruvate hydrolase and maleylpyruvate isomerase. Disruption of *genR* resulted in loss of these enzyme activities and inability to grow on gentisate or 3-hydroxybenzoate (Shen *et al.*, 2005b), indicating that GenR is an activator of gentisate pathway in *C. glutamicum*. In addition, it was proposed that *genK* and *genH* were also under the positive control of GenR (Schroder and Tauch, 2010). The IclR-type regulator GenR was further identified as a dual-function protein that acts as activator of the transcription of *genDFM* and *genKH* operons for 3-hydroxybenzoate and gentisate catabolism in response to 3-hydroxybenzoate and gentisate and as repressor of its own expression in *C. glutamicum*. Footprinting analyses demonstrated that GenR bound to four sites with different affinities. In addition, GenR was found to bind three sites upstream of *genDFM* and *genKH* and to activate transcription of these genes according to transcriptional activation type Class I and Class II. The results reveal a new regulatory function of the IclR family in the catabolism of aromatic compounds (Chao and Zhou, 2013).

Other IclR type regulators that are potentially involved in regulation of aromatic metabolism are NCgl0531 and NCgl2954. NCgl0531 is deduced to be regulator for 4-cresol degradation.

Figure 7.2 (opposite) Regulation mechanisms and genetic organizations involved in aromatic compound metabolism in *Corynebacterium glutamicum*. (a) The IclR family regulators, including PcaR for protocatechuate metabolism, GenR for gentisate metabolism, NCgl2954 for putative hydroxyquinol metabolism and NCgl0531 for putative 4-cresol metabolism. (b) The TetR family regulators, including RolR for resorcinol metabolism and PaaR for phenylacetic acid metabolism. (c) The LuxR family regulators, including PcaO for protocatechuate metabolism and BenR for benzoate metabolism. (d) The LysR-type regulator QsuR for quinate metabolism. (e) The PadR-type regulator VanR for vanillate metabolism. Regulatory genes and proteins are shown in black while structural genes are shown in white. Arrows indicate activation effect and blunt bars indicate repression effect. Question marks indicate unknown regulation type. Functions of regulators confirmed experimentally are indicated in solid lines, whereas functions only assumed are shown in dash lines. The '+' symbols indicate up-regulation of expression while the '–' symbols indicate down-regulation of expression, and 'ND' indicates no significant difference when the regulatory genes are disrupted, as verified by real-time reverse transcription PCR or enzyme activity assays. The (putative) operators for each regulator are also shown below the clusters (Shen *et al.*, 2005b; Brinkrolf *et al.*, 2006; Huang *et al.*, 2006; Qi *et al.*, 2007; Haußmann *et al.*, 2009; Teramoto *et al.*, 2009; Zhao *et al.*, 2010; Li *et al.*, 2011, 2012; Chen *et al.*, 2012; Chao and Zhou, 2013).

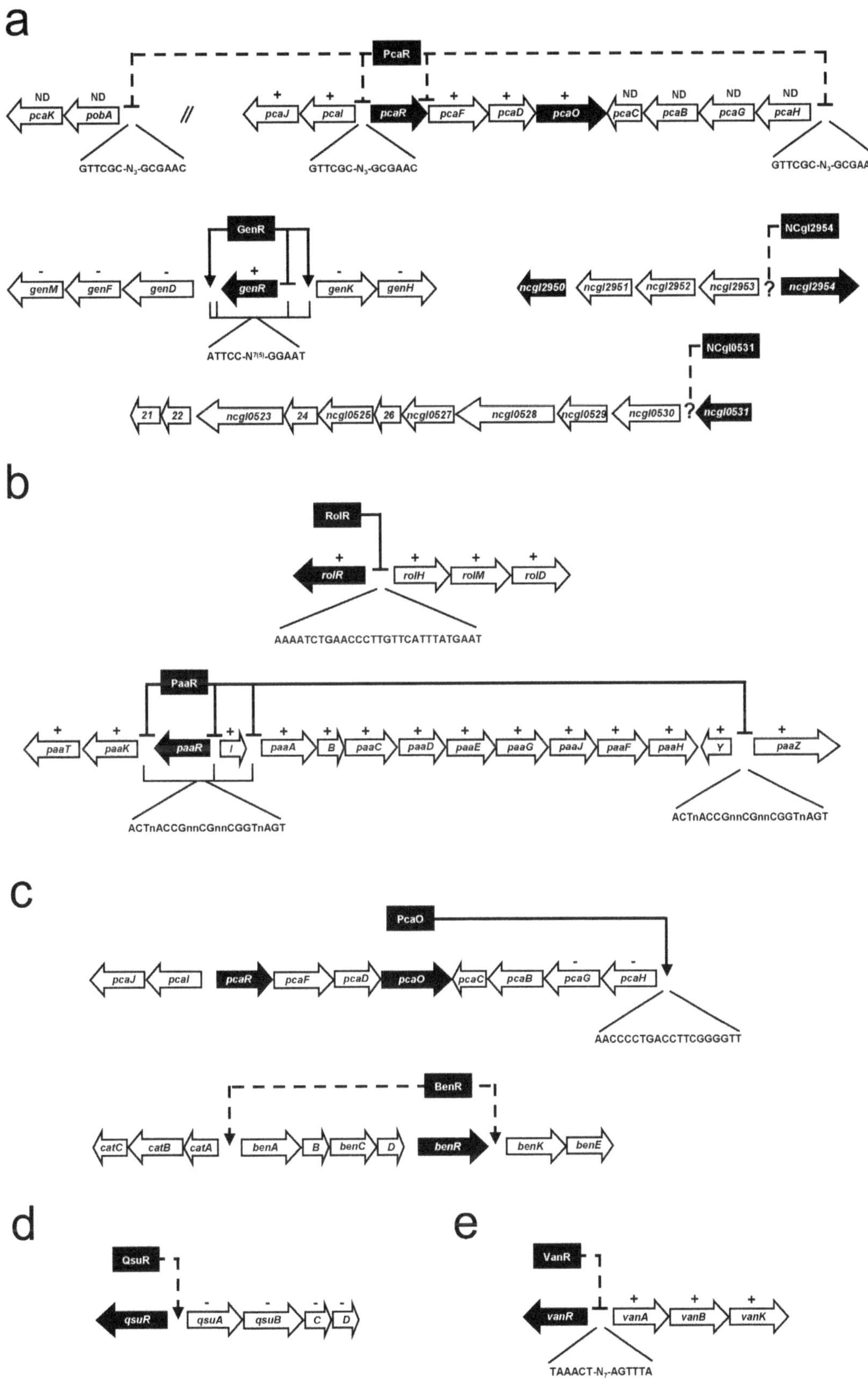
a
PcaR
ND ND
pcaK pobA
GTTCGC-N3-GCGAAC
pcaJ pcaI pcaR pcaF pcaD pcaO pcaC pcaB pcaG pcaH
GenR
genM genF genD genR genK genH
ATTCC-N7(5)-GGAAT
NCgl2954
ncgl2950 ncgl2951 ncgl2952 ncgl2953 ncgl2954
NCgl0531
21 22 ncgl0523 24 ncgl0525 26 ncgl0527 ncgl0528 ncgl0529 ncgl0530 ncgl0531
b
RolR
rolR rolH rolM rolD
AAAATCTGAACCCTTGTTCATTTATGAAT
PaaR
paaT paaK paaR I paaA B paaC paaD paaE paaG paaJ paaF paaH Y paaZ
ACTnACCGnnCGnnCGGTnAGT
c
PcaO
AACCCCTGACCTTCGGGGTT
BenR
catC catB catA benA B benC D benR benK benE
d
QsuR
qsuR qsuA qsuB C D
e
VanR
vanR vanA vanB vanK
TAAACT-N7-AGTTTA

Comparative proteomic analysis revealed that NCgl0524, NCgl0525, NCgl0527 and NCgl0530 were specifically synthesized when 4-cresol was used as sole carbon and energy source for growth (Qi *et al.*, 2007). The regulatory gene *ncgl0531* was adjacent to *ncgl0524-ncgl0530* genes. Thus, *ncgl0531* was likely to encode the regulator controlling 4-cresol metabolism. Similarly, the IclR regulator NCgl2954 is deduced to be involved in regulation of hydroxyquinol pathway. The *ncgl2954* located at the neighbour and in the opposite direction of the gene cluster *ncgl2951–2953*, which encodes the enzymes for the hydroxyquinol pathway. Thus, NCgl2954 is a good candidate to regulate this hydroxyquinol pathway (Brinkrolf *et al.*, 2006).

The TetR-type regulators: RolR and PaaR

The TetR-type regulator was named after the classical transcriptional regulator in transposon Tn*10* that conferred resistance against tetracycline (Beck *et al.*, 1982). TetR-type regulators have been identified in many Gram-negative bacteria and have been extensively investigated (Hinrichs *et al.*, 1994; Orth *et al.*, 2000; Saenger *et al.*, 2000). To date, a total of 16 TetR regulators were identified in *C. glutamicum* ATCC 13032, which were composed to the largest family of DNA-binding transcriptional regulators in this bacterium (Schroder and Tauch, 2010). Two TetR-type regulators, RolR and PaaR, were identified to be involved in the regulation of aromatic degradation in *C. glutamicum*.

The RolR negatively regulates resorcinol metabolism in *C. glutamicum* (Huang *et al.*, 2006). RolR and some other TetR regulators usually function as repressors and share a common regulatory mechanism: In state I (without effector) they bind to target DNA region and thus block the transcription of target gene(s). In state II, an effector binds to the TetR protein and induces conformational change of the TetR protein. This stops the binding of TetR protein to its targeted DNA binding site and the repression of gene transcription is thus released (Ramos *et al.*, 2005). RolR repressed the transcription of *rolHMD* and its own gene as a form of homodimer by binding to a unique operator *rolO*, which was a 29-bp sequence (Brinkrolf *et al.*, 2006; Li *et al.*, 2012). The operator *rolO* was located in the intragenic region of both *rolR* and *rolHMD*. A roadblock model was proposed to explain the repression of the transcription of *rolHMD* and of its own gene by RolR (Fig. 7.3) (Li *et al.*, 2012). Resorcinol and its metabolic intermediate hydroxyquinol were identified as effectors for RolR and can dissociate the RolR–*rolO* complex (Li *et al.*, 2012). The crystal structure of RolR in its effector-bound (with resorcinol) and *apo*-form revealed a decrease of spatial distance between two recognition helices when binding with resorcinol, which made RolR unsuitable for DNA binding (Li *et al.*, 2011). The *rolO*-like motifs and similar genetic organizations comprising *rolO*-like motifs and homologues of regulator TetR were identified in the genomes of *Burkholderia*, *Mycobacterium*, *Rhodococcus* and *Streptomyces* species (Li *et al.*, 2012), suggesting that these bacteria might share a common regulatory mechanism.

PaaR is another TetR-type regulator involved in regulation of aromatic compound metabolism in *C. glutamicum* (Chen *et al.*, 2012). Deletion of *paaR* gene resulted in significant up-regulation of mRNA levels of all *paa* genes, indicating that PaaR was a negative regulator of phenylacetic acid metabolism. The binding site of PaaR was identified to be a 22-bp palindromic motif. Phenylacetyl-CoA (PA-CoA) was the effector of PaaR. By genome data-mining, orthologous PaaR regulons were identified in taxonomic groups including actinobacteria, *Alpha*- and *Betaproteobacteria*, *Deinococcus-Thermus* and flavobacteria (Chen *et al.*, 2012).

The LuxR-type regulators: PcaO and BenR

The LuxR family regulators are widespread in bacteria and are known for their involvement in quorum sensing (QS) and other physiological adaptation to environments such as biofilm formation, plasmid transfer and antibiotics biosynthesis (Chen and Xie, 2011; Santos *et al.*, 2012). In *C. glutamicum*, PcaO is an atypical large ATP-binding LuxR (LAL) family regulator and regulates the protocatechuate branch of β-ketoadipate pathway. PcaO positively regulated the *pcaHG*

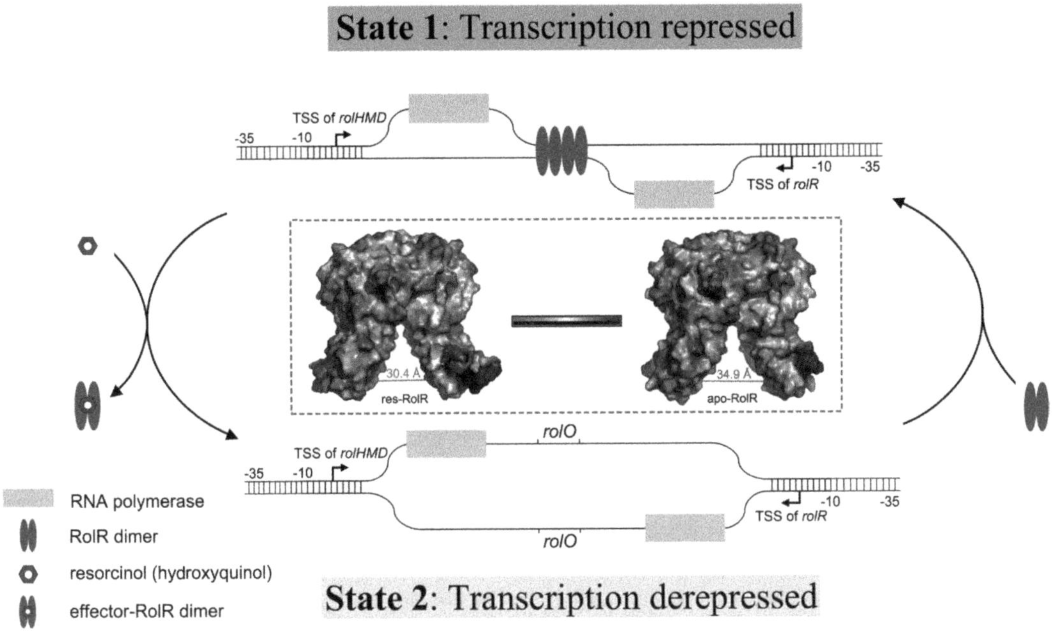

Figure 7.3 One stone two birds: a roadblock model to explain the regulation by RolR on resorcinol degradation. The crystal structures of RolR in its effector-bound (with resorcinol, res-RolR) and apo- (apo-RolR) forms are shown in the red box. With resorcinol-binding, the centre-to-centre separation of the recognition helices is decreased from 34.9Å to 30.4Å, thus not suitable for DNA binding (Li *et al.*, 2011, 2012). A colour version of this figure is available in the plate section at the back of the book.

encoding protocatechuate 3,4-dioxygenase by binding to a 20-bp operator located upstream of the −35 region of the *pcaHG* promoter. Gel shift data showed that ATP weakened the binding between PcaO and operator, whereas ADP strengthened the binding. Protocatechuate was identified as an effector for PcaO and its effect is concentration-dependent: the binding was reduced at low concentrations but increased at high concentrations. It was proposed that both protocatechuate molecule and the ratio of ATP to ADP in cells affected the precise control of expression of *pcaHG* (Zhao *et al.*, 2010).

BenR, another LuxR-type regulator, is proposed that BenR positively regulates the *ben* and/or *cat* genes required for utilization of benzyl alcohol, benzoate, phenol and catechol (Brinkrolf *et al.*, 2006). When *C. glutamicum* grew on benzoate as sole carbon source, the BenR protein was significantly up-regulated (Haußmann *et al.*, 2009). However, the role of BenR in regulation of catechol pathway remains to be experimentally validated.

The LysR-type regulator in hydroaromatic compound degradation: QsuR

The LysR-type regulators constitute the largest known family of DNA-binding transcriptional regulators in prokaryotes (Maddocks and Oyston, 2008). The first characterized LysR regulated the decarboxylation of diaminopimelate (Stragier and Patte, 1983; Stragier *et al.*, 1983). LysR-type regulators act as either activators or repressors to control the expression of genes involved in multiple physiological processes such as metabolism, cell division, quorum sensing and virulence (Maddocks and Oyston, 2008). In *C. glutamicum*, a LysR-type protein QsuR is involved in regulation of the catabolism of hydroaromatic compounds, quinate and shikimate (Teramoto *et al.*, 2009). The expression levels of the *qsuABCD* operon was significantly elevated when cells were grown on quinate or shikimate compared with growth on glucose as sole carbon source. Disruption of *qsuR* resulted in a decreased expression levels of *qsuABCD* genes, even in the presence of shikimate. Thus, it was proposed that QsuR was an activator of quinate/shikimate catabolic pathway (Teramoto *et al.*, 2009).

The PadR-type regulator: VanR

The operon *vanABK* was proved to be involved in vanillate catabolism and uptake in *C. glutamicum* (Merkens *et al.*, 2005; Chaudhry *et al.*, 2007). The regulator-encoding gene *vanR* is located in the opposite direction of *vanABK* and it encoding a PadR-type regulator (Brinkrolf *et al.*, 2006). The expression level of *vanABK* was remarkably increased in the *vanR* mutant when compared with that in the parent strain of *C. glutamicum*, indicating that VanR negatively regulates vanillate catabolism. Downstream of the mapped *vanABK* promoter, a putative operator sequence with an imperfect palindromic structure was identified and a similar structure was also found in *C. efficiens* (Brinkrolf *et al.*, 2006).

The global regulator GlxR

Besides pathway-specific regulators, global regulators are also involved in regulation of aromatic compound degradation, although the knowledge of those global regulations is very limited. The GlxR, a cyclic adenosine monophosphate (cAMP) receptor protein, was first identified by screening repressors for *aceB* gene of glyoxylate bypass of *C. glutamicum* (Kim *et al.*, 2004a). Mutation of *glxR* gene resulted in severe growth defect and differential expression of over 100 proteins (Moon *et al.*, 2007), suggesting that it controls a core regulon and acts as a regulatory hub. Besides *aceB* of glyoxylate bypass, GlxR also regulates other genes involved in various physiological processes and metabolic pathways (Kimura, 2002; Letek *et al.*, 2006; Han *et al.*, 2008; Jungwirth *et al.*, 2008). By using *in silico* analysis combined with an *in vitro* detection approach, 215 potential GlxR operators, which might be involved in control of 281 operons comprising 439 genes, were identified (Kohl *et al.*, 2008; Kohl and Tauch, 2009). Thus, about 14% of the annotated genes are under control of GlxR in *C. glutamicum*. As for aromatic compound degradation, putative GlxR binding sites were located at the genetic clusters involved in metabolism of benzoate, 4-hydroxybenzoate, protocatechuate, ferulate, vanillin, vanillate, resorcinol, hydroxyquinol, 3-hydroxybenzoate and gentisate (Kohl *et al.*, 2008; Schroder and Tauch, 2010). Recently, GlxR binding sites were identified within the upstream regions of *paaR* as well as other *paa* genes. Therefore, it may be possible that GlxR regulated simultaneously the transcriptions of *paaR* and other structural *paa* genes during phenylacetic acid metabolism in *C. glutamicum* (Chen *et al.*, 2012).

Perspectives

The genetics and physiology of aromatic degradation and assimilation have been advanced significantly in *C. glutamicum*, since the first genetic and biochemical identification of a catechol 1,2-dioxygenase from this bacterium a decade ago (Shen *et al.*, 2004). However, as summarized in a recent review (Shen *et al.*, 2012), knowledge of aromatic compound tolerance and cross-talk among aromatic catabolism and other cellular physiological processes is still largely unexplored in *C. glutamicum*. With the advent of multiple novel technologies and advanced systems biology tools for *C. glutamicum*, advancements in the following aspects are expected.

New information on physiology of *C. glutamicum* and its genetic and biochemical basis will continue to emerge. This will not only advance the understanding of aromatic compound degradation and assimilation with *C. glutamicum*, but also shed lights on the understanding of biodegradation of aromatic compounds in other bacterial groups. Examples to support this expectation are (1) the discovery of a mycothiol-depended gentisate pathway in *C. glutamicum*, which promoted the identification of similar metabolism in *Rhodococcus* species (Liu *et al.*, 2011) and (2) the discovery of a link between aromatic degradation and gluconeogenesis in *C. glutamicum*, which lead to the advancement of understanding that gluconeogenesis is an essential physiological process in bacteria growing on aromatic compounds as sole carbon sources (Qi *et al.*, 2007).

Reactions of aromatic degradation may be used for engineering novel pathways in metabolic network for *C. glutamicum*. Newly defined amino acid production processes using aromatic compounds such as phenol and naphthalene as sole carbon source were reported (Lee *et al.*, 2010a,b). We consider that it is more industrially relevant if aromatic metabolism is explored in a way that promotes the currently used carbon sources (such as glucose and fructose) is converted into the key metabolic intermediates for amino acids or other fermentative products.

New fermentation processes may be tested and developed based on understanding of assimilation of aromatic compounds with *C. glutamicum*. In particular, modifications of fermentative media may be expected that are oriented to accelerate certain metabolic fluxes and optimize product yield with *C. glutamicum* in the future.

References

Beck, C. F., Mutzel, R., Barbé, J., and Müller, W. (1982) A multifunctional gene (*tetR*) controls Tn*10*-encoded tetracycline resistance. *J. Bacteriol.* **150**, 633–642.

Binder, S., Schendzielorz, G., Stabler, N., Krumbach, K., Hoffmann, K., Bott, M., and Eggeling, L. (2012) A high-throughput approach to identify genomic variants of bacterial metabolite producers at the single-cell level. *Genome Biol.* **13**, R40.

Blombach, B., and Seibold, G. M. (2010) Carbohydrate metabolism in *Corynebacterium glutamicum* and applications for the metabolic engineering of L-lysine production strains. *Appl. Microbiol. Biotechnol.* **86**, 1313–1322.

Brinkrolf, K., Brune, I., and Tauch, A. (2006) Transcriptional regulation of catabolic pathways for aromatic compounds in *Corynebacterium glutamicum*. *Genet. Mol. Res.* **5**, 773–789.

Brinkrolf, K., Brune, I., and Tauch, A. (2007) The transcriptional regulatory network of the amino acid producer *Corynebacterium glutamicum*. *J. Biotechnol.* **129**, 191–211.

Brown, E. D., Vivas, E. I., Walsh, C. T., and Kolter, R. (1995) MurA (MurZ), the enzyme that catalyzes the first committed step in peptidoglycan biosynthesis, is essential in *Escherichia coli*. *J. Bacteriol.* **177**, 4194–4197.

Cao, B., Nagarajan, K., and Loh, K. C. (2009) Biodegradation of aromatic compounds: current status and opportunities for biomolecular approaches. *Appl. Microbiol. Biotechnol.* **85**, 207–228.

Caplan, A. J., Cyr, D. M., and Douglas, M. G. (1993) Eukaryotic homologues of *Escherichia coli dnaJ*: a diverse protein family that functions with hsp70 stress proteins. *Mol. Biol. Cell* **4**, 555–563.

Chain, P. S., Denef, V. J., Konstantinidis, K. T., Vergez, L. M., Agullo, L., Reyes, V. L., Hauser, L., Cordova, M., Gomez, L., Gonzalez, M., *et al.* (2006) *Burkholderia xenovorans* LB400 harbors a multi-replicon, 9.73-Mbp genome shaped for versatility. *Proc. Natl. Acad. Sci. U.S.A.* **103**, 15280–15287.

Chao, H., and Zhou, N. Y. (2013) GenR, an IclR-type regulator, activates and represses the transcription of *gen* genes involved in 3-hydroxybenzoate and gentisate catabolism in *Corynebacterium glutamicum*. *J. Bacteriol.* **195**, 1598–1609.

Chapman, P. J., and Ribbons, D. W. (1976a) Metabolism of resorcinylic compounds by bacteria: alternative pathways for resorcinol catabolism in *Pseudomonas putida*. *J. Bacteriol.* **125**, 985–998.

Chapman, P. J., and Ribbons, D. W. (1976b) Metabolism of resorcinylic compounds by bacteria: orcinol pathway in *Pseudomonas putida*. *J. Bacteriol.* **125**, 975–984.

Chaudhry, M. T., Huang, Y., Shen, X.-H., Poetsch, A., Jiang, C.-Y., and Liu, S.-J. (2007) Genome-wide investigation of aromatic acid transporters in *Corynebacterium glutamicum*. *Microbiology* **153**, 857–865.

Chen, J., and Xie, J. (2011) Role and regulation of bacterial LuxR-like regulators. *J. Cell. Biochem.* **112**, 2694–2702.

Chen, X., Kohl, T. A., Ruckert, C., Rodionov, D. A., Li, L. H., Ding, J. Y., Kalinowski, J., and Liu, S. J. (2012) Phenylacetic acid catabolism and its transcriptional regulation in *Corynebacterium glutamicum*. *Appl. Environ. Microbiol.* **78**, 5796–5804.

Collier, L. S., Nichols, N. N., and Neidle, E. L. (1997) *BenK* encodes a hydrophobic permease-like protein involved in benzoate degradation by *Acinetobacter* sp. strain ADP1. *J. Bacteriol.* **179**, 5943–5946.

Davis, J. R., and Sello, J. K. (2010) Regulation of genes in *Streptomyces* bacteria required for catabolism of lignin-derived aromatic compounds. *Appl. Microbiol. Biotechnol.* **86**, 921–929.

Dayan, F. E., Watson, S. B., and Nanayakkara, N. P. (2007) Biosynthesis of lipid resorcinols and benzoquinones in isolated secretory plant root hairs. *J. Exp. Bot.* **58**, 3263–3272.

Díaz, E. (2004) Bacterial degradation of aromatic pollutants: a paradigm of metabolic versatility. *Int. Microbiol.* **7**, 173–180.

El-Said Mohamed, M. (2000) Biochemical and molecular characterization of phenylacetate-coenzyme A ligase, an enzyme catalyzing the first step in aerobic metabolism of phenylacetic acid in *Azoarcus evansii*. *J. Bacteriol.* **182**, 286–294.

Elsemore, D. A., and Ornston, L. N. (1994) The *pca-pob* supraoperonic cluster of *Acinetobacter calcoaceticus* contains *quiA*, the structural gene for quinate-shikimate dehydrogenase. *J. Bacteriol.* **176**, 7659–7666.

Erb, T. J., Ismail, W., and Fuchs, G. (2008) Phenylacetate metabolism in thermophiles: characterization of phenylacetate-CoA ligase, the initial enzyme of the hybrid pathway in *Thermus thermophilus*. *Curr. Microbiol.* **57**, 27–32.

Eulberg, D., Lakner, S., Golovleva, L. A., and Schlomann, M. (1998) Characterization of a protocatechuate catabolic gene cluster from *Rhodococcus opacus* 1CP: evidence for a merged enzyme with 4-carboxymuconolactone-decarboxylating and 3-oxoadipate enol-lactone-hydrolyzing activity. *J. Bacteriol.* **180**, 1072–1081.

Feng, J., Che, Y., Milse, J., Yin, Y. J., Liu, L., Ruckert, C., Shen, X. H., Qi, S. W., Kalinowski, J., and Liu, S. J. (2006)

The gene *ncgl2918* encodes a novel maleylpyruvate isomerase that needs mycothiol as cofactor and links mycothiol biosynthesis and gentisate assimilation in *Corynebacterium glutamicum*. *J. Biol. Chem.* **281**, 10778–10785.

Ferrández, A., Miñambres, B., García, B., Olivera, E. a. R., Luengo, J. M., García, J. L., and Díaz, E. (1998) Catabolism of phenylacetic acid in *Escherichia coli*. *J. Biol. Chem.* **273**, 25974–25986.

Ferreira, M. I., Marchesi, J. R., and Janssen, D. B. (2008) Degradation of 4-fluorophenol by *Arthrobacter* sp. strain IF1. *Appl. Microbiol. Biotechnol.* **78**, 709–717.

Franco, A. V., Liu, D., and Reeves, P. R. (1998) The wzz (cld) protein in *Escherichia coli*: amino acid sequence variation determines O-antigen chain length specificity. *J. Bacteriol.* **180**, 2670–2675.

Fuchs, G., Boll, M., and Heider, J. (2011) Microbial degradation of aromatic compounds – from one strategy to four. *Nat. Rev. Microbiol.* **9**, 803–816.

Gerischer, U. (2002) Specific and global regulation of genes associated with the degradation of aromatic compounds in bacteria. *J. Mol. Microbiol. Biotechnol.* **4**, 111–121.

Groseclose, E. E., and Ribbons, D. W. (1981) Metabolism of resorcinylic compounds by bacteria: new pathway for resorcinol catabolism in *Azotobacter vinelandii*. *J. Bacteriol.* **146**, 460–466.

Han, S. O., Inui, M., and Yukawa, H. (2008) Effect of carbon source availability and growth phase on expression of *Corynebacterium glutamicum* genes involved in the tricarboxylic acid cycle and glyoxylate bypass. *Microbiology* **154**, 3073–3083.

Harwood, C. S., and Parales, R. E. (1996) The β-ketoadipate pathway and the biology of self-identity. *Annu. Rev. Microbiol.* **50**, 553–590.

Haußmann, U., and Poetsch, A. (2012) Global proteome survey of protocatechuate- and glucose-grown *Corynebacterium glutamicum* reveals multiple physiological differences. *J. Proteomics* **75**, 2649–2659.

Haußmann, U., Qi, S. W., Wolters, D., Rogner, M., Liu, S. J., and Poetsch, A. (2009) Physiological adaptation of *Corynebacterium glutamicum* to benzoate as alternative carbon source – a membrane proteome-centric view. *Proteomics* **9**, 3635–3651.

Hawkins, A. R., Lamb, H. K., Moore, J. D., Charles, I. G., and Roberts, C. F. (1993) The pre-chorismate (shikimate) and quinate pathways in filamentous fungi: theoretical and practical aspects. *J. Gen. Microbiol.* **139**, 2891–2899.

Hinrichs, W., Kisker, C., Duvel, M., Muller, A., Tovar, K., Hillen, W., and Saenger, W. (1994) Structure of the Tet repressor–tetracycline complex and regulation of antibiotic resistance. *Science* **264**, 418–420.

Huang, Y., Zhao, K. X., Shen, X. H., Chaudhry, M. T., Jiang, C. Y., and Liu, S. J. (2006) Genetic characterization of the resorcinol catabolic pathway in *Corynebacterium glutamicum*. *Appl. Environ. Microbiol.* **72**, 7238–7245.

Huang, Y., Zhao, K. X., Shen, X. H., Jiang, C. Y., and Liu, S. J. (2008) Genetic and biochemical characterization of a 4-hydroxybenzoate hydroxylase from *Corynebacterium glutamicum*. *Appl. Microbiol. Biotechnol.* **78**, 75–83.

Ikeda, M., and Nakagawa, S. (2003) The *Corynebacterium glutamicum* genome: features and impacts on biotechnological processes. *Appl. Microbiol. Biotechnol.* **62**, 99–109.

Iwagami, S. G., Yang, K., and Davies, J. (2000) Characterization of the protocatechuic acid catabolic gene cluster from *Streptomyces* sp. strain 2065. *Appl. Environ. Microbiol.* **66**, 1499–1508.

Jiménez, J. I., Minambres, B., Garcia, J. L., and Díaz, E. (2002) Genomic analysis of the aromatic catabolic pathways from *Pseudomonas putida* KT2440. *Environ. Microbiol.* **4**, 824–841.

Jungwirth, B., Emer, D., Brune, I., Hansmeier, N., Pühler, A., Eikmanns, B. J., and Tauch, A. (2008) Triple transcriptional control of the resuscitation promoting factor 2 (*rpf2*) gene of *Corynebacterium glutamicum* by the regulators of acetate metabolism RamA and RamB and the cAMP-dependent regulator GlxR. *FEMS Microbiol. Lett.* **281**, 190–197.

Kalinowski, J., Bathe, B., Bartels, D., Bischoff, N., Bott, M., Burkovski, A., Dusch, N., Eggeling, L., Eikmanns, B. J., Gaigalat, L., *et al.* (2003) The complete *Corynebacterium glutamicum* ATCC 13032 genome sequence and its impact on the production of L-aspartate-derived amino acids and vitamins. *J. Biotechnol.* **104**, 5–25.

Kim, H. J., Kim, T. H., Kim, Y., and Lee, H. S. (2004a) Identification and characterization of *glxR*, a gene involved in regulation of glyoxylate bypass in *Corynebacterium glutamicum*. *J. Bacteriol.* **186**, 3453–3460.

Kim, S. J., Jones, R. C., Cha, C. J., Kweon, O., Edmondson, R. D., and Cerniglia, C. E. (2004b) Identification of proteins induced by polycyclic aromatic hydrocarbon in *Mycobacterium vanbaalenii* PYR-1 using two-dimensional polyacrylamide gel electrophoresis and *de novo* sequencing methods. *Proteomics* **4**, 3899–3908.

Kim, Y. H., Cho, K., Yun, S. H., Kim, J. Y., Kwon, K. H., Yoo, J. S., and Kim, S. I. (2006) Analysis of aromatic catabolic pathways in *Pseudomonas putida* KT 2440 using a combined proteomic approach: 2-DE/MS and cleavable isotope-coded affinity tag analysis. *Proteomics* **6**, 1301–1318.

Kimura, E. (2002) Triggering mechanism of L-glutamate overproduction by DtsR1 in coryneform bacteria. *J. Biosci. Bioeng.* **94**, 545–551.

Kitagawa, W., Kimura, N., and Kamagata, Y. (2004) A novel *p*-nitrophenol degradation gene cluster from a gram-positive bacterium, *Rhodococcus opacus* SAO101. *J. Bacteriol.* **186**, 4894–4902.

Kohl, T. A., and Tauch, A. (2009) The GlxR regulon of the amino acid producer *Corynebacterium glutamicum*: detection of the corynebacterial core regulon and integration into the transcriptional regulatory network model. *J. Biotechnol.* **143**, 239–246.

Kohl, T. A., Baumbach, J., Jungwirth, B., Puhler, A., and Tauch, A. (2008) The GlxR regulon of the amino acid producer *Corynebacterium glutamicum*: *in silico* and *in vitro* detection of DNA binding sites of a global transcription regulator. *J. Biotechnol.* **135**, 340–350.

Krogh, A., Larsson, B., von Heijne, G., and Sonnhammer, E. L. (2001) Predicting transmembrane protein topology with a hidden Markov model: application to complete genomes. *J. Mol. Biol.* **305**, 567–580.

Krüger, E., Völker, U., and Hecker, M. (1994) Stress induction of *clpC* in *Bacillus subtilis* and its involvement in stress tolerance. *J. Bacteriol.* **176**, 3360–3367.

Kvint, K., Nachin, L., Diez, A., and Nystrom, T. (2003) The bacterial universal stress protein: function and regulation. *Curr. Opin. Microbiol.* **6**, 140–145.

Lack, L. (1959) The enzymic oxidation of gentisic acid. *Biochim. Biophys. Acta* **34**, 117–123.

Lee, S. Y., Kim, Y. H., and Min, J. (2010a) Conversion of phenol to glutamate and proline in *Corynebacterium glutamicum* is regulated by transcriptional regulator ArgR. *Appl. Microbiol. Biotechnol.* **85**, 713–720.

Lee, S. Y., Le, T. H., Chang, S. T., Park, J. S., Kim, Y. H., and Min, J. (2010b) Utilization of phenol and naphthalene affects synthesis of various amino acids in *Corynebacterium glutamicum*. *Curr. Microbiol.* **61**, 596–600.

Letek, M., Valbuena, N., Ramos, A., Ordonez, E., Gil, J. A., and Mateos, L. M. (2006) Characterization and use of catabolite-repressed promoters from gluconate genes in *Corynebacterium glutamicum*. *J. Bacteriol.* **188**, 409–423.

Leveau, J. H., Zehnder, A. J., and van der Meer, J. R. (1998) The *tfdK* gene product facilitates uptake of 2,4-dichlorophenoxyacetate by *Ralstonia eutropha* JMP134(pJP4). *J. Bacteriol.* **180**, 2237–2243.

Levin, D. A. (1971) Plant phenolics: an ecological perspective. *Am Nat* **105**, 157–181.

Li, D. F., Zhang, N., Hou, Y. J., Huang, Y., Hu, Y., Zhang, Y., Liu, S. J., and Wang, D. C. (2011) Crystal structures of the transcriptional repressor RolR reveals a novel recognition mechanism between inducer and regulator. *PLoS One* **6**, e19529.

Li, T., Zhao, K., Huang, Y., Li, D., Jiang, C. Y., Zhou, N., Fan, Z., and Liu, S. J. (2012) The TetR-type transcriptional repressor RolR from *Corynebacterium glutamicum* regulates resorcinol catabolism by binding to a unique operator, *rolO*. *Appl. Environ. Microbiol.* **78**, 6009–6016.

Li, T., Chen, X., Chaudhry, M. T., Zhang, B., Jiang, C. Y., Liu, S. J. (2014) Genetic characterization of 4-cresol catabolism in *Corynebacterium glutamicum*. *J Biotechnol.* **192** (Pt B), 355–365.

Liu, T. T., Xu, Y., Liu, H., Luo, S., Yin, Y. J., Liu, S. J., and Zhou, N. Y. (2011) Functional characterization of a gene cluster involved in gentisate catabolism in *Rhodococcus* sp. strain NCIMB 12038. *Appl. Microbiol. Biotechnol.* **90**, 671–678.

Luengo, J. M., Garcia, J. L., and Olivera, E. R. (2001) The phenylacetyl-CoA catabolon: a complex catabolic unit with broad biotechnological applications. *Mol. Microbiol.* **39**, 1434–1442.

Luengo, J. M., Arias, S., Arcos, M., and Olivera, E. R. (2007) The Catabolism of Phenylacetic Acid and Other Related Molecules in *Pseudomonas putida* U. Ramos, J.-L., and Filloux, A. (eds.) Springer, Dortrecht, The Netherlands, pp. 147–192.

Lv, Y., Wu, Z., Han, S., Lin, Y., and Zheng, S. (2011) Genome sequence of *Corynebacterium glutamicum* S9114, a strain for industrial production of glutamate. *J. Bacteriol.* **193**, 6096–6097.

Lv, Y., Liao, J., Wu, Z., Han, S., Lin, Y., and Zheng, S. (2012) Genome sequence of *Corynebacterium glutamicum* ATCC 14067, which provides insight into amino acid biosynthesis in coryneform bacteria. *J. Bacteriol.* **194**, 742–743.

Lykidis, A., Perez-Pantoja, D., Ledger, T., Mavromatis, K., Anderson, I. J., Ivanova, N. N., Hooper, S. D., Lapidus, A., Lucas, S., Gonzalez, B., *et al.* (2010) The complete multipartite genome sequence of *Cupriavidus necator* JMP134, a versatile pollutant degrader. *PLoS One* **5**, e9729.

Lynch, B. S., Delzell, E. S., and Bechtel, D. H. (2002) Toxicology review and risk assessment of resorcinol: thyroid effects. *Regul. Toxicol. Pharmacol.* **36**, 198–210.

McLeod, M. P., Warren, R. L., Hsiao, W. W., Araki, N., Myhre, M., Fernandes, C., Miyazawa, D., Wong, W., Lillquist, A. L., Wang, D., *et al.* (2006) The complete genome of *Rhodococcus* sp. RHA1 provides insights into a catabolic powerhouse. *Proc. Natl. Acad. Sci. U.S.A.* **103**, 15582–15587.

Maddocks, S. E., and Oyston, P. C. (2008) Structure and function of the LysR-type transcriptional regulator (LTTR) family proteins. *Microbiology* **154**, 3609–3623.

Marshak-Rothstein, A., Margolies, M. N., Benedetto, J. D., and Gefter, M. L. (1981) Two structurally distinct and independently regulated idiotypic families associated with the A/J response to azophenylarsonate. *Eur. J. Immunol.* **11**, 565–572.

Martínez-Blanco, H., Reglero, A., Rodriguez-Aparicio, L. B., and Luengo, J. M. (1990) Purification and biochemical characterization of phenylacetyl-CoA ligase from *Pseudomonas putida*. A specific enzyme for the catabolism of phenylacetic acid. *J. Biol. Chem.* **265**, 7084–7090.

Meiswinkel, T. M., Gopinath, V., Lindner, S. N., Nampoothiri, K. M., and Wendisch, V. F. (2013) Accelerated pentose utilization by *Corynebacterium glutamicum* for accelerated production of lysine, glutamate, ornithine and putrescine. *Microb. Biotechnol.* **6**, 131–140.

Merkens, H., Beckers, G., Wirtz, A., and Burkovski, A. (2005) Vanillate metabolism in *Corynebacterium glutamicum*. *Curr. Microbiol.* **51**, 59–65.

Mitra, A., Kitamura, Y., Gasson, M. J., Narbad, A., Parr, A. J., Payne, J., Rhodes, M. J., Sewter, C., and Walton, N. J. (1999) 4-Hydroxycinnamoyl-CoA hydratase/lyase (HCHL) – an enzyme of phenylpropanoid chain cleavage from *Pseudomonas*. *Arch. Biochem. Biophys.* **365**, 10–16.

Mohamed, M., Ismail, W., Heider, J., and Fuchs, G. (2002) Aerobic metabolism of phenylacetic acids in *Azoarcus evansii*. *Arch. Microbiol.* **178**, 180–192.

Molina-Henares, A. J., Krell, T., Eugenia Guazzaroni, M., Segura, A., and Ramos, J. L. (2006) Members of the IclR family of bacterial transcriptional regulators function as activators and/or repressors. *FEMS Microbiol. Rev.* **30**, 157–186.

Moon, M. W., Park, S. Y., Choi, S. K., and Lee, J. K. (2007) The phosphotransferase system of *Corynebacterium glutamicum*: features of sugar transport and carbon regulation. *J. Mol. Microbiol. Biotechnol.* **12**, 43–50.

Nichols, N. N., and Harwood, C. S. (1997) PcaK, a high-affinity permease for the aromatic compounds 4-hydroxybenzoate and protocatechuate from *Pseudomonas putida*. *J. Bacteriol.* **179**, 5056–5061.

Nogales, J., Macchi, R., Franchi, F., Barzaghi, D., Fernandez, C., Garcia, J. L., Bertoni, G., and Díaz, E. (2007) Characterization of the last step of the aerobic phenylacetic acid degradation pathway. *Microbiology* **153**, 357–365.

Nordin, K., Unell, M., and Jansson, J. K. (2005) Novel 4-chlorophenol degradation gene cluster and degradation route via hydroxyquinol in *Arthrobacter chlorophenolicus* A6. *Appl. Environ. Microbiol.* **71**, 6538–6544.

Orth, P., Schnappinger, D., Hillen, W., Saenger, W., and Hinrichs, W. (2000) Structural basis of gene regulation by the tetracycline inducible Tet repressor-operator system. *Nat. Struct. Biol.* **7**, 215–219.

Pao, S. S., Paulsen, I. T., and Saier, M. H., Jr. (1998) Major facilitator superfamily. *Microbiol. Mol. Biol. Rev.* **62**, 1–34.

Patrauchan, M. A., Florizone, C., Dosanjh, M., Mohn, W. W., Davies, J., and Eltis, L. D. (2005) Catabolism of benzoate and phthalate in *Rhodococcus* sp. strain RHA1: redundancies and convergence. *J. Bacteriol.* **187**, 4050–4063.

Perez-Pantoja, D., Donoso, R., Agullo, L., Cordova, M., Seeger, M., Pieper, D. H., and Gonzalez, B. (2012) Genomic analysis of the potential for aromatic compounds biodegradation in *Burkholderiales*. *Environ. Microbiol.* **14**, 1091–1117.

Pessione, E., Giuffrida, M. G., Prunotto, L., Barello, C., Mazzoli, R., Fortunato, D., Conti, A., and Giunta, C. (2003) Membrane proteome of *Acinetobacter radioresistens* S13 during aromatic exposure. *Proteomics* **3**, 1070–1076.

Pieper, D. H., Pollmann, K., Nikodem, P., Gonzalez, B., and Wray, V. (2002) Monitoring key reactions in degradation of chloroaromatics by *in situ* ^{1}H nuclear magnetic resonance: solution structures of metabolites formed from *cis*-dienelactone. *J. Bacteriol.* **184**, 1466–1470.

Qi, S. W., Chaudhry, M. T., Zhang, Y., Meng, B., Huang, Y., Zhao, K. X., Poetsch, A., Jiang, C. Y., Liu, S., and Liu, S. J. (2007) Comparative proteomes of *Corynebacterium glutamicum* grown on aromatic compounds revealed novel proteins involved in aromatic degradation and a clear link between aromatic catabolism and gluconeogenesis via fructose-1,6-bisphosphatase. *Proteomics* **7**, 3775–3787.

Ramos, J. L., Martinez-Bueno, M., Molina-Henares, A. J., Teran, W., Watanabe, K., Zhang, X., Gallegos, M. T., Brennan, R., and Tobes, R. (2005) The TetR family of transcriptional repressors. *Microbiol. Mol. Biol. Rev.* **69**, 326–356.

Ren, Q., Kang, K. H., and Paulsen, I. T. (2004) TransportDB: a relational database of cellular membrane transport systems. *Nucleic Acids Res.* **32**, D284–288.

Rey, D. A., Nentwich, S. S., Koch, D. J., Ruckert, C., Puhler, A., Tauch, A., and Kalinowski, J. (2005) The McbR repressor modulated by the effector substance S-adenosylhomocysteine controls directly the transcription of a regulon involved in sulphur metabolism of *Corynebacterium glutamicum* ATCC 13032. *Mol. Microbiol.* **56**, 871–887.

Saenger, W., Orth, P., Kisker, C., Hillen, W., and Hinrichs, W. (2000) The tetracycline repressor – a paradigm for a biological switch. *Angew. Chem. Int. Ed. Engl.* **39**, 2042–2052.

Santos, C. L., Correia-Neves, M., Moradas-Ferreira, P., and Mendes, M. V. (2012) A walk into the LuxR regulators of *Actinobacteria*: phylogenomic distribution and functional diversity. *PLoS One* **7**, e46758.

Schroder, J., and Tauch, A. (2010) Transcriptional regulation of gene expression in *Corynebacterium glutamicum*: the role of global, master and local regulators in the modular and hierarchical gene regulatory network. *FEMS Microbiol. Rev.* **34**, 685–737.

Shen, W., Liu, W., Zhang, J., Tao, J., Deng, H., Cao, H., and Cui, Z. (2010) Cloning and characterization of a gene cluster involved in the catabolism of *p*-nitrophenol from *Pseudomonas putida* DLL-E4. *Bioresour. Technol.* **101**, 7516–7522.

Shen, X., and Liu, S. (2005) Key enzymes of the protocatechuate branch of the β-ketoadipate pathway for aromatic degradation in *Corynebacterium glutamicum*. *Sci. China C. Life Sci.* **48**, 241–249.

Shen, X. H., Liu, Z. P., and Liu, S. J. (2004) Functional identification of the gene locus (*ncgl2319*) and characterization of catechol 1,2-dioxygenase in *Corynebacterium glutamicum*. *Biotechnol. Lett.* **26**, 575–580.

Shen, X. H., Huang, Y., and Liu, S. J. (2005a) Genomic analysis and identification of catabolic pathways for aromatic compounds in *Corynebacterium glutamicum*. *Microbes Environ.* **20**, 160–167.

Shen, X. H., Jiang, C. Y., Huang, Y., Liu, Z. P., and Liu, S. J. (2005b) Functional identification of novel genes involved in the glutathione-independent gentisate pathway in *Corynebacterium glutamicum*. *Appl. Environ. Microbiol.* **71**, 3442–3452.

Shen, X. H., Zhou, N. Y., and Liu, S. J. (2012) Degradation and assimilation of aromatic compounds by *Corynebacterium glutamicum*: another potential for applications for this bacterium? *Appl. Microbiol. Biotechnol.* **95**, 77–89.

Shin, N. R., Whon, T. W., Roh, S. W., Kim, M. S., Jung, M. J., Lee, J., and Bae, J. W. (2011) Genome sequence of *Corynebacterium nuruki* S6-4^T, isolated from alcohol fermentation starter. *J. Bacteriol.* **193**, 4257.

Sikkema, J., de Bont, J. A., and Poolman, B. (1994) Interactions of cyclic hydrocarbons with biological membranes. *J. Biol. Chem.* **269**, 8022–8028.

Stragier, P., and Patte, J. C. (1983) Regulation of diaminopimelate decarboxylase synthesis in *Escherichia coli*. III. Nucleotide sequence and regulation of the *lysR* gene. *J. Mol. Biol.* **168**, 333–350.

Stragier, P., Richaud, F., Borne, F., and Patte, J.-C. (1983) Regulation of diaminopimelate decarboxylase synthesis in *Escherichia coli*: I. Identification of a *lysR* gene encoding an activator of the *lysA* gene. *J. Mol. Biol.* **168**, 307–320.

Sunnarborg, A., Klumpp, D., Chung, T., and LaPorte, D. C. (1990) Regulation of the glyoxylate bypass operon: cloning and characterization of *iclR*. *J. Bacteriol.* **172**, 2642–2649.

Takenaka, S., Okugawa, S., Kadowaki, M., Murakami, S., and Aoki, K. (2003) The metabolic pathway of 4-aminophenol in *Burkholderia* sp. strain AK-5 differs from that of aniline and aniline with C-4 substituents. *Appl. Environ. Microbiol.* **69**, 5410–5413.

Tam le, T., Eymann, C., Albrecht, D., Sietmann, R., Schauer, F., Hecker, M., and Antelmann, H. (2006) Differential gene expression in response to phenol and catechol reveals different metabolic activities for the degradation of aromatic compounds in *Bacillus subtilis*. *Environ. Microbiol.* **8**, 1408–1427.

Teramoto, H., Inui, M., and Yukawa, H. (2009) Regulation of expression of genes involved in quinate and shikimate utilization in *Corynebacterium glutamicum*. *Appl. Environ. Microbiol.* **75**, 3461–3468.

Teufel, R., Mascaraque, V., Ismail, W., Voss, M., Perera, J., Eisenreich, W., Haehnel, W., and Fuchs, G. (2010) Bacterial phenylalanine and phenylacetate catabolic pathway revealed. *Proc. Natl. Acad. Sci. U.S.A.* **107**, 14390–14395.

Teufel, R., Gantert, C., Voss, M., Eisenreich, W., Haehnel, W., and Fuchs, G. (2011) Studies on the mechanism of ring hydrolysis in phenylacetate degradation: a metabolic branching point. *J. Biol. Chem.* **286**, 11021–11034.

Tomás-Gallardo, L., Canosa, I., Santero, E., Camafeita, E., Calvo, E., López, J. A., and Floriano, B. (2006) Proteomic and transcriptional characterization of aromatic degradation pathways in *Rhodoccocus* sp. strain TFB. *Proteomics* **6**, S119–S132.

Tuck, C. O., Perez, E., Horvath, I. T., Sheldon, R. A., and Poliakoff, M. (2012) Valorization of biomass: deriving more value from waste. *Science* **337**, 695–699.

Udaka, S. (1960) Screening method for microorganisms accumulating metabolites and its use in the isolation of *Micrococcus glutamicus*. *J. Bacteriol.* **79**, 754–755.

Venturi, V., Zennaro, F., Degrassi, G., Okeke, B. C., and Bruschi, C. V. (1998) Genetics of ferulic acid bioconversion to protocatechuic acid in plant-growth-promoting *Pseudomonas putida* WCS358. *Microbiology* **144**, 965–973.

Wang, S. H., Y. Xu, S. J. Liu, and N. Y. Zhou (2011) Conserved residues in the aromatic acid/H^+ symporter family are important for benzoate uptake by NCgl2325 in *Corynebacterium glutamicum*. *Int. Biodeterior. Biodegrad.* **65**, 527–532.

Wehrmann, A., Morakkabati, S., Kramer, R., Sahm, H., and Eggeling, L. (1995) Functional analysis of sequences adjacent to *dapE* of *Corynebacterium glutamicum* reveals the presence of *aroP*, which encodes the aromatic amino acid transporter. *J. Bacteriol.* **177**, 5991–5993.

Wei, M., Zhang, J. J., Liu, H., and Zhou, N. Y. (2010) para-Nitrophenol 4-monooxygenase and hydroxyquinol 1,2-dioxygenase catalyze sequential transformation of 4-nitrocatechol in *Pseudomonas* sp. strain WBC-3. *Biodegradation* **21**, 915–921.

Xu, Y., Yan, D. Z., and Zhou, N. Y. (2006) Heterologous expression and localization of gentisate transporter Ncg12922 from *Corynebacterium glutamicum* ATCC 13032. *Biochem. Biophys. Res. Commun.* **346**, 555–561.

Xu, Y., Gao, X., Wang, S. H., Liu, H., Williams, P. A., and Zhou, N. Y. (2012a) MhbT is a specific transporter for 3-hydroxybenzoate uptake by Gram-negative bacteria. *Appl. Environ. Microbiol.* **78**, 6113–6120.

Xu, Y., Wang, S. H., Chao, H. J., Liu, S. J., and Zhou, N. Y. (2012b) Biochemical and molecular characterization of the gentisate transporter GenK in *Corynebacterium glutamicum*. *PLoS One* **7**, e38701.

Yang, Y. F., Zhang, J. J., Wang, S. H., and Zhou, N. Y. (2010) Purification and characterization of the *ncgl2923*-encoded 3-hydroxybenzoate 6-hydroxylase from *Corynebacterium glutamicum*. *J. Basic Microbiol.* **50**, 599–604.

Yoshida, M., Oikawa, T., Obata, H., Abe, K., Mihara, H., and Esaki, N. (2007) Biochemical and genetic analysis of the gamma-resorcylate (2,6-dihydroxybenzoate) catabolic pathway in *Rhizobium* sp. strain MTP-10005: identification and functional analysis of its gene cluster. *J. Bacteriol.* **189**, 1573–1581.

Yukawa, H., Omumasaba, C. A., Nonaka, H., Kos, P., Okai, N., Suzuki, N., Suda, M., Tsuge, Y., Watanabe, J., Ikeda, Y., *et al.* (2007) Comparative analysis of the *Corynebacterium glutamicum* group and complete genome sequence of strain R. *Microbiology* **153**, 1042–1058.

Zhang, S., Sun, W., Xu, L., Zheng, X., Chu, X., Tian, J., Wu, N., and Fan, Y. (2012) Identification of the para-nitrophenol catabolic pathway, and characterization of three enzymes involved in the hydroquinone pathway, in *Peudomonas* sp. 1-7. *BMC Microbiol.* **12**, 27.

Zhao, B., Yeo, C. C., Lee, C. C., Geng, A., Chew, F. T., and Poh, C. L. (2004) Proteome analysis of gentisate-induced response in *Pseudomonas alcaligenes* NCIB 9867. *Proteomics* **4**, 2028–2036.

Zhao, K. X., Huang, Y., Chen, X., Wang, N. X., and Liu, S. J. (2010) PcaO positively regulates *pcaHG* of the β-ketoadipate pathway in *Corynebacterium glutamicum*. *J. Bacteriol.* **192**, 1565–1572.

Zhao, Z., Ding, J. Y., Li, T., Zhou, N. Y., and Liu, S. J. (2011) The *ncgl1108* (*PheP*$_{Cg}$) gene encodes a new L-Phe transporter in *Corynebacterium glutamicum*. *Appl. Microbiol. Biotechnol.* **90**, 2005–2013.

Zhou, N. Y., Fuenmayor, S. L., and Williams, P. A. (2001) *nag* genes of *Ralstonia* (formerly *Pseudomonas*) sp. strain U2 encoding enzymes for gentisate catabolism. *J. Bacteriol.* **183**, 700–708.

Engineering *Corynebacterium glutamicum* for the Production of Organic Acids and Alcohols

8

Bernhard J. Eikmanns and Michael Bott

Abstract

Owing to its success in the large-scale production of L-glutamate and L-lysine and the tremendous increase in knowledge on its metabolism and regulatory networks as well as the establishment of reliable tools for genetic engineering, *Corynebacterium glutamicum* has become a favourite model organism in white biotechnology. Within the past decade, its potential for the production of other amino acids, but also of a variety of other metabolites of commercial interest has been explored extensively. A major driving force of these studies is the aim to establish a bio-based economy based on renewable carbon sources, which enables to reduce our dependency on fossil carbon sources and to establish sustainable production processes. Important targets in this direction are organic acids, which are employed e.g. as precursors for bulk chemicals and bio-based polymers, and alcohols, which serve, for example, as biofuels. In this review, we will summarize the current knowledge on *C. glutamicum* strains developed for the production of the organic acids succinate, L- and D-lactate, pyruvate, the 2-keto acids ketoisovalerate, ketoisocaproate and ketoglutarate and of the alcohols ethanol and isobutanol. From the data presented it becomes obvious that *C. glutamicum* has an enormous potential for the biotechnological synthesis of many of these compounds, which is comparable and often even better than that described for alternative production hosts.

Introduction

Traditionally, *Corynebacterium glutamicum* is employed for the large scale fermentative production of the amino acids L-glutamate (worldwide > 2.9 million t/a) and L-lysine (> 1.9 million t/a) (Ajinomoto Co., 2013a,b). To improve the production performance by metabolic engineering approaches, the respective biosynthetic pathways, the central carbon metabolism, and the regulation of the relevant pathways in *C. glutamicum* were analysed in detail and genetic tools as well as systems biology approaches on the 'omics' level have been developed and employed (overviews in Eggeling and Bott, 2005; Sauer and Eikmanns, 2005; Wendisch *et al.*, 2006a,b; Bott, 2007; Takors *et al.*, 2007; Burkovski, 2008; Brinkrolf *et al.*, 2010; Schröder and Tauch, 2010; Teramoto *et al.*, 2011; Becker and Wittmann, 2012; Bott and Brocker, 2012; Vertes *et al.*, 2012; Yukawa and Inui, 2013). However, recent studies also explored the usefulness of *C. glutamicum* for the production of other commodity or specialty chemicals, such as the diamines cadaverine and putrescine (Mimitsuka *et al.*, 2007; Kind and Wittmann, 2011; Schneider and Wendisch, 2011), the sugar alcohol xylitol (Sasaki *et al.*, 2010), γ-amino butyric acid (Takahashi *et al.*, 2012), polyhydroxybutyrate (Song *et al.*, 2013), the chaperone ectoine (Becker *et al.*, 2013), carotenoids (Heider *et al.*, 2012, 2014) and in particular, several organic mono- or dicarboxylic acids and the biofuels ethanol and isobutanol. In this review, we summarize the current knowledge and recent achievements on metabolic engineering of *C. glutamicum* for the production of organic acids and alcohols from renewable carbon sources. We present the homologous and/or heterologous, aerobic and anaerobic catabolic pathways from relevant carbon sources leading to succinate, L- and D-lactate, pyruvate, the 2-keto acids ketoisovalerate, ketoisocaproate and ketoglutarate and to the alcohols ethanol and isobutanol, and discuss their regulation. Special emphasis is placed on genetic strain engineering and on the central metabolic network of the respective production strains under aerobic and anaerobic conditions. The aerobic and the anaerobic catabolism of selected carbon sources is depicted in Figs. 8.1 and 8.2, respectively.

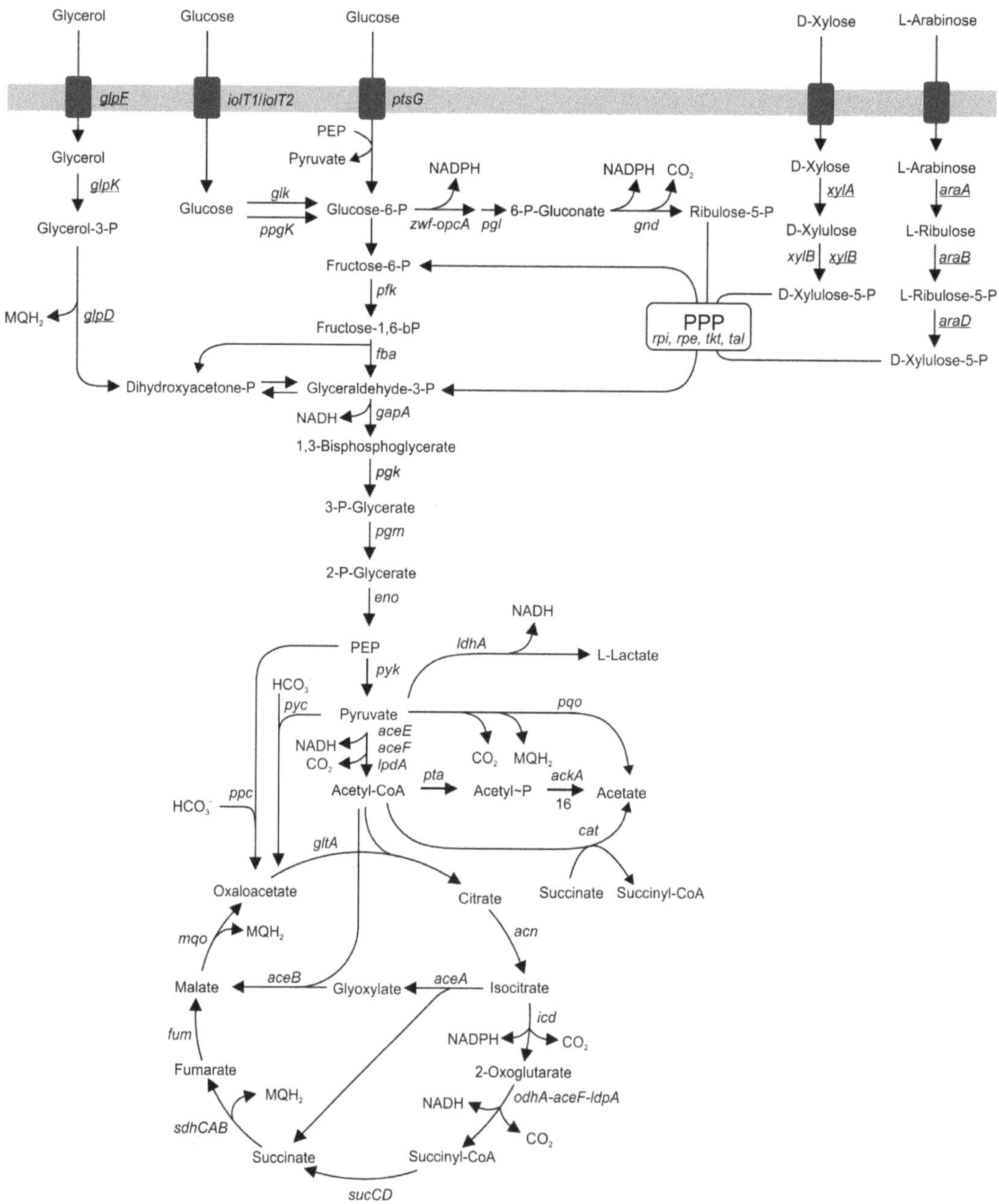

Figure 8.1 Aerobic catabolism of selected carbon sources by *Corynebacterium glutamicum*. The genes encoding the corresponding enzymes are indicated as well as reducing equivalents (NADH, MQH_2) formed or consumed. In contrast to glucose, D-xylose, L-arabinose, and glycerol cannot be utilized by *C. glutamicum* wild type and require the introduction of genes from other species, such as *Escherichia coli*, which are underlined.

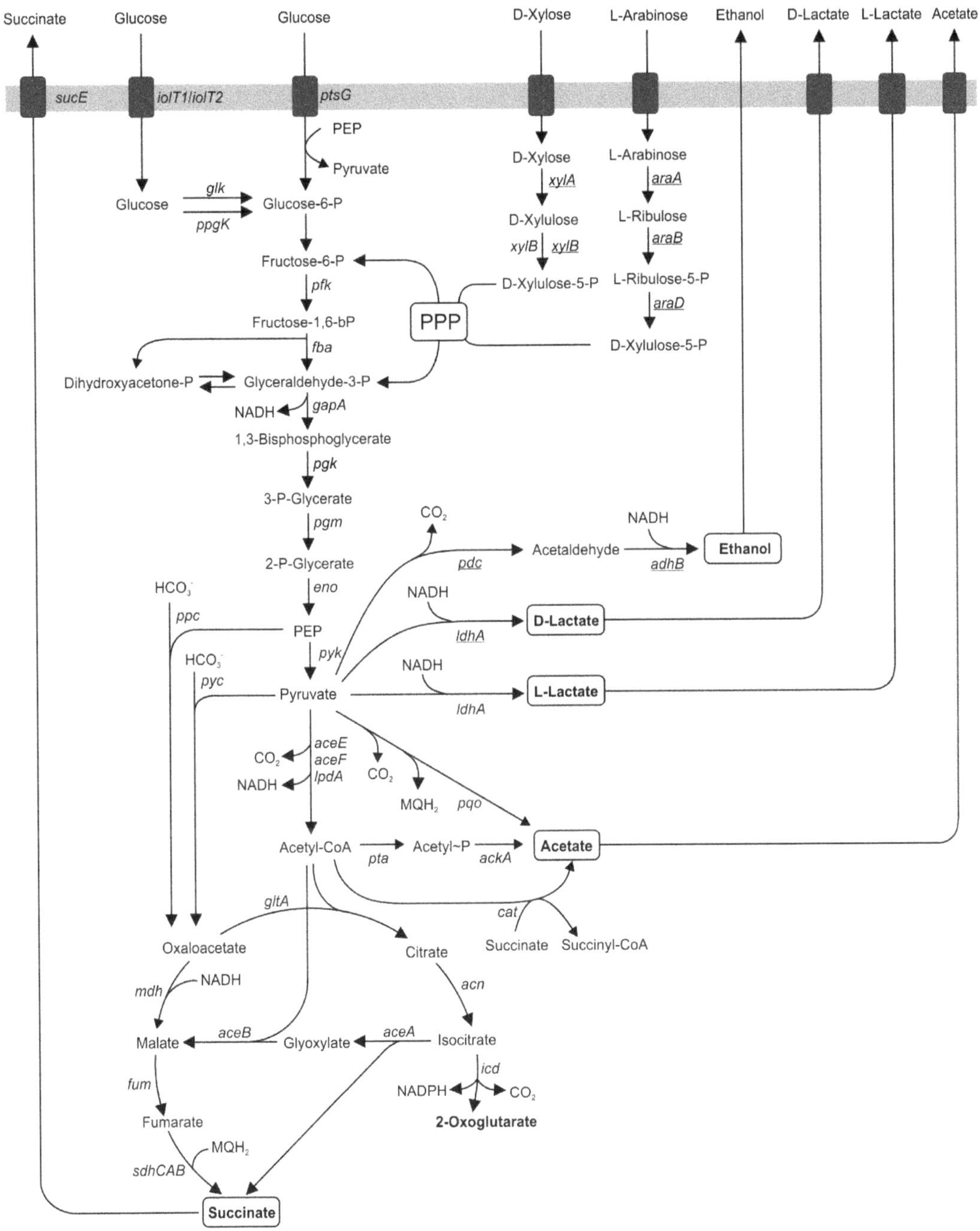

Figure 8.2 Anaerobic sugar catabolism of *Corynebacterium glutamicum* forming L-lactate, succinate, and acetate as natural products or D-lactate or ethanol as non-natural products. The genes encoding the corresponding enzymes are indicated as well as reducing equivalents (NADH, MQH_2) formed or consumed. Genes derived from other species for D-xylose and L-arabinose utilization and for ethanol formation are underlined.

Metabolic engineering of *Corynebacterium glutamicum* for organic acid production

Succinate

Succinate has been identified as an important platform chemical that can be produced from renewable carbon sources. The products that can be derived from succinate are commodity chemicals such as 1,4-butandiol, 2-pyrrolidone, and tetrahydrofuran, bio-based polymers such as polybutylene succinate, and other derivatives such succinate esters and salts, sulfosuccinates, or sulfosuccinamates (Zeikus *et al.*, 1999; Werpy and Peterson, 2004; McKinlay *et al.*, 2007; Cukalovic and Stevens, 2008; Sauer *et al.*, 2008). The estimated market size for succinate and its derived products varies 100-fold between 0.25 and 25 million tons per year (Bozell and Peterson, 2010). Currently succinic acid is produced at a scale of around 20,000–40,000 tons per year mainly from petroleum-derived maleic acid (Cukalovic and Stevens, 2008). Owing to the broad range of applications mentioned above, microbial succinate production has been intensively studied in the past decade with various organisms, such as *Anaerobiospirillum succiniciproducens, Actinobacillus succinogenes, Mannheimia succiniciproducens, Escherichia coli, C. glutamicum* and yeasts. An overview of these studies as well on production conditions, downstream processing, and companies involved in the bio-based succinate business can be found in a recent review article (Litsanov *et al.*, 2014).

Succinate formation from glucose under oxygen-limited or anaerobic conditions

The capability of *C. glutamicum* to form succinate as an end product of glucose metabolism was first noticed in a study with strain ATCC 17965 analysing carbon flux during batch cultivation in a 20-l bioreactor, in which cells were first grown under oxygen excess and thereafter under oxygen limitation. In the latter phase, lactate was formed as major product and succinate and acetate as minor products (Dominguez *et al.*, 1993) (see Fig. 8.2). The authors suggested that succinate is formed via the reductive branch of the tricarboxylic acid (TCA) cycle, with oxaloacetate being synthesized by phosphoenolpyruvate (PEP) carboxylase (PEPCx) (Dominguez *et al.*, 1993). In a later study with *C. glutamicum* strain R, cells were first cultivated aerobically and after harvesting incubated in glucose minimal medium under oxygen-deprived conditions (dissolved oxygen (DO) <0.01 ppm) at a high cell density (50 g wet weight/l). Under these conditions, the cells did not grow, but metabolized glucose to L-lactate (1.55 mol/mol), succinate (0.22 mol/mol) and acetate (0.04 mol/mol) (Inui *et al.*, 2004b). The addition of bicarbonate or pyruvate stimulated glucose consumption, which was explained by an increase of the NAD^+/NADH ratio (from 1.7 to 2.9 and 3.5), leading to an increase in activity of glyceraldehyde-3-phosphate dehydrogenase (GAPDH). This enzyme is known to be strongly inhibited by high NADH/NAD^+ ratios (Dominguez *et al.*, 1998). Bicarbonate increased succinate formation, which is coupled to NADH consumption, by improved PEP or pyruvate carboxylation to oxaloacetate. Pyruvate on the other hand serves as substrate for lactate formation, which is also coupled to NADH oxidation. Studies with defined mutants showed that lactate formation is strictly dependent on the *ldhA* gene and that succinate formation is strongly dependent on PEPCx, whereas pyruvate carboxylase (PCx) plays a minor role in the wild type (Inui *et al.*, 2004b). In contrast, in a Δ*ldhA* mutant, overexpression of PCx, but not of PEPCx, caused a doubling of succinate formation and a corresponding increase in glucose consumption. As almost no succinate was found in a Δ*fum* mutant, it must be formed via the reductive branch of the TCA cycle (Fig. 8.2). Analysis of a Δ*aceA* mutant lacking isocitrate lyase argued against a role of the glyoxylate shunt in succinate formation (Inui *et al.*, 2004b). In conclusion, the results with strain R were in full agreement with those of strain ATCC 17965 (Dominguez *et al.*, 1993). The role of bicarbonate for anaerobic glucose metabolism by *C. glutamicum* strain R was later analysed in more detail, revealing a profound influence on the lactate/succinate ratio, as described in 'L- and D-Lactate formation under oxygen-deprived conditions' below (Okino *et al.*, 2005).

A study similar to the one described above for the wild type (Okino *et al.*, 2005) was subsequently performed for the Δ*ldhA* mutant of *C. glutamicum* strain R overexpressing PCx using plasmid pCRA717 (Okino *et al.*, 2008a). Whereas the specific succinate formation rate and the substrate-specific product yield ($Y_{P/S}$) were independent of the glucose concentration in the range of 20–420 mM, these parameters were strongly dependent on the bicarbonate concentration, reaching maximal values (2.6 mmol/h × g cell dry weight, 1.42 mol/mol) at about 200 mM bicarbonate. With increasing cell density, the succinate production rate increased linearly up to 40 g cell dry weight/l and reached a value of 142 mM/h (16.8 g/l × h) at 60 g cell dry weight/l. In an experiment with intermittent addition of glucose and bicarbonate, a succinate concentration of 1.24 M (146 g/l) with a rate of 44 mM/h (3.2 g/l × h) was achieved after 46 hours at 50 g cell dry weight/l. Besides succinate, 0.27 M acetate (16 g/l) and trace amounts of lactate, malate, and pyruvate as well as some amino acids were formed. The $Y_{P/S}$ were 1.40 mol succinate/mol glucose (0.92 g/g) and 0.29 mol acetate/mol glucose (0.10 g/g). These results demonstrated an outstanding potential of *C. glutamicum* for succinate production.

A non-optimal characteristic of the succinate production strain described above (Okino *et al.*, 2008a) was the formation of significant amounts of acetate as a by-product. This problem was addressed in a study with the type strain ATCC 13032 (Litsanov *et al.*, 2012a). In contrast to the previous studies, the aerobically pre-cultivated cells were not suspended in minimal medium for anaerobic glucose conversion, but in 0.9% (wt/vol) NaCl solution containing glucose and bicarbonate and the pH was kept at 6.9. The parent strain Δ*ldhA* overexpressing a PCx gene carrying a Pro458Ser exchange (pAN6-*pyc*^{P458S}) formed per mol glucose 1.15 mol succinate, 0.46 mol acetate, and traces of pyruvate, malate, fumarate, and α-ketoglutarate. After deletion of the genes known to be involved in acetate formation, i.e. phosphotransacetylase (PTA) and acetate kinase (AK) (Δ*pta-ackA*), pyruvate:quinone oxidoreductase (PQO) (Δ*pqo*), and acetyl-CoA:CoA transferase (Δ*cat*) (see Figs. 8.1 and 8.2), the acetate $Y_{P/S}$ of the resulting strain BOL-1/pAN6-*pyc*^{P458S} was in fact reduced from 0.46 to 0.04 mol/mol glucose. The 90% reduction of acetate formation was accompanied by a reduced succinate yield (1.01 mol/mol) and increased formation of pyruvate (0.21 mol/mol) and α-ketoglutarate (0.13 mol/mol). The decreased succinate yield was due to the reduced formation of NADH by the pyruvate dehydrogenase complex and/or reduced formation of MQH_2 by pyruvate:quinone oxidoreductase. Comparable values were obtained when the *pyc*^{P458S} gene was integrated into the chromosome under the control of the strong *tuf* promoter (*tuf* encodes elongation factor Tu) from *C. glutamicum* rather than provided by plasmid pAN6-*pyc*^{P458S}. The glucose consumption and succinate production rates of the resulting strain BOL-2 were in fact higher than for strain BOL-1/pAN6-*pyc*^{P458S} (Litsanov *et al.*, 2012a).

The succinate yield from glucose is limited by the redox balance, as the two NADH formed by glycolytic conversion of glucose to pyruvate are sufficient for conversion of only one molecule oxaloacetate to succinate. To overcome this limitation, the *fdh* gene for an NAD^+-dependent formate dehydrogenase from *Mycobacterium vaccae* was integrated under control of the *tuf* promoter into the genome of strain BOL-2 (Litsanov *et al.*, 2012a). In the absence of formate, the resulting strain BOL-3 behaved like its parent, but when supplemented with formate, the succinate $Y_{P/S}$ increased from 1.00 mol/mol to 1.27 mol/mol. The BOL-3 strain suffered from a temporary inhibition of glucose catabolism, presumably caused by an inhibition of GAPDH (encoded by *gapA*) due to excess NADH produced by formate oxidation. This inhibition could be overcome by plasmid-driven overexpression of the *gapA* gene and led to a succinate $Y_{P/S}$ of 1.41 mol/mol. In a fed-batch experiment of strain BOL-3/pAN6-*gap* (12.5 g cell dry weight/l) with intermittent addition of glucose and formate, 1134 mM succinate (134 g/l) were formed within 53 hours from 679 mM glucose (122 g/l) and 772 mM formate (36 g/l), corresponding to a succinate $Y_{P/S}$ of 1.67 mol/mol glucose (1.1 g/g). The by-products α-ketoglutarate, malate, acetate, fumarate, and pyruvate corresponded to 10% of the total glucose carbon consumed. The overall specific rates for glucose consumption, formate consumption, and succinate formation were 0.95, 1.08, 1.59 mmol/h × g cell dry weight, respectively. Thus, using a proper strain and a suitable supplementary electron and CO_2 donor, a homosuccinate fermentation becomes feasible: Glucose + 2 Formate$^-$ → 2 Succinate^{2-} + 2 H^+ + 2 H_2O ($\Delta G^{0\prime}$ = −315 kJ/mol glucose) (Litsanov *et al.*, 2012a).

The use of formate as co-substrate for succinate formation besides glucose as described by (Litsanov *et al.*, 2012a) might increase the fermentation costs and impede downstream processing. An alternative approach to increase the succinate yield from glucose is the simultaneous use of the reductive branch of the TCA cycle and the glyoxylate shunt as reported for *E. coli* (Vemuri *et al.*, 2002). This combination allows a maximal $Y_{P/S}$ of 1.71 mol succinate/mol glucose when carbon dioxide is provided in addition: 7 Glucose + 6 CO_2 → 12 Succinate^{2-} + 24 H^+ + 6 H_2O (ΔG^0 = −275 kJ/mol glucose). To engineer a *C. glutamicum* strain for this type of conversion and with eliminated lactate and reduced acetate formation, the genes for *ldhA*, *pta*, *pqo* and *cat* were deleted, the native promoter of the PEPCx gene *ppc* was replaced by the promoter of the native *sod* gene (superoxide dismutase), the PCx gene *pyc* was overexpressed using plasmid pEpyc and the genes of the glyoxylate shunt, *aceA* (isocitrate lyase) and *aceB* (malate synthase), were overexpressed from a second compatible plasmid (pXaceAB) (Zhu *et al.*, 2014). However, despite high specific activities of isocitrate lyase (0.68 U/mg protein) and malate synthase (0.41 U/mg protein), the succinate $Y_{P/S}$ of 1.23 mol/mol of strain SA3 was comparable to that of the reference strain SA2 without *aceAB* overexpression (1.21 mol/mol). Additional overexpression of the citrate synthase gene *gltA* causing a 9-fold increased specific activity (1.69 U/mg protein) led to a 9% increase of the succinate $Y_{P/S}$ (1.34 mol/mol) and lowered by-product levels in strain SA4. A critical issue for the application of the glyoxylate shunt is the inhibition of isocitrate lyase activity by various metabolites including succinate, for which an inhibition constant of 1.48 mM was determined (Reinscheid *et al.*, 1994). Accordingly, a low intracellular succinate concentration is very important for an active glyoxylate shunt. For that purpose, the succinate exporter gene *sucE* (see below) was overexpressed and the resulting strain SA5 showed in fact an increased succinate $Y_{P/S}$ of 1.43 mol/mol (Zhu *et al.*, 2014). In a fed-batch cultivation of strain SA5 with 27.5 g cell dry weight/l, 926 mM succinate, 104 mM α-ketoglutarate, 51 mM acetate, 19 mM fumarate, 7 mM malate and 8 mM pyruvate were formed from 701 mM glucose (and carbon dioxide) within 98 hours. The resulting succinate $Y_{P/S}$ was 1.32 mol/mol and the overall volumetric productivity was 9.4 mM/h. These values indicate that there is still room for improvement of this process.

Another approach for succinate production by the reductive branch of the TCA cycle employed the *C. glutamicum* strain ELB-P, an ATCC 13032 derivative with deletions of *aceE* (encoding the E1 subunit of the pyruvate dehydrogenase complex PDHC), *pqo*, *ldhA*, *alaT* and *avtA* (alanine aminotransferases) and a modification of *ilvN* (ΔC-T, leaky variant of acetohydroxyacid synthase) (Wieschalka *et al.*, 2012, 2013). Owing to the lack of pyruvate dehydrogenase activity, this strain is acetate-auxotrophic and was cultivated in fed-batch in three phases: (i) an aerobic growth phase on glucose plus acetate, (ii) a self-induced microaerobic phase at the end of the exponential growth phase triggered by minimal aeration, and (iii) an anaerobic production phase including gassing with CO_2. In this process, 330 mM succinate (39 g/l) were formed with a $Y_{P/S}$ of 1.02 mol/mol glucose and formation of 30 mM pyruvate as major by-product (Wieschalka *et al.*, 2013).

Succinate formation from pentoses under oxygen-limited or anaerobic conditions

When using lignocellulosic biomass as feedstock for the industrial production of chemicals such as succinate, the utilization of pentoses derived from hemicellulose is an important issue. The major pentoses present in hemicellulose are xylose and arabinose, both of which cannot be used by the currently studied *C. glutamicum* wild-type strains. However, recombinant strains were constructed which overcome this deficiency (Figs. 8.1 and 8.2).

For xylose consumption, expression of the *E. coli xylA* gene encoding xylose isomerase was sufficient to allow aerobic growth on xylose as sole carbon source (Kawaguchi *et al.*, 2006).

XylA converts xylose to xylulose, which then is phosphorylated by the endogenous xylulokinase XylB to xylulose-5-phosphate, an intermediate of the pentose phosphate cycle. Simultaneous overexpression of the *E. coli xylA* and *xylB* genes in strain CRX2 improved growth compared with overexpression of *xylA* alone, but the growth rate was still 1.4-fold higher on glucose than on xylose (Kawaguchi *et al.*, 2006). Under oxygen deprivation conditions, xylose was converted by strain CRX2 to lactate (53–54%), succinate (21–25%) and acetate (6–8%). The lactate $Y_{P/S}$ was lower on xylose than on glucose (64–68%), correlating with a lower L-lactate dehydrogenase (LDH) activity, whereas the succinate $Y_{P/S}$ was higher than on glucose (14%). Pre-cultivation on xylose increased the specific xylose consumption rate by 40% to 4.6 mmol/h × g cell dry weight, which corresponds to 80% of the glucose consumption rate (Kawaguchi *et al.*, 2006). When a 2:1 mixture of glucose and xylose was provided, both sugars were consumed in parallel, but glucose clearly had a negative influence of on xylose consumption, the molecular basis of which is not known yet.

In order to establish a pathway for arabinose utilization, the *E. coli* genes *araA*, *araB* and *araD* encoding L-arabinose isomerase, L-ribulokinase and L-ribulose-5-phosphate 4-epimerase were expressed on plasmid pCRA820 in *C. glutamicum* under the control of a constitutive promoter (Kawaguchi *et al.*, 2008). These enzymes convert L-arabinose via L-ribulose and L-ribulose 5-phosphate to D-xylulose 5-phosphate (see Figs. 8.1 and 8.2) and allowed aerobic growth of the recombinant *C. glutamicum* strain on L-arabinose as sole carbon and energy source with a growth rate of 0.27 h^{-1}. Under oxygen deprivation, L-arabinose was converted predominantly to lactate (37%), succinate (33–40%) and small amounts of acetate. Cells pre-cultured on L-arabinose showed a higher succinate $Y_{P/S}$ (40%) and 1.3- to 1.4-fold increased specific rates of arabinose consumption (3.4 mmol/h × g cell dry weight), lactate production, and succinate production. When supplied with a mixture of 50 g/l glucose and 10 g/l arabinose, both sugars were metabolized simultaneously, glucose with a rate of about 3.9 mmol/h × g cell dry weight, arabinose with a rate of 0.4 mmol/h × g cell dry weight. After glucose consumption, the lactate $Y_{P/S}$ significantly decreased (from 62% to 11%), whereas the succinate and acetate $Y_{P/S}$ increased (from 20% and 7% to 47% and 25%, respectively). As in the case of xylose, glucose had a negative influence on arabinose consumption. Xylose and arabinose utilization by recombinant *C. glutamicum* strains was possible although no heterologous transport genes were introduced, showing that not yet identified endogenous transporters catalyse the uptake of these pentoses.

Whereas the two studies summarized above used pure sugars as carbon sources, a recent study reported on the use of acid hydrolysates of corn cobs as source of anaerobic succinate production (Wang *et al.*, 2014). The recombinant ATCC 13032 derivative NC-2 contained an *ldhA* deletion and the *E. coli xylAB* genes were integrated chromosomally into the *pta-ackA* locus under the control of the *trc* promoter. The corn cobs were treated with 2% (vol/vol) sulfuric acid at a ratio of 1:5 (wt/vol) for 2 hours at 105°C, then brought to pH 6 by calcium hydroxide at 50°C and filtered. Optimally, a charcoal treatment was included to remove inhibitory substances. The major components of the resulting hydrolysates were xylose (55 g/l), glucose (4 g/l), arabinose (1 g/l) and acetate (6 g/l). Under oxygen-deprived conditions (12 g cell dry weight/l) in the presence of 300 mM sodium carbonate the cells formed 41 g/l succinate within 48 hours with a volumetric productivity of 0.85 g/l × h. Xylose (55 g/l) and glucose (4 g/l) were completely consumed. The yield of succinate from total sugar was 0.69 g/g. This example demonstrates that corn cob hydrolysates and possibly also other hydrolysates derived from hemicellulose-containing biomass can serve as substrate for anaerobic succinate production by suitable strains of *C. glutamicum*.

Succinate formation from glucose, glycerol and arabinose under aerobic conditions

Besides the anaerobic processes described above, also aerobic succinate production with *C. glutamicum* was studied in recent years (Fig. 8.3). Under aerobic conditions, succinate is an intermediate of the TCA cycle (Bott, 2007). By deleting the *sdhCAB* genes encoding succinate:menaquinone

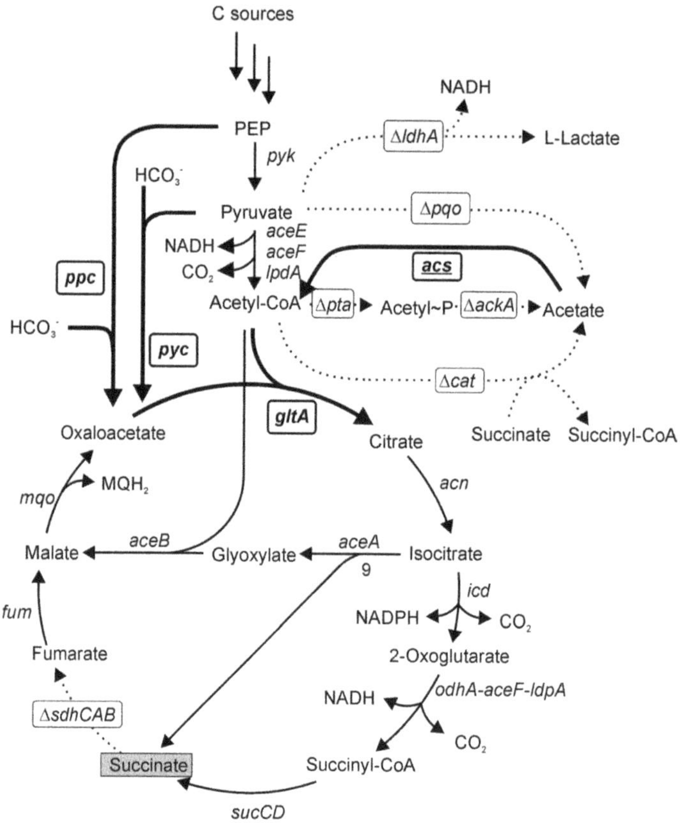

Figure 8.3 Metabolic engineering for aerobic succinate formation with *Corynebacterium glutamicum*. The reactions abolished are indicated by dotted lines, the reactions enhanced are indicated by bold lines. The genes deleted (Δ*sdhCAB*, Δ*pta*, Δ*ackA*, Δ*pqo*, Δ*cat*, Δ*ldhA*), enhanced (*ppc*, *pyc*, *gltA*), or introduced from *Bacillus subtilis* (acs) to improve the production of succinate and to reduce the formation of by-products are boxed. For further details, see text.

oxidoreductase (succinate dehydrogenase), the TCA cycle is interrupted and succinate becomes an end product (Litsanov *et al.*, 2012b). In fact, a Δ*sdhCAB* mutant produced significant amounts of succinate (40 mM) from glucose; however, the major product formed was acetate (125 mM). By deleting the genes for the known acetate-forming pathways, which are PTA plus AK (*pta-ackA*), PQO (*pqo*), and acetyl-CoA:CoA transferase (*cat*), acetate formation was reduced by more than 80% and succinate formation increased by 65% (66 mM). When interrupting the TCA cycle by deletion of the *sdhCAB* genes, oxaloacetate can no longer be regenerated within the cycle but has to be provided by anaplerotic reactions. Whereas overexpression of the pyc^{P458S} gene for a pyruvate carboxylase variant improved the succinate titre only marginally by 3% (68 mM), simultaneous plasmid-based overexpression of pyc^{P458S} and *ppc*, encoding PEPCx, increased the succinate concentration by 24% (82 mM), the $Y_{P/S}$ by 29% (0.36 mol/mol), and the specific production rate by 15% (1.60 mmol/h × g cell dry weight) (Litsanov *et al.*, 2012b). Thus, an increased PEPCx activity is advantageous for aerobic succinate production by strain BL1.

An undesirable feature of aerobic succinate production via the TCA cycle is the loss of two substrate carbon atoms as CO_2 in the isocitrate dehydrogenase and the 2-oxoglutarate dehydrogenase reactions. Use of the glyoxylate shunt could avoid this carbon loss; however, overexpression of the corresponding genes *aceA* (isocitrate lyase) and *aceB* (malate synthase) in strain BL1 (Δ*sdhCAB*, Δ*ack-pta*, Δ*pqo*, Δ*cat*) led to only a 9% increase of the succinate titre (72 mM) and to a 7% increase in succinate $Y_{P/S}$ (0.30 mol/mol) (Litsanov *et al.*, 2012b). The minor effect of *aceAB* overexpression was presumably due to the inhibition of isocitrate lyase by succinate, as described above. A carbon balance analysis of the strains described above show that the majority of carbon from glucose is converted into biomass and CO_2 (31% and 21%, respectively) and only 21% into succinate. An improvement is possible by growth-decoupled succinate production, which was realized by nitrogen limitation, causing a 45% decreased biomass formation and a 43% increase in the succinate titre (90 mM after 117 hours) with strain BL1 pAN6-*pyc*^{P458S}-*ppc*. Whereas acetate formation was decreased by 50% (15 mM), pyruvate (25 mM) and 2-oxoglutarate (18 mM) were formed as additional by-products. The succinate $Y_{P/S}$ reached 0.45 mol/mol, corresponding to a 40% increase compared with non-limited cultures (Litsanov *et al.*, 2012b). This first study on aerobic succinate production with *C. glutamicum* revealed that the organism is suitable for that purpose, but that the yields were much lower than under anaerobic conditions due to the growth of the cells and CO_2 formation.

A subsequent study addressed the problem of by-product formation during aerobic succinate production (Zhu *et al.*, 2013). Like strain BL1 described above, strain ZX1 contained deletions of *sdhCAB*, *pqo*, *cat* and *pta*, and in addition a deletion of *ldhA*. Furthermore, the native promoters of *pyc* and *ppc* were replaced by the *sod* promoter of the superoxide dismutase gene in order to increase their expression. The strain produced 25 mM succinate from 55 mM glucose (0.43 mol/mol) and formed pyruvate (7 mM) and acetate (9 mM) as by-products. To reduce the acetate accumulation, an acetate assimilation pathway from *Bacillus subtilis* was introduced into strain ZX1 using plasmid pEacsA, which contains the *acsA* gene under the control of the *trc* promoter. *acsA* encodes acetyl-CoA synthetase, an enzyme converting acetate and CoA-SH to acetyl-CoA with concomitant hydrolysis of ATP to AMP. *C. glutamicum* wild type does not possess such an enzyme. Plasmid-based *acsA* expression in *C. glutamicum* led to a specific ACS activity of 8 mU/mg protein and completely suppressed acetate formation, but at the cost of a 50% increased pyruvate accumulation (15 mM) (Zhu *et al.*, 2013). A further improvement could be obtained by simultaneous overexpression of the endogenous *gltA* gene for citrate synthase. Strain ZX1 (pEacsAgltA) showed a 6-fold increase citrate synthase activity (4.5 U/mg protein) and formed only 6 mM pyruvate, but 35 mM succinate (0.60 mol/mol) with a specific productivity of 1.26 mmol/h × g cell dry weight. This improvement was at least in part due to a decrease in the cellular acetyl-CoA level, which was increased by expression of *acsA*. Additional overexpression of the succinate exporter gene *sucE* (see below) using plasmid pEacsAgltAsucE did not improve succinate titre, yield, and specific productivity, suggesting that export was not limiting under the tested conditions. In a fed-batch experiment with strain ZX1 (pEacsAgltA) 240 mM succinate were produced within 68 hours with a $Y_{P/S}$ of 0.63 mol/mol and a volumetric productivity of 3.55 mM/h (Zhu *et al.*, 2013). As in this study a medium containing 10 g/l tryptone and 10 g/l yeast extract was used, an exact comparison to the previous study in which a minimal medium without any complex components was employed (Litsanov *et al.*, 2012b) is not possible.

Two further studies on aerobic succinate production deal with the use of alternative substrates, glycerol and arabinose. Glycerol is an interesting substrate as it is a by-product of biodiesel production (Wendisch *et al.*, 2011) whereas arabinose is a component of hemicellulose, as described above. *C. glutamicum* wild type is unable to grow with glycerol as sole carbon and energy source, but can do so when transformed with plasmid pVWEx1-*glpFKD* encoding the *E. coli* genes for the glycerol facilitator (*glpF*), glycerol kinase (*glpK*) and glycerol 3-phosphate dehydrogenase (*glpD*) (Rittmann *et al.*, 2008). When the succinate-producing *C. glutamicum* strain BL1 was transformed with this plasmid and cultivated in a minimal medium with 400 mM glycerol, it produced 79 mM succinate within 22 hours with a specific rate of 1.55 mmol/h × g cell dry weight, a succinate $Y_{P/S}$ of 0.21 mol/mol glycerol and a volumetric productivity of 3.59 mM/h (Litsanov *et al.*, 2013). 35 mM acetate was formed as by-product. Thus, glycerol was shown to be a suitable substrate for aerobic succinate production. For arabinose utilization, strain ZX1 (pEacsAgltA) was transformed with a second compatible plasmid pXaraBAD encoding the *E. coli* genes previously shown to enable arabinose utilization (Kawaguchi *et al.*, 2008). The resulting strain grew with arabinose and produced 74 mM succinate with a $Y_{P/S}$ of 0.58 mol/mol arabinose and low by-product titres (6 mM acetate) (Chen *et al.*, 2014). Additionally, mixtures of arabinose and glucose were used with a similar yield and titres up to 110 mM succinate (Chen *et al.*, 2014). Thus, arabinose can also be used as substrate for aerobic succinate production.

Succinate export

Although succinate is formed as an end product of anaerobic catabolism by many bacteria, knowledge about the export of succinate is rare. In *E. coli*, the secondary carriers DcuA, DcuB and DcuC were shown to be capable of catalysing succinate export (for review see Janausch *et al.*, 2002), but homologues were not found in *C. glutamicum*. Recently, the succinate exporter SucE, a secondary carrier, was identified in *C. glutamicum* (Huhn *et al.*, 2011). A bioinformatic approach based on the comparison of membrane proteins of *C. glutamicum* and two rumen bacteria forming succinate as a fermentation end product, *A. succinogenes* and *M. succiniciproducens*, led to the identification of 20 succinate exporter candidates. Inactivation of 12 of the corresponding genes and testing for anaerobic succinate production revealed a diminished production for the cg2425 mutant. Further studies confirmed that the cg2425 gene product, termed SucE, is involved in succinate export, but not in import and that besides SucE additional succinate export systems must be present in *C. glutamicum*. In the presence of SucE, the intracellular succinate concentration was lower than in its absence (Huhn *et al.*, 2011). The SucE protein consists of 539 amino acid residues and contains 10–12 transmembrane helices and a large central cytoplasmic loop of 180 residues. Within the transporter classification system, SucE belongs to the AAE (aspartate:alanine exchanger) family, TC No 2.A.81 (Saier *et al.*, 2006). In another study published shortly after (Huhn *et al.*, 2011), the succinate exporter was searched for by comparing global gene expression of a Δ*ldhA* mutant of *C. glutamicum* under aerobic and microaerobic conditions using DNA microarrays (Fukui *et al.*, 2011). The gene NCgl2130 (corresponding to cg2425, termed *sucE1*) showed a 3-fold increased mRNA level under microaerobic conditions. Deletion of *sucE1* in the Δ*ldhA* background prevented succinate formation from glucose under anaerobic conditions, whereas plasmid-driven overexpression of NCgl2130 increased succinate formation by 50%. Experiments in a bioreactor revealed that under microaerobic conditions succinate formation and glucose consumption were reduced by 30% and 20% in the Δ*ldhA*Δ*sucE1* mutant compared with the Δ*ldhA* mutant, whereas glucose consumption and succinate formation were completely abolished in the double mutant under anaerobic conditions. This suggests that SucE1 could be essential for succinate export under anaerobic conditions, whereas alternative export systems can replace SucE1 under microaerobic conditions. In agreement with the proposed function of SucE1, the cytoplasmic succinate concentration was found to be significantly higher in the Δ*ldhA*Δ*sucE1* mutant (8.1 mM) than in the Δ*ldhA* mutant (4.9 mM) when measured under microaerobic conditions (Fukui *et al.*, 2011). The activity of SucE1 was also tested *in vitro* by reconstitution of partially purified SucE1 protein with a C-terminal hexahistidine tag into liposomes made of *E. coli* lipids. The resulting proteoliposomes loaded with 100 mM unlabelled succinate and incubated with 100 μM [^{3}H]-succinate catalysed accumulation of [^{3}H]-succinate with an exchange rate of 2.4 nmol/min × mg protein (Fukui *et al.*, 2011). The results of the two studies clearly demonstrate an important role of SucE (SucE1) in succinate export by *C. glutamicum*; however, the mechanism of transport remains unclear. At an intracellular pH of 7.5, succinate is predominantly present as dianion ($pK_{a1} = 4.16$, $pK_{a2} = 5.61$) and the membrane potential could serve as driving force. However, the establishment proton-motive force under anaerobic conditions in the absence of a respiratory chain has not been studied yet in *C. glutamicum*.

Lactate

Similar to succinate, lactate, which exists as L- and D-isomer, can be used for a variety of applications, e.g. as a preservative in the food, pharmaceutical, leather and textile industries, as a precursor for the synthesis of base chemicals and as precursor for polymer production (Hofvendahl and Hahn-Hägerdal, 2000; Sauer *et al.*, 2008; Bozell and Peterson, 2010; Gao *et al.*, 2011; Okano *et al.*, 2010). Polylactic acid (PLA) is a biodegradable polyester, which has similar mechanical properties as polyethylene terephthalate. Stereocomplex PLA, which is composed of both D- and L-lactate, is especially interesting as it has a 40–50°C higher melting point compared with the PLA composed of only L- or D-lactate and a higher mechanical strength (Pang *et al.*, 2010). A variety of microbes produce lactate naturally or after metabolic engineering, including lactic acid bacteria, *E. coli*, *C. glutamicum* and yeast. Although lactic acid bacteria are highly efficient lactate producers, their complex nutritional requirements impose higher costs for cultivation and downstream processing compared with less fastidious microorganisms (Okano *et al.*, 2010).

L- and D-Lactate formation under oxygen-deprived conditions

As already described in 'Succinate formation from pentoses under oxygen-limited or anaerobic conditions' above, *C. glutamicum* forms L-lactate as major product of glucose catabolism under oxygen-limited or anoxic conditions (Dominguez *et al.*, 1993; Inui *et al.*, 2004b; Okino *et al.*, 2005) (Fig. 8.2). In the absence of bicarbonate, 1.79 mol lactate/mol glucose and 0.09 mol succinate/mol glucose were formed, whereas in the presence of 400 mM bicarbonate, 1.44 mol lactate/mol glucose and 0.44 mol succinate/mol glucose were synthesized, corresponding to lactate:succinate ratios of 13.1:1 and 3.2:1. Furthermore, the glucose consumption rate was increased about 2-fold in the presence of bicarbonate concentrations above 100 mM and was independent of the initial glucose concentration between 10 mM and 400 mM. On the other hand, the rates of lactate and succinate formation were linearly correlated to the cell concentration and reached values up to 477 mM/h (42.9 g/l/h) and 99 mM/h (11.7 g/l/h) at 60 g cell dry weight/l, which correspond to specific rates of 133 and 28 nmol/min/mg cell dry weight, respectively. In a fed-batch experiment with intermittent addition of glucose and an initial bicarbonate concentration of 400 mM, 1061 mM lactate (95 g/l), 192 mM succinate (23 g/l), and 50 mM acetate (3 g/l) were formed after 6 hours with 30 g cell dry weight/l (Okino *et al.*, 2005). Using a continuous cell-recycling system, the rates of glucose conversion to lactate and succinate remained constant for 360 hours (Okino *et al.*, 2005).

In order to produce D-lactate instead of L-lactate with *C. glutamicum* (Fig. 8.2), the *ldhA* genes encoding the fermentative D-LDH from either *E. coli* or *Lactobacillus delbrueckii* were overexpressed in the Δ*ldhA* mutant of *C. glutamicum* strain R using suitable plasmids (Okino *et al.*, 2008b). Cell extracts of the recombinant strains showed specific D-LDH activities of 6.5 μmol/min × mg protein for the *E. coli* enzyme expressed from the *C. glutamicum ldhA* promoter (pCRB201), 9.0 μmol/min × mg protein for the *L. delbrueckii* enzyme expressed from the same promoter (pCRB204) and 30.5 μmol/min × mg protein for the *L. delbrueckii* enzyme expressed from the constitutive *tac* promoter (pCRB203). For D-lactate production, aerobically grown cells were incubated under oxygen-deprived conditions in mineral salts medium at the desired cell density. The pH was maintained at 7.0 by ammonia addition. At a cell density of 10 g cell dry weight/l, the lactic acid production rates of the two strains with the *L. delbrueckii* D-LDH (Δ*ldhA*/pCRB203 and Δ*ldhA*/pCRB204) were 5-fold higher (74 mM/h) than those of strain Δ*ldhA*/pCRB201 with the *E. coli* D-LDH (14 mM/h). The specific production rates of strain Δ*ldhA*/pCRB204 were constant up to 20 g cell dry weight/l (6–7 mol/h × g cell dry weight) and slightly decreased at higher cell densities. At 60 g cell dry weight/l, a volumetric productivity of 256 mM/h (23 g/l × h) was reached. Under fed-batch conditions with 60 g cell dry weight/l, within 30 hours 1336 mM D-lactate (120 g/l) with an optical purity of >99.9% were formed with 146 mM succinate (17 g/l) and 52 mM acetate (3 g/l) as by-products. The molar yields with respect to glucose were 1.73 for D-lactate, 0.19 for succinate and 0.07 for acetate.

In a further study on D-lactate production with the *L. delbrueckii* D-LDH, a parent strain lacking *ldhA* and *ppc* (encoding PEPCx) was used carrying plasmid pCRB215 which led to a very high

D-LDH activity of 370 µmol/min × mg protein (Tsuge *et al.*, 2013). Furthermore, genes encoding ten glycolytic enzymes (*pgm, pgi, pfkA, fba, tpi, gapA, pgk, pgm, eno, pyk*) were introduced individually in a second (or second plus third) copy into the chromosome under the control of the *tac* promoter. It has to be mentioned that the *glk* gene encoding glucokinase is not required for glycolysis in *C. glutamicum*, as glucose is taken up and phosphorylated via the PEP-dependent phosphotransferase system. The enzymes encoded by the overexpressed genes showed activities that were increased between 1.7-fold in the case of enolase and 82-fold in the case of glucokinase and the strains showed similar growth rates under aerobic conditions as the parent strain. Interestingly, under oxygen-deprived conditions, overexpression of glucokinase, GAPDH, phosphofructokinase, fructose-bisphosphate aldolase and triosephosphate isomerase increased D-lactate productivity by 98%, 39%, 15%, 13% and 10% compared with the parent strain (20 mmol/l × h). Overexpression of pyruvate kinase gene decreased the productivity by 70%, those of the other genes (*pgi, pgk, pgm, eno*) was without effect. Thus, only increased enzyme activities upstream of 1,3-bisphosphoglycerate had a positive effect. Overexpression of the individual genes had differential effects on the yields of D-lactate, succinate, dihydroxyacetone and glycerol. For example, glucokinase overexpression decreased the D-lactate $Y_{P/S}$, but increased the yields of dihydroxyacetone and glycerol. Glucokinase is required for phosphorylating glucose that is not taken up via the PEP-dependent phosphotransferase system, but via alternative secondary transporters, in particular *iolT1* and *iolT2* (Ikeda *et al.*, 2011; Lindner *et al.*, 2011). These transporters are usually transcriptionally repressed by IolR, which controls genes involved in *myo*-inositol metabolism (Klaffl *et al.*, 2013). An increased glucose uptake capacity by *iolR* deletion was recently shown to be beneficial for overproduction of L-leucine and might also be helpful for other products derived from glycolytic intermediates (Vogt *et al.*, 2014). As expression of the *iolT* genes was not deliberately altered, the positive effect of glucokinase overexpression on D-lactate productivity remains unclear. In fed-batch experiments, up to 1445 mM D-lactate were formed within 80 hours by the strain overexpressing glucokinase, which was 65% higher than that of the parent strain (Tsuge *et al.*, 2013). Thus, overexpression of certain glycolytic genes offers the potential to improve D-lactate production with *C. glutamicum*.

The values described above demonstrate that anaerobic L- and D-lactate production with *C. glutamicum* is highly competitive compared with other microorganisms such as *E. coli* (Zhou *et al.*, 2003). Further improvements might be possible by reducing the formation of the by-products and by a further increase in productivity.

Lactate-based polyester production with *Corynebacterium glutamicum*

Besides metabolic engineering aiming at D-lactate production, another study reported on the direct production of D-lactate-based polyester by *C. glutamicum* (Song *et al.*, 2012). Plasmid-based expression of the genes encoding the *E. coli* D-LDH (*ldhA*), the *Megasphaera elsdenii* propionyl-CoA transferase (*pct*) and mutated polyhydroxyalkanoate synthase from *Pseudomonas* sp. 61-3 (PhaC1-S325T/Q481K) termed lactate polymerizing enzyme (*phaC1*(ST/QK) in *C. glutamicum* ATCC13803) led to the synthesis of a polymer (1% wt/wt) composed of 99.3% lactate and 0.7% 3-hydroxybutyrate (Song *et al.*, 2012). Biological synthesis of lactate-based polymers could be an alternative to the chemical synthesis, which involves heavy metal-catalysed polymerization of biologically produced lactate (Auras *et al.*, 2004).

Pyruvate

Pyruvate has a broad application range as the building block for the synthesis of various chemicals and polymers and as an ingredient or additive in food, cosmetics and pharmaceuticals (Li *et al.*, 2001; Zhu *et al.*, 2008). Further applications are its use in weight loss diets (Roufs, 1996; Stanko *et al.*, 1992a,b), in anti-oxidants (Deboer *et al.*, 1993) and in anti-acne and anti-ageing skin treatment (Cotellessa *et al.*, 2004; Ghersetich *et al.*, 2004). Chemical synthesis and production of pyruvate by dehydration and decarboxylation of tartaric acid is feasible; however, it is relatively cost-ineffective (Howard and Fraser, 1932; Li *et al.*, 2001). Therefore, there was a considerable interest in development of cost-effective fermentative pyruvate production processes and in fact, several approaches have been made for pyruvate production with microorganisms, such as multi-auxotrophic yeasts and recombinant *E. coli* strains [reviewed in Li *et al.* (2001), Wendisch *et al.* (2006a) and Zhu *et al.* (2008)]. However, recently, *C. glutamicum* has also been successfully engineered to produce pyruvate with high yields and productivities (see below).

In nearly all organisms, pyruvate is a central intermediate in the carbon and energy metabolism, connecting glycolysis and/or the pentose phosphate pathway with the oxidative or reductive TCA cycle (Figs. 8.1 and 8.2, respectively) and serving as a precursor for the biosynthesis of a variety of amino and organic acids. Thus, it could be expected that the major pyruvate-drawing reaction, i.e. the pyruvate dehydrogenase complex- (PDHC-) reaction, had to be down-regulated or even eliminated for the construction of an efficient pyruvate-producing *C. glutamicum* strain. Schreiner *et al.* (2005) functionally characterized the PDHC in *C. glutamicum* and inactivated it by deletion of the *aceE* gene, encoding the E1p subunit of the complex. The resulting strain, *C. glutamicum* Δ*aceE*, showed no PDHC activity and required acetate (or ethanol) as an additional carbon source for growth on glucose (Blombach *et al.*, 2009; Schreiner *et al.*, 2005). However, when acetate was exhausted from the medium and growth stopped, the mutant showed a more than 10-fold higher intracellular concentration of pyruvate than the parental wild type (about 26 mM vs. 2 mM) and it secreted significant amounts of L-alanine (up to 30 mM), L-valine (up to 30 mM) and pyruvate (up to 35 mM) from glucose (Blombach *et al.*, 2007).

Pyruvate formation from glucose under limited oxygen supply

The extracellular accumulation of L-alanine and L-valine, both amino acids derived from pyruvate, and in particular the intracellular and extracellular accumulation of pyruvate itself by *C. glutamicum* Δ*aceE* indicated an intracellular surplus of pyruvate and prompted the authors to employ this strain as a basis for development of more efficient strains for the overproduction of pyruvate and of pyruvate-derived products (see next paragraphs). Deletion of the pyruvate:quinone oxidoreductase (PQO) gene *pqo* led to a significant increase of pyruvate formation (50–55 mM) and a $Y_{P/S}$ of 0.48 mol/mol of glucose (Wieschalka *et al.*, 2012). Additional deletion of the L-LDH gene *ldhA* and substitution of the native acetohydroxyacid synthase (AHAS, the first enzyme in the pathway from pyruvate to L-valine) by an attenuated derivative (ΔC-T IlvN) nearly abolished L-valine formation (< 1 mM) and shifted the product spectrum to pyruvate (about 190 mM with a $Y_{P/S}$ of 1.36 mol/mol of glucose). The undesired residual L-alanine by-product formation (about 10 mM) was reduced by about 50% by successive deletion of the *alaT* and *avtA* genes encoding alanine aminotransferase and valine-pyruvate aminotransferase, respectively (Wieschalka *et al.*, 2012). The final strain, designated as *C. glutamicum* ELB-P, produced about 200 mM pyruvate in shake-flasks with a $Y_{P/S}$ of 1.49 mol/mol of glucose. In fed-batch fermentations with limited oxygen supply (0–5% DO) at the end of the growth phase and in the production phase, the cells accumulated more than 500 mM (45 g/l) pyruvate with a $Y_{P/S}$ of about 0.95 mol/mol of glucose (Wieschalka *et al.*, 2012).

The best pyruvate-producing *E. coli* strains YYC202 and ALS1059 formed under optimal conditions 720 mM (63 g/l) and 1022 mM (90 g/l) pyruvate with $Y_{P/S}$ of 1.74 and 1.39 mol/mol of glucose, respectively (Zelic *et al.*, 2003; Zhu *et al.*, 2008). These impressive production characteristics were obtained on metabolic engineering of the pyruvate-drain-off, reduction of by-product formation and additionally improvement of the glycolytic flux. However, as the pyruvate yield of *C. glutamicum* ELB-P in shake-flask experiments is in the same range as in the *E. coli* strains, further metabolic engineering (e.g. on glycolysis) and process engineering approaches (e.g. on sophisticated oxygen supply) with *C. glutamicum* ELB-P might lead to a superior pyruvate production system.

2-Ketoisovalerate, 2-ketoisocaproate and 2-ketoglutarate

The 2-keto acids 2-ketoisovalerate (KIV), 2-ketoisocaproate (KIC) and 2-ketoglutarate (KG) are the direct precursors of L-valine, L-leucine and L-glutamate, respectively, and they are used many-fold as therapeutic agents. KIV and KIC serve as nitrogen-free substitutes for their corresponding branched-chain amino acids in chronic kidney disease patients (Teschan *et al.*, 1998; Feiten *et al.*, 2005; Aparicio *et al.*, 2009, 2012a,b; Chang *et al.*, 2009). Owing to catabolic properties, KIC is also used as supplement in so-called 'functional food' (Zanchi *et al.*, 2011) and it has been discussed to provide benefits in the recovery of muscle function after exercise-induced muscle damage (van Someren *et al.*, 2005; Nunan *et al.*, 2010). KG has been reported to reduce uraemia and to supplement endoperitoneal function to treat chronic renal insufficiency in haemodialysis patients (Riedel *et al.*, 1996; Teplan *et al.*, 2001; Jo *et al.*, 2012). All three 2-keto acids are precursors or intermediates for the synthesis of a variety of further compounds, e.g. L-valine, pantothenate synthesis or 3-methyl-1-butanol (potential biofuel) in case of KIC and L-glutamate and derivatives in case of KG (Krill, 2003; Connor and Liao, 2008; Connor *et al.*, 2010; Jo *et al.*, 2012). Additionally, KG is used for flavour generation in dairy products and during sausage fermentation (Banks *et al.*, 2001; Tjener *et al.*, 2004; Gutiérrez-Méndez *et al.*, 2008). KIV, KIC and KG can be synthesized by chemical synthesis (Cooper *et al.*, 1983), for KG, an enzymatic process using glutamate dehydrogenase (GDH) and NADH oxidase has been established (Ödman *et al.*, 2004). However, very recently also fermentative production of all three 2-keto acids have been established by metabolic engineering of *C. glutamicum* (see below).

As in many microorganisms and plants, KIV and KIC are synthesized in *C. glutamicum* from two molecules of pyruvate in three or six reactions, respectively (Fig. 8.4). These are catalysed by AHAS (encoded by *ilvBN*), acetohydroxyacid isomeroreductase (AHAIR; encoded by *ilvC*) and dihydroxyacid dehydratase (DHAD; encoded by *ilvD*), which are common for KIV and KIC synthesis, and subsequently by isopropylmalate (IPM) synthase (IPMS; encoded by *leuA*), IPM dehydratase (IPMD; encoded by *leuCD*) and IPM dehydrogenase (IPMDH; encoded by *leuB*), which represent the so-called IPM pathway and which are specific for KIC formation. Under physiological conditions, KIV and KIC are further metabolized (aminated) to L-valine and L-leucine, respectively, by transaminase B (TAB, encoded by *ilvE*) (Fig. 8.4).

A key enzyme for L-valine biosynthesis in *C. glutamicum*, and thus also for the production of KIV and KIC with this organism, is the AHAS enzyme, which is feedback-inhibited by L-valine, L-leucine and L-isoleucine to a residual activity of about 50% (Eggeling *et al.*, 1987) and which is competitively inhibited by physiological concentrations of KIV (Krause *et al.*, 2010b). In addition, the expression of the *C. glutamicum ilvBN* genes (and thus, the synthesis of AHAS) is attenuated about 50% by the presence of the three branched-chain amino acids and expression of *ilvBNC* is induced by a so far unknown mechanism in the presence of 2-ketobutyrate, which (aside from pyruvate) is a precursor of L-isoleucine biosynthesis (Eggeling *et al.*, 1987; Morbach *et al.*, 2000). A further key enzyme for KIC (and L-leucine) synthesis in *C. glutamicum* is the first enzyme of the IPM pathway, i.e. IPMS. This enzyme catalyses the condensation of KIV and acetyl-CoA to IPM and is strongly feed-back-inhibited by low concentrations of L-leucine ($K_i = 0.4$ mM; (Pátek *et al.*, 1994)) and competitively inhibited by KIC ($K_i = 8$ mM at 0.5 mM KIV; (Bückle-Vallant *et al.*, 2014)). Expression of the respective *leuA* gene has been suggested to be subject to 'LEU element-dependent' transcription attenuation (Pátek *et al.*, 1994; Seliverstov *et al.*, 2005) and to a rather moderate repression by the IclR-type transcriptional regulator LtbR (Brune *et al.*, 2007). In contrast, LtbR represses the *leuCD* and *leuB* genes (and thus, the synthesis of IPMD and IPMDH) more than 10-fold; however, the effector has not been identified so far (Brune *et al.*, 2007)). The IPMD enzyme was non-competitively inhibited by KIC with a K_i of about 5 mM (Bückle-Vallant *et al.*, 2014).

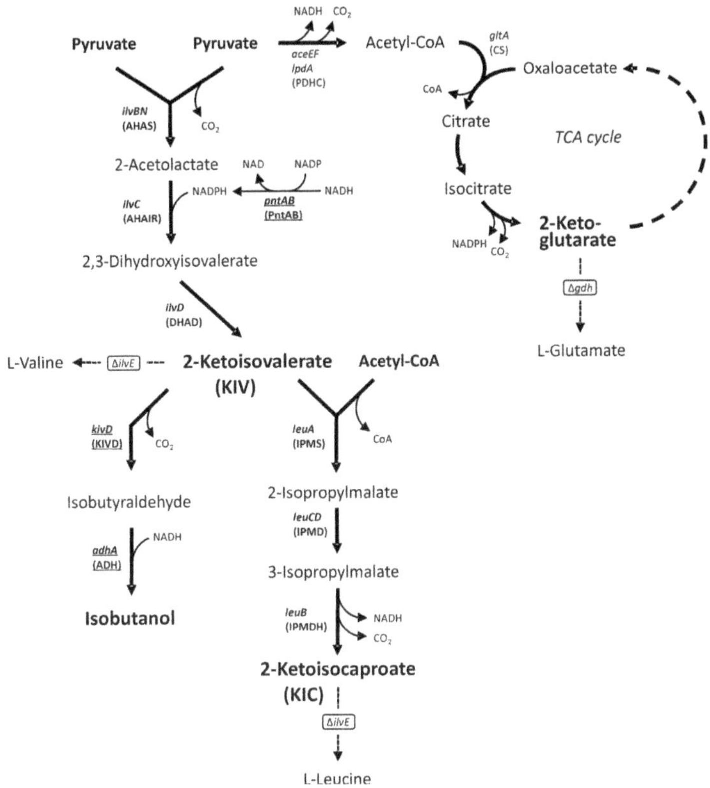

Figure 8.4 Metabolic engineering for the formation of 2-ketoisovalerate (KIV), 2-ketoisocaproate (KIC), 2-ketoglutarate and isobutanol with *Corynebacterium glutamicum*. Homologous genes overexpressed and the encoded enzymes are indicated bold, heterologous genes overexpressed and the corresponding enzymes are shown in bold and underlined, and deleted genes are boxed. For KIV production, the PDHC reaction was abolished by deleting *aceE* or reduced by *aceE* promoter exchange. For KIC production, the CS reaction was reduced by *gltA* promoter exchange. Abbreviations: ADH, alcohol dehydrogenase; AHAS, acetohydroxyacid synthase; AHAIR, acetohydroxyacid isomeroreductase; CS, citrate synthase; DHAD, dihydroxyacid dehydratase; IPMS, isopropylmalate synthase; IMPD, isopropylmalate dehydratase; IPMDH, isopropylmalate dehydrogenase; KivD, ketoacid decarboxylase; PDHC, pyruvate dehydrogenase complex; PntAB, transhydrogenase. The respective gene names are given in brackets. The pathway from KIV and acetyl-CoA to KIC is designated isopropylmalate pathway (IPM).

2-Ketoisovalerate formation from glucose under aerobic conditions

Since the PDHC-deficient strain *C. glutamicum* Δ*aceE* was observed to accumulate L-alanine, L-valine and pyruvate (Schreiner *et al.*, 2005; Blombach *et al.*, 2007; see above) and since KIV originates from two molecules of pyruvate, the Δ*aceE* strain was used as a platform for KIV production (Krause *et al.*, 2010a). Deletion of the TAB gene *ilvE*, overexpression of the AHAS, AHAIR and DHAD genes *ilvBNCD* and subsequent deletion of the PQO gene *pqo* resulted in *C. glutamicum* Δ*aceE* Δ*ilvE* Δ*pqo* (pJC4*ilvBNCD*). Owing to the *aceE* and *ilvE* deletions, this strain required acetate and the branched-chain amino acids L-valine, L-leucine and L-isoleucine as supplements for growth. However, in the production phase, i.e. when acetate was completely exhausted and growth has stopped, the strain produced up to about 220 mM (25.5 g/l) KIV with a volumetric productivity of about 5 mM/h and showed a maximal $Y_{P/S}$ of 0.56 mol/mol of glucose in the production phase (Krause *et al.*, 2010a). In a further approach to optimize KIV production, Buchholz *et al.* (2013) recently constructed a *C. glutamicum* strain with reduced instead of completely abolished PDHC activity. This strain with about 20% of the wild-type PDHC activity was constructed by chromosomal replacement of the native promoter of the *aceE* gene by a mutated variant of the *dapA* promoter, which was generated by Vašicová *et al.* (1999). Upon additional deletion of *ilvE*, *pqo* and *ppc* (the latter encoding the anaplerotic enzyme PEPCx) and plasmid-bound overexpression of *ilvBNCD*, the strain produced about 230 mM (26 g/l) KIV in the growth phase and at the end of the fed-batch fermentation, it accumulated even more than 300 mM (34.8 g/l) KIV with a $Y_{P/S}$ of 0.24 mol/mol of glucose and a volumetric productivity of about 6.9 mM/h (Buchholz *et al.*, 2013). However, the additional formation of 30 mM pyruvate and 33 mM L-valine as undesired by-products indicate that even this optimized strain or the process conditions for KIV production can be further improved.

2-Ketoisocaproate formation from glucose under aerobic conditions

The successful generation of an efficient KIV producer of *C. glutamicum* (see above) prompted Bückle-Vallant *et al.* (2014) to approach the construction of a *C. glutamicum* strain for the fermentative production of KIC. However, although KIV is a precursor of KIC, the PDHC-deficient KIV-producing strain (see above) was not suitable as a basis for development of a KIC-producing strain since (due to the *aceE* deletion) the former strain lacks PDHC activity and therefore is not able to form acetyl-CoA from glucose. Acetyl-CoA on the other side, is the second substrate of the IPMS (Fig. 8.4) and therefore essential for KIC production. Starting with *C. glutamicum* wild type, a basic KIC-producing strain was constructed by deletion of the *ltbR* and *ilvE* genes (i.e. inactivation of LtbR and TAB), overexpression of the *ilvBNCD* genes (i.e. enhancement of the specific activities of AHAS, AHAIR and DHAD) and reduction of citrate synthase (CS) activity to about 30% of the parental activity by exchange of the native promoter of the CS gene *gltA* (Bückle-Vallant *et al.*, 2014). To avoid suppressor mutations complementing the lower CS activity by activation or derepression of the methylcitrate synthase genes *prpC1* and *prpC2*, both of them were also deleted. The resulting strain *C. glutamicum* VB (pJC4*ilvBNCD*) produced about 31 mM (4.0 g/l) KIC with a $Y_{P/S}$ of about 0.26 mol/mol of glucose; however, it also produced about 33 mM KIV and about 6 mM of L-valine and of L-leucine (Bückle-Vallant *et al.*, 2014). The appearance of these undesired by-product in combination with the (known) sensitivity of the IPMS towards L-leucine and KIV indicated a limitation of the carbon flux in the IPM pathway from KIV to KIC. This hypothesis turned out to be valid since additional plasmid-bound expression of a *leuA* allele (*leuA*$^{EC\text{-}G462D}$), encoding a feedback-inhibition-resistant IPMS from *E. coli*, led to an about 2-fold higher KIC production (about 54 mM, i.e. 7 g/l with a $Y_{P/S}$ of about 0.2 mol/mol of glucose) and to a strong reduction of KIV formation (< 2 mM) (Bückle-Vallant *et al.*, 2014). The cultivation of *C. glutamicum* VB (pJC4*ilvBNCD leuA*$^{EC\text{-}G462D}$) with glucose plus acetate resulted in even higher concentrations of KIC (about 71 mM, i.e. 9.2 g/l), indicating that the intracellular acetyl-CoA concentration is still not high enough to allow maximum specific activity of IPMS.

Regarding the performance of the KIV-producing *C. glutamicum* strain (see above) and also having in mind that Vogt *et al.* (2014) recently reported on a metabolically engineered *C. glutamicum* strain that produced about 180 mM of L-leucine, there clearly is room for further improvement of KIC production with this organism. Possible targets include the (feedback-) inhibition of AHAS by the branched-chain amino acids and by KIV, respectively, the KIC inhibition of IPMS and of IPMD, and the relatively low specific activity of the *leuA*$^{EC\text{-}G462D}$ encoded IPMS variant. Other optimization strategies would include a further reduction of the CS activity, an increase of glucose uptake and/or glycolysis and the optimization of process parameters, such as fed-batch cultivation with elaborated substrate feeding, pH and/or oxygen control.

2-Ketoglutarate formation from complex substrates under aerobic conditions

For construction of a *C. glutamicum* KG producer strain, Jo *et al.* (2012) used an undefined glutamate-overproducing mutant, obtained by classical methyl-nitro-nitrosoguanidine mutagenesis. KG is an intermediate of the TCA cycle and a precursor of L-glutamate and the glutamate family of amino acids. Accordingly, the authors disrupted the genes for GDH and glutamate synthase (*gdhA* and *gltB* gene, respectively), shutting down the drain-off of KG to glutamate and afterwards, they inactivated the isocitrate lyase gene (*aceA*) to knock out the competing glyoxylate cycle. It turned out that inactivation of the *gdhA* and *aceA* genes is most relevant for KG production while the disruption of *gltB* seems to be of minor importance. In shake-flask batch fermentations, the *gdhA-gltB-aceA* triple mutant *C. glutamicum* JH110 produced up to about 85 mM (12.4 g/l), in a preliminary experiment using a 5-l fermenter, the strain produced even 325 mM (47.5 g/l) KG after 120 hours of cultivation in a complex medium containing sugar cane molasses, soybean protein hydrolysate and glucose as main carbon sources (Jo *et al.*, 2012). However, the authors did not give data on productivity or yield, or on other details of the KG production process.

Metabolic engineering of *Corynebacterium glutamicum* for alcohol production

Several alcohols (i.e. organic compounds characterized by one or more hydroxyl groups attached to a carbon atom of an alkyl group, such as ethanol, 1-butanol or 1,3-propanediol) are well-known as natural fermentation products of bacteria or yeasts. While ethanol production by fermentation with *Saccharomyces cerevisiae* has a long lasting history over the centuries, the directed microbial production of some higher-chain alcohols (e.g. 1-butanol or 1,3-propanediol) goes back only to the beginning of the twentieth century. Nowadays, the fermentative production of ethanol, 1-butanol and 1,3-propanediol, using *Saccharomyces*, *Clostridium* and recombinant *E. coli* strains, respectively (Nakamura and Whited, 2003; Dürre, 2007; Nevoigt, 2008; Geddes *et al.*, 2011; Jang *et al.*, 2012; Schiel-Bengelsdorf *et al.*, 2013), is of major industrial relevance. In recent years, also the fermentative production of non-natural branched-chain alcohols, such as 2-methyl-1-butanol, 3-methyl-1-butanol and 2-methyl-1-propanol (isobutanol), gained more and more attention, since these alcohols possess higher energy contents and less hygroscopy than (bio)ethanol and thus are regarded superior as biofuels (Li *et al.*, 2010). In accordance, metabolic engineering and synthetic pathway engineering has been applied in several host organisms (*E. coli, B. subtilis, Ralstonia eutropha, Clostridium cellulolyticum, S. cerevisiae*) to increase respective productivities and, thus, to come to economical production processes (reviewed in (Atsumi *et al.*, 2008) and in (Eikmanns and Blombach, 2014)). Owing to a long tradition as robust, user-friendly and industrially employed organism and based on the well-known central carbon metabolism and physiology of *C. glutamicum*, metabolic engineering approaches have recently also been initialized to develop efficient *C. glutamicum* strains for the production of alcohols, in particular of ethanol and isobutanol (see below).

Ethanol

Fermentative ethanol production using so-called first-generation feedstocks (starch hydrolysate, sucrose) amounts worldwide to more than 80 million m^3 per year, about 80% being used as biofuel (http://www.cropenergies.com), the residual 20% serve for chemical (e.g. as solvent, disinfectant, cleaning agent, perfume and anti-aspirant additive, precursor for ethene production by chemical catalysis) or for food purposes (alcoholic drinks and acetic acid). *S. cerevisiae* is the main production organism (Jeffries and Jin, 2004; Mussatto *et al.*, 2010; Amorim *et al.*, 2011; Abreu-Cavalheiro and Monteiro, 2013; Della-Bianca *et al.*, 2013). However, in the last decades, aside from *S. cerevisiae* a variety of other micro-organisms have successfully been engineered to efficiently produce (bio)ethanol (e.g. *Zymomonas mobilis, E. coli, Klebsiella oxytoca, Neurospora crassa*) with high yields and productivities, not only from starch, sucrose, and glucose, but also from second generation feedstocks such as cellulose, hemicellulose and lignocellulose (Dien *et al.*, 2003; Nevoigt, 2008; Matsushita *et al.*, 2009; Mussatto *et al.*, 2010; Jang *et al.*, 2012; Dogaris *et al.*, 2013; Nielsen *et al.*, 2013).

Ethanol formation from glucose or from lignocellulose under oxygen-limited conditions

Inui *et al.* (2004a) engineered *C. glutamicum* for fuel ethanol production by high-level expression of the *Z. mobilis* pyruvate decarboxylase and alcohol dehydrogenase genes (*pdc* and *adhB*, respectively) from the homologous *ldhA* promoter (Fig. 8.2). Resting cells of the resulting strain already produced significant amounts of ethanol under oxygen-deprivation conditions; however, with acetate, lactate and succinate as by-products. Subsequent inactivation of the NAD-dependent L-LDH gene *ldhA* and of the PEPCx gene *ppc* abolished lactate and dramatically decreased succinate formation, and increased the ethanol yield more than 3-fold and ethanol productivity 1.4-fold. Surprisingly, the addition of trace amounts of pyruvate or of acetaldehyde accelerated the ethanol production rate 2-fold (Inui *et al.*, 2004a). It is noteworthy to mention that the productivity increased in proportion to the cell density and using their system (*C. glutamicum* R-*ldhA*-pCRA723) at very high densities, the authors achieved maximal volumetric productivities of more than 640 mmol/l (=29.5 g/l)/h and a final ethanol concentration of 1.7 M (78.2 g/l) (Inui *et al.*, 2004a; Sakai *et al.*, 2007). This final concentration easily competes with the maximal final ethanol concentrations achieved with other bacterial systems, such as *E. coli* LY01 (1.4 M), *K. oxytoca* M5A1(pLO1555) (1.0 M) or *Z. mobilis* AX101 (0.9 M) (Dien *et al.*, 2003).

Having in mind the use of lignocellulose biomass as substrate for ethanol production with *C. glutamicum* R-*ldhA*-pCRA723, the effects of microbial inhibitors usually generated during pretreatment of lignocellulose (Canilha *et al.*, 2012) on ethanol formation was investigated (Sakai *et al.*, 2007). Probably because of the growth arrest under the oxygen-deprivation conditions applied, *C. glutamicum* showed high tolerance to all organic acids, furans and phenolic inhibitors tested. Additionally, different combinations of the inhibitors only slightly affected the ethanol productivity of *C. glutamicum* R-*ldhA*-pCRA723 (Sakai *et al.*, 2007). These results indicate that *C. glutamicum* may be an ideal platform for ethanol production from pre-treated second generation feedstocks without detoxification steps.

Isobutanol

Owing to several chemical properties, such as high water miscibility, high vapour pressure, high octane number and relatively low energy density, ethanol is not considered as optimal substitute for gasoline (Li *et al.*, 2010). In contrast, higher alcohols such as 1-butanol or the branched-chain C_4 alcohol isobutanol possess properties much more similar to gasoline, they are more easy and safer to handle, fully compatible with engines and pipelines, and therefore considered as economically attractive next generation biofuels. At present, both 1-butanol and isobutanol are used many-fold and widely as solvents and as precursor for bulk chemicals, such as butylacrylate or butylacetate (from 1-butanol) or isobutene (from isobutanol). The latter nowadays is produced by petroleum refining in large quantities (> 10 million tons per year) and it is used as gasoline additive and for the production of butyl rubber, isoprene and other speciality chemicals (van Leeuwen *et al.*, 2012).

Isobutanol formation from glucose under oxygen-limited or anaerobic conditions

Several bacteria (mainly *E. coli*) and yeasts (*Saccharomyces*) have successfully been engineered to produce isobutanol, some strains with exceptional high yields (Atsumi *et al.*, 2008, 2009; Smith *et al.*, 2010; Baez *et al.*, 2011; Bastian *et al.*, 2011; Chen *et al.*, 2011; Li *et al.*, 2011; Lamsen and Atsumi, 2012; Buijs *et al.*, 2013; Wen *et al.*, 2013; Eikmanns and Blombach, 2014). Very recently, *C. glutamicum* has also been metabolically engineered for isobutanol production. In a first approach, Smith *et al.* (2010) harnessed the 2-ketoisovalerate pathway (physiologically leading to L-valine) (see Fig. 8.4) for isobutanol production by overexpression of the *B. subtilis alsS* gene (encoding AHAS), the *C. glutamicum ilvC*, *ilvD* and *adhA* genes (encoding AHAIR, DHAD and ADH, respectively), and the *Lactococcus lactis kivD* gene (encoding ketoacid decarboxylase) in *C. glutamicum*. This approach led to the formation of 56 mM isobutanol within 48 hours and 87 mM within 96 hours. Additional inactivation of the L-LDH and PCx genes (*ldhA* and *pyc*, respectively) resulted in the production of 107 mM (4.9 g/l) isobutanol.

Based on the KIV-producing strain *C. glutamicum* Δ*aceE* Δ*ilvE* Δ*pqo* (pJC4*ilvBNCD*) (Krause *et al.*, 2010a) (see the section 'KIV formation from glucose under aerobic conditions' above), Blombach *et al.* (2011) constructed several *C. glutamicum* strains producing isobutanol from glucose under oxygen-deprivation conditions by deletion of the L-LDH and malate dehydrogenase (MDH) genes *ldhA* and *mdh*, respectively, and additional overexpression of the *L. lactis kivD*, the *C. glutamicum adhA* and the *pntAB* genes, the latter encoding the membrane-bound transhydrogenase from *E. coli* (Kabus *et al.*, 2007). The resulting strain *C. glutamicum* Iso7 produced about 82 mM (6.1 g/l) isobutanol with a $Y_{P/S}$ of about 0.77 mol/mol of glucose in shaken bottles under oxygen-deprivation. It is noteworthy to mention that NADPH supply is a critical factor for isobutanol production with *C. glutamicum*. Although *C. glutamicum* is able to convert $NADH + H^+$ to $NADPH + H^+$ by a metabolic cycle consisting of the combined reactions of PCx or PEPCx, MDH and malic enzyme (Sauer and Eikmanns, 2005), both this cycle and the heterologous transhydrogenase enzyme was essential for efficient isobutanol production (Blombach *et al.*, 2011).

To transfer the process conditions of isobutanol production with *C. glutamicum* Iso7 from the sealed bottle to a bioreactor, Blombach *et al.* (2011) developed a two-stage (two-phase) fermentation. In the first stage under fully aerobic conditions, *C. glutamicum* Iso7 grew to high cell densities with glucose and acetate as substrates and produced no isobutanol. When acetate was completely consumed and cell proliferation stopped, the second stage was started by switching off the aeration. After rapidly consuming the residual oxygen, *C. glutamicum* Iso7 produced up to 180 mM (13.3 g/l) Isobutanol with a $Y_{P/S}$ of about 0.48 mol/mol of glucose and a volumetric productivity of 4.4 mmol/l/h. The production stop at about 180 mM was discussed to be due to isobutanol toxicity and Blombach and Eikmanns (2011) suggested to overcome the negative effect by an integrated gas (N_2) stripping process, as successfully applied for 1-butanol production with clostridia (Ezeji *et al.*, 2005).

Very recently, isobutanol production with *C. glutamicum* was further improved by strain optimization and prohibiting product (isobutanol) inhibition (Yamamoto *et al.*, 2013). As a basis strain, these authors used *C. glutamicum* IBU1, an L-LDH-deficient *C. glutamicum* strain carrying plasmid-bound *ilvBNCD* genes expressed from the *tac* promoter and plasmid-bound *L. lactis kivD* and *S. cerevisiae adh2* genes, both expressed from the corynebacterial *ldhA* promoter. Improvement of isobutanol production was obtained by adjusting the expression of *kivD* by promoter exchange and substituting *adh2* by *adhP*, encoding an NAD^+-dependent isobutanol dehydrogenase from *E. coli*. A further improvement of the isobutanol yield by strain optimization was possible by disruption of the PEPCx gene, plasmid-bound expression of a NAD^+-specific AHAIR and overexpression of several genes encoding glycolytic enzymes (GAPDH, pyruvate kinase, phosphofructokinase, phosphoglucose isomerase). Under process optimization, i.e. continuous solvent extraction with oleyl alcohol, *C. glutamicum* IBU8 produced up to 503 mM (37.2 g/l) isobutanol with a $Y_{P/S}$ of 0.78 mol/mol of glucose (Yamamoto *et al.*, 2013). About 80% of the formed isobutanol was found in the extraction layer and it was nearly impurity-free. After prolonged reaction time and with high-density cell suspensions, the final isobutanol titre even reached 981 mM (72.6 g/l), the volumetric productivity 35.7 mmol/l/h. Even though these production data are very impressive and the final titres by far exceeded those of the best *E. coli* isobutanol producer (i.e. 50.8 g/l with *E. coli* JCL7pSA55/pSA69 (Atsumi *et al.*, 2008; Baez *et al.*, 2011)), the authors speculate that even higher productivity gains might be possible on further optimization of the extraction procedure (Wieschalka *et al.*, 2013; Yamamoto *et al.*, 2013).

It is noteworthy to mention that isobutanol production with *C. glutamicum* obviously is more efficient in two stage fermentations (aerobic growth in complex or minimal media and, after harvest of the cells and resuspension in new medium, transfer to sealed bottles) than in one-stage fermentations (growth and production in a single bioreactor). With *C. glutamicum* Iso7 and *C. glutamicum* IBU8, the $Y_{P/S}$ were 0.77 and 0.78 mol/mol of glucose, respectively, in the

Table 8.1 Maximal substrate specific yields ($Y_{P/S}$), productivities, titres, and the respective references of the so far most efficient processes for organic acid and alcohol production with *Corynebacterium glutamicum* strains under different process conditions

Product and strain	Process	Substrate(s)	$Y_{P/S}$ $mol_{product}/mol_{substrate}$ (g/g)	Productivity,[a] mM/h (g/l h)	Maximal titre, mM (g/l)	Reference
Succinate						
C. glutamicum R *ΔldhA* pCRA717	Anaerobic, closed bottle, repeated substrate addition, mineral medium	Glucose, bicarbonate	1.40 (0.92)	44 (3.18)	1240 (146.3)	Okino *et al.* (2008a)
C. glutamicum BOL-3 pAN6-*gap*	Anaerobic, fed-batch fermentation in bioreactor, 0.9% NaCl solution	Glucose, bicarbonate, formate	1.67 (1.09)	21 (2.48)	1134 (133.8)	Litsanov *et al.* (2012a)
C. glutamicum SA5	Anaerobic, fed-batch fermentation in bioreactor, mineral medium	Glucose, bicarbonate	1.32 (0.87)	9.4 (1.11)	926 (109.3)	Zhu *et al.* (2014)
C. glutamicum BL-1 pAN6-*pyc*^{P458S}*ppc*	Aerobic, nitrogen limitation, batch fermentation, mineral medium	Glucose	0.45 (0.30)	0.8 (0.09)	90 (10.6)	Litsanov *et al.* (2012b)
C. glutamicum ZX1 pEacsAgltA	Aerobic, fed-batch fermentation, rich medium	Glucose, tryptone, yeast extract	0.63 (0.41)	3.6 (0.42)	241 (28.4)	Zhu *et al.* (2013)
C. glutamicum BL-1 pVWEx1-*glpFKD*	Aerobic, batch fermentation, mineral medium	Glycerol	0.21 (0.27)	3.6 (0.42)	79 (9.3)	Litsanov *et al.* (2013)
C. glutamicum ZX1 pEacsAgltA pXaraBAD	Aerobic, batch fermentation, mineral medium	Arabinose	0.58 (0.45)	0.9 (0.10)	74 (8.7)	Chen *et al.* (2014)
L-Lactate						
C. glutamicum R	Anaerobic, closed bottle, fed-batch, mineral medium	Glucose	1.79 (0.89)	62 (5.59)	574 (51.7)	Okino *et al.* (2005)
C. glutamicum R	Anaerobic, closed bottle, fed-batch, mineral medium	Glucose, bicarbonate	1.42 (0.71)	Up to 477 (42.97)	1061 (95.6)	Okino *et al.* (2005)
D-Lactate						
C. glutamicum R *ΔldhA* pCRB204	Anaerobic, closed bottle, mineral medium	Glucose	1.73 (0.87)	44.5 (4.01)	1336 (120.3)	Okino *et al.* (2008b)
C. glutamicum R LPglc192 pCRB215	Anaerobic, closed bottle, fed-batch, mineral medium	Glucose	1.66 (0.83)	18.1 (1.63)	1446 (130.2)	Tsuge *et al.* (2013)
Pyruvate						
C. glutamicum ELB-P	Aerobic, shake flask	Glucose	1.49 (0.72)	n.s.	200 (17.6)	Wieschalka *et al.* (2012, 2013)
C. glutamicum ELB-P	Low oxygen tension, fed-batch fermentation	Glucose	0.97 (0.47)	5.6 (0.49)	512 (44.5)	Wieschalka *et al.* (2012, 2013)

Product and strain	Process	Substrate(s)	$Y_{P/S}$ $mol_{product}/mol_{substrate}$ (g/g)	Productivity,[a] mM/h (g/l h)	Maximal titre, mM (g/l)	Reference
2-Ketoisovalerate						
C. glutamicum Δ*aceE* Δ*pqo* Δ*ilvE* (pJC4ilvBNCD)	Aerobic, fed-batch fermentation	Glucose	0.56 (0.36)	5.0 (0.58)	220 (25.5)	Krause *et al.* (2010a)
C. glutamicum aceE A16 Δ*pqo* Δ*ppc* Δ*ilvE* (pJC4ilvBNCD)	Aerobic, fed-batch fermentation	Glucose	0.24 (0.15)	6.9 (0.8)	303 (35.1)	Buchholz *et al.* (2013)
2-Ketoisocaproate						
C. glutamicum VB (pJC4ilvBNCDleuA$^{EC\text{-}G462D}$)	Aerobic, shake flask	Glucose	0.22 (0.16)	3.4 (0.44)	54 (7.0)	Bückle-Vallant *et al.* (2014)
C. glutamicum VB (pJC4ilvBNCDleuA$^{EC\text{-}G462D}$)	Aerobic, shake flask	Glucose, acetate	n.s.	3.4 (0.44)	71 (9.2)	Bückle-Vallant *et al.* (2014)
2-Ketoglutarate						
C. glutamicum JH110	Aerobic, fed-batch fermentation	Glucose, molasses, soybean hydrolysate	n.s.	2.7 (0.39)	325 (47.5)	Jo *et al.* (2012)
Ethanol						
C. glutamicum R-*ldhA*-pCRA723	Anaerobic, closed bottle, fed-batch	Glucose	1.48 (0.38)	642 (29.5)	1700 (78.2)	Inui *et al.* (2004a), Sakai *et al.* (2007)
Isobutanol						
C. glutamicum Iso7	Anaerobic, fed-batch fermentation	Glucose	0.48 (0.20)	4.4 (0.33)	180 (13.3)	Blombach *et al.* (2011)
C. glutamicum IBU8	Anaerobic, fed-batch fermentation, continuous solvent extraction	Glucose	0.78 (0.32)	35.7 (2.64)	981 (72.6)	Yamamoto *et al.* (2013)

[a]During production phase.
n.s., not specified.

two-stage process, whereas it was significantly lower (0.48 mol isobutanol/mol of glucose) when growth and production of strain Iso7 occurred in the same bioreactor. These results indicate that the transition from fully aerobic to anaerobic conditions are of major relevance for the isobutanol production process. This hypothesis is corroborated by similar findings for succinate production with *C. glutamicum* ELB-P (see the section 'Succinate formation from glucose under oxygen-limited or anaerobic conditions' above) (Wieschalka *et al.*, 2013; own unpublished results). It will be interesting to study the physiological consequences and the adjustment of key enzymatic steps in *C. glutamicum* during, before and after slow or fast shifts from aerobic to anaerobic conditions and understanding of the metabolic adaptation certainly will help to further optimize the production of isobutanol and/or products formed under anaerobic conditions.

Concluding remarks

As summarized in Table 8.1, studies in the past decade have disclosed an outstanding potential of *C. glutamicum* for the production of various organic acids and the alcohols ethanol and isobutanol. For several of these products, the $Y_{P/S}$ values, the productivities, and the maximal titres achieved with *C. glutamicum* strains are comparable and often even better than the ones described for alternative production hosts. The efficiency in the development of these strains was possible due to the reliable protocols available for genetic engineering and to the extensive knowledge on the metabolism of *C. glutamicum* that has been obtained predominantly by studies on amino acid production. Furthermore, the robustness of *C. glutamicum* based on its high tolerance to various stresses, such as osmotic stress, is an important prerequisite for obtaining the high product titres shown in Table 8.1 for several products. Another remarkable property of *C. glutamicum* that is key for the production of many organic acids, ethanol, and isobutanol is its ability for anaerobic catabolism without growth. Although anaerobic production processes with *C. glutamicum* require aerobic pre-cultivation for biomass generation, the possibility to perform the conversion of substrate into product at very high cell densities and over long time periods allows exceptional productivities and titres that can compensate the hindrance of a two-step process. The next step that is now desired is the industrial implementation of the production strains described in this chapter. The future will show whether the production of organic acids and alcohols will become another success story for *C. glutamicum*.

References

Abreu-Cavalheiro, A., and Monteiro, G. (2013) Solving ethanol production problems with genetically modified yeast strains. *Braz. J. Microbiol.* **44**, 665–671.

Ajinomoto Co., Inc. (2013a) Feed-use amino acids business. Available at: http://www.ajinomoto.com/en/ir/pdf/Feed-useAA-Oct2013.pdf.

Ajinomoto Co., Inc. (2013b) First quarter-FY2013 Market and other information. Available at: http://www.ajinomoto.com/en/ir/pdf/FY13_Data_E.pdf.

Amorim, H. V., Lopes, M. L., de Castro Oliveira, J. V., Buckeridge, M. S., and Goldman, G. H. (2011) Scientific challenges of bioethanol production in Brazil. *Appl. Microbiol. Biotechnol.* **91**, 1267–1275.

Aparicio, M., Cano, N. J., Cupisti, A., Ecder, T., Fouque, D., Garneata, L., Liou, H. H., Lin, S., Schober-Halstenberg, H. J., Teplan, V., and Zakar, G. (2009) Keto-acid therapy in predialysis chronic kidney disease patients: consensus statements. *J. Ren. Nutr.* **19**, S33–S35.

Aparicio, M., Bellizzi, V., Chauveau, P., Cupisti, A., Ecder, T., Fouque, D., Garneata, L., Lin, S., Mitch, W. E., Teplan, V., Zakar, G., and Yu, X. (2012a) Keto acid therapy in predialysis chronic kidney disease patients: final consensus. *J. Ren. Nutr.* **22**, S22–S24.

Aparicio, M., Bellizzi, V., Chauveau, P., Cupisti, A., Ecder, T., Fouque, D., Garneata, L., Lin, S., Mitch, W. E., Teplan, V., Zakar, G., and Yu, X. (2012b) Protein-restricted diets plus keto/amino acids – a valid therapeutic approach for chronic kidney disease patients. *J. Ren. Nutr.* **22**, S1–S21.

Atsumi, S., Hanai, T., and Liao, J. C. (2008) Non-fermentative pathways for synthesis of branched-chain higher alcohols as biofuels. *Nature* **451**, 86–89.

Atsumi, S., Li, Z., and Liao, J. C. (2009) Acetolactate synthase from *Bacillus subtilis* serves as a 2-ketoisovalerate decarboxylase for isobutanol biosynthesis in *Escherichia coli*. *Appl. Environm. Microbiol.* **75**, 6306–6311.

Auras, R., Harte, B., and Selke, S. (2004) An overview of polylactides as packaging materials. *Macromol. Biosci.* **4**, 835–864.

Baez, A., Cho, K. M., and Liao, J. C. (2011) High-flux isobutanol production using engineered *Escherichia coli*: a bioreactor study with *in situ* product removal. *Appl. Microbiol. Biotechnol.* **90**, 1681–1690.

Banks, J. M., Yvon, M., Gripon, J. C., de la Fuente, M. A., Brechany, E. Y., Williams, A. G., and Muir, D. D. (2001) Enhancement of amino acid catabolism in Cheddar cheese using α-ketoglutarate: amino acid degradation in relation to volatile compounds and aroma character. *Int. Dairy J.* **11**, 235–243.

Bastian, S., Liu, X., Meyerowitz, J. T., Snow, C. D., Chen, M. M., and Arnold, F. H. (2011) Engineered ketol-acid reductoisomerase and alcohol dehydrogenase enable anaerobic 2-methylpropan-1-ol production at theoretical yield in *Escherichia coli*. *Metab. Eng.* **13**, 345–352.

Becker, J., and Wittmann, C. (2012) Bio-based production of chemicals, materials and fuels -*Corynebacterium glutamicum* as versatile cell factory. *Curr. Opin. Biotechnol.* **23**, 631–640.

Becker, J., Schäfer, R., Kohlstedt, M., Harder, B. J., Borchert, N. S., Stoveken, N., Bremer, E., and Wittmann, C. (2013) Systems metabolic engineering of *Corynebacterium glutamicum* for production of the chemical chaperone ectoine. *Microb. Cell Fact.* **12**, 110.

Blombach, B., and Eikmanns, B. J. (2011) Current knowledge on isobutanol production with *Escherichia coli*, *Bacillus subtilis* and *Corynebacterium glutamicum*. *Bioeng. Bugs* **2**, 346–350.

Blombach, B., Schreiner, M. E., Holatko, J., Bartek, T., Oldiges, M., and Eikmanns, B. J. (2007) L-Valine production with pyruvate dehydrogenase complex-deficient *Corynebacterium glutamicum*. *Appl. Environm. Microbiol.* **73**, 2079–2084.

Blombach, B., Arndt, A., Auchter, M., and Eikmanns, B. J. (2009) L-valine production during growth of pyruvate dehydrogenase complex-deficient *Corynebacterium glutamicum* in the presence of ethanol or by inactivation of the transcriptional regulator SugR. *Appl. Environ. Microbiol.* **75**, 1197–1200.

Blombach, B., Riester, T., Wieschalka, S., Ziert, C., Youn, J. W., Wendisch, V. F., and Eikmanns, B. J. (2011) *Corynebacterium glutamicum* tailored for efficient isobutanol production. *Appl. Environ. Microbiol.* **77**, 3300–3310.

Bott, M. (2007) Offering surprises: TCA cycle regulation in *Corynebacterium glutamicum*. *Trends Microbiol.* **15**, 417–425.

Bott, M., and Brocker, M. (2012) Two-component signal transduction in *Corynebacterium glutamicum* and other corynebacteria: on the way towards stimuli and targets. *Appl. Microbiol. Biotechnol.* **94**, 1131–1150.

Bozell, J. J., and Peterson, G. (2010) Technology development for the production of biobased products from biorefinery carbohydrates – the US Department of Energy's "Top 10" revisited. *Green Chem.* **12**, 539–554.

Brinkrolf, K., Schröder, J., Pühler, A., and Tauch, A. (2010) The transcriptional regulatory repertoire of *Corynebacterium glutamicum*: reconstruction of the network controlling pathways involved in lysine and glutamate production. *J. Biotechnol.* **149**, 173–182.

Brune, I., Jochmann, N., Brinkrolf, K., Hüser, A. T., Gerstmeir, R., Eikmanns, B. J., Kalinowski, J., Pühler, A., and Tauch, A. (2007) The IclR-type transcriptional repressor LtbR regulates the expression of leucine and tryptophan biosynthesis genes in the amino acid producer *Corynebacterium glutamicum*. *J. Bacteriol.* **189**, 2720–2733.

Buchholz, J., Schwentner, A., Brunnenkan, B., Gabris, C., Grimm, S., Gerstmeir, R., Takors, R., Eikmanns, B. J., and Blombach, B. (2013) Platform engineering of *Corynebacterium glutamicum* with reduced pyruvate dehydrogenase complex activity for improved production of L-lysine, L-valine, and 2-ketoisovalerate. *Appl. Environ. Microbiol.* **79**, 5566–5575.

Bückle-Vallant, V., Krause, F. S., Messerschmidt, S., and Eikmanns, B. J. (2014) Metabolic engineering of *Corynebacterium glutamicum* for 2-ketoisocaproate production. *Appl. Microbiol. Biotechnol.* **98**, 297–311.

Buijs, N. A., Siewers, V., and Nielsen, J. (2013) Advanced biofuel production by the yeast *Saccharomyces cerevisiae*. *Curr. Opin. Chem. Biol.* **17**, 480–488.

Burkovski, A. (ed.) (2008) Corynebacteria: Genomics and Molecular Biology. Caister Academic Press, Norfolk, UK.

Canilha, L., Kumar Chandel, A., dos Santos Milessi, T. S., Fernandes Antunes, F. A., da Costa Freitas, W. L., das Gracas Almeida Felipe, M., and da Silva, S. S. (2012) Bioconversion of sugarcane biomass into ethanol: an overview about composition, pretreatment methods, detoxification of hydrolysates, enzymatic saccharification, and ethanol fermentation. *J. Biomed. Biotechnol.* **2012**, 989572.

Chang, J. H., Kim, D. K., Park, J. T., Kang, E. W., Yoo, T. H., Kim, B. S., Choi, K. H., Lee, H. Y., Han, D. S., and Shin, S. K. (2009) Influence of ketoanalogs supplementation on the progression in chronic kidney disease patients who had training on low-protein diet. *Nephrology* **14**, 750–757.

Chen, T., Zhu, N. Q., and Xia, H. H. (2014) Aerobic production of succinate from arabinose by metabolically engineered *Corynebacterium glutamicum*. *Bioresource Technol.* **151**, 411–414.

Chen, X., Nielsen, K. F., Borodina, I., Kielland-Brandt, M. C., and Karhumaa, K. (2011) Increased isobutanol production in *Saccharomyces cerevisiae* by overexpression of genes in valine metabolism. *Biotechnol. Biofuels* **4**, 21.

Connor, M. R., and Liao, J. C. (2008) Engineering of an *Escherichia coli* strain for the production of 3-methyl-1-butanol. *Appl. Environm. Microbiol.* **74**, 5769–5775.

Connor, M. R., Cann, A. F., and Liao, J. C. (2010) 3-Methyl-1-butanol production in *Escherichia coli*: random mutagenesis and two-phase fermentation. *Appl. Microbiol. Biotechnol.* **86**, 1155–1164.

Cooper, A. J., Ginos, J. Z., and Meister, A. (1983) Synthesis and properties of the α-keto acids. *Chem. Rev.* **83**, 321–358.

Cotellessa, C., Manunta, T., Ghersetich, I., Brazzini, B., and Peris, K. (2004) The use of pyruvic acid in the treatment of acne. *J. Eur. Acad. Dermatol. Venereol.* **18**, 275–278.

Cukalovic, A., and Stevens, C. V. (2008) Feasibility of production methods for succinic acid derivatives: a marriage of renewable resources and chemical technology. *Biofuels Bioprod. Bioref.* **2**, 505–529.

Deboer, L. W. V., Bekx, P. A., Han, L. H., and Steinke, L. (1993) Pyruvate enhances recovery of rat hearts after ischemia and reperfusion by preventing free-radical generation. *Am. J. Physiol.* **265**, H1571–H1576.

Della-Bianca, B. E., Basso, T. O., Stambuk, B. U., Basso, L. C., and Gombert, A. K. (2013) What do we know about the yeast strains from the Brazilian fuel ethanol industry? *Appl. Microbiol. Biotechnol.* **97**, 979–991.

Dien, B. S., Cotta, M. A., and Jeffries, T. W. (2003) Bacteria engineered for fuel ethanol production: current status. *Appl. Microbiol. Biotechnol.* **63**, 258–266.

Dogaris, I., Mamma, D., and Kekos, D. (2013) Biotechnological production of ethanol from renewable resources by *Neurospora crassa*: an alternative to conventional yeast fermentations? *Appl. Microbiol. Biotechnol.* **97**, 1457–1473.

Dominguez, H., Nezondet, C., Lindley, N. D., and Cocaign, M. (1993) Modified carbon flux during oxygen limited growth of *Corynebacterium glutamicum* and the consequences for amino acid overproduction. *Biotechnol. Lett.* **15**, 449–454.

Dominguez, H., Rollin, C., Guyonvarch, A., Guerquin-Kern, J. L., Cocaign-Bousquet, M., and Lindley, N. D. (1998) Carbon-flux distribution in the central metabolic pathways of *Corynebacterium glutamicum* during growth on fructose. *Eur. J. Biochem.* **254**, 96–102.

Dürre, P. (2007) Biobutanol: an attractive biofuel. *Biotechnol. J.* **2**, 1525–1534.

Eggeling, L., and Bott, M. (eds.) (2005) Handbook of *Corynebacterium glutamicum*. CRC Press, Taylor & Francis Group, Boca Raton, FL, USA.

Eggeling, I., Cordes, C., Eggeling, L., and Sahm, H. (1987) Regulation of acetohydroxy acid synthase in *Corynebacterium glutamicum* during fermentation of α-ketobutyrate to L-isoleucine. *Appl. Microbiol. Biotechnol.* **25**, 346–351.

Eikmanns, B. J., and Blombach, B. (2014) Isobutanol. In: Bisaria V. S., and Kondo A. (eds.) Bioprocessing of Renewable Resources to Commodity Bioproducts. John Wiley & Sons, Hoboken, NJ, USA, pp. 329–354.

Ezeji, T. C., Karcher, P. M., Qureshi, N., and Blaschek, H. P. (2005) Improving performance of a gas stripping-based recovery system to remove butanol from *Clostridium beijerinckii* fermentation. *Bioprocess Biosyst. Eng.* **27**, 207–214.

Feiten, S. F., Draibe, S. A., Watanabe, R., Duenhas, M. R., Baxmann, A. C., Nerbass, F. B., and Cuppari, L. (2005) Short-term effects of a very-low-protein diet supplemented with ketoacids in nondialyzed chronic kidney disease patients. *Eur. J. Clin. Nutr.* **59**, 129–136.

Fukui, K., Koseki, C., Yamamoto, Y., Nakamura, J., Sasahara, A., Yuji, R., Hashiguchi, K., Usuda, Y., Matsui, K., Kojima, H., and Abe, K. (2011) Identification of succinate exporter in *Corynebacterium glutamicum* and its physiological roles under anaerobic conditions. *J. Biotechnol.* **154**, 25–34.

Gao, C., Ma, C. Q., and Xu, P. (2011) Biotechnological routes based on lactic acid production from biomass. *Biotechnol. Adv.* **29**, 930–939.

Geddes, C. C., Mullinnix, M. T., Nieves, I. U., Peterson, J. J., Hoffman, R. W., York, S. W., Yomano, L. P., Miller, E. N., Shanmugam, K. T., and Ingram, L. O. (2011) Simplified process for ethanol production from sugarcane bagasse using hydrolysate-resistant *Escherichia coli* strain MM160. *Bioresour. Technol.* **102**, 2702–2711.

Ghersetich, I., Brazzini, B., Peris, K., Cotellessa, C., Manunta, T., and Lotti, T. (2004) Pyruvic acid peels for the treatment of photoaging. *Dermatol. Surg.* **30**, 32–36.

Gutiérrez-Méndez, N., Vallejo-Cordoba, B., González-Córdova, A. F., Nevárez-Moorillón, G. V., and Rivera-Chavira, B. (2008) Evaluation of aroma generation of *Lactococcus lactis* with an electronic nose and sensory analysis. *J. Dairy Sci.* **91**, 49–57.

Heider, S. A., Peters-Wendisch, P., and Wendisch, V. F. (2012) Carotenoid biosynthesis and overproduction in *Corynebacterium glutamicum*. *BMC Microbiol.* **12**, 198.

Heider, S. A., Peters-Wendisch, P., Netzer, R., Stafnes, M., Brautaset, T., and Wendisch, V. F. (2014) Production and glucosylation of C50 and C 40 carotenoids by metabolically engineered *Corynebacterium glutamicum*. *Appl. Microbiol. Biotechnol.* **98**, 1223–1235.

Hofvendahl, K., and Hahn-Hägerdal, B. (2000) Factors affecting the fermentative lactic acid production from renewable resources. *Enzyme Microb. Technol.* **26**, 87–107.

Howard, J. W., and Fraser, W. A. (1932) Preparation of pyruvic acid. *Org. Synth. Coll.* **1**, 475–480.

Huhn, S., Jolkver, E., Krämer, R., and Marin, K. (2011) Identification of the membrane protein SucE and its role in succinate transport in *Corynebacterium glutamicum*. *Appl. Microbiol. Biotechnol.* **89**, 327–335.

Ikeda, M., Mizuno, Y., Awane, S., Hayashi, M., Mitsuhashi, S., and Takeno, S. (2011) Identification and application of a different glucose uptake system that functions as an alternative to the phosphotransferase system in *Corynebacterium glutamicum*. *Appl. Microbiol. Biotechnol.* **90**, 1443–1451.

Inui, M., Kawaguchi, H., Murakami, S., Vertes, A. A., and Yukawa, H. (2004a) Metabolic engineering of *Corynebacterium glutamicum* for fuel ethanol production under oxygen-deprivation conditions. *J. Mol. Microbiol. Biotechnol.* **8**, 243–254.

Inui, M., Murakami, S., Okino, S., Kawaguchi, H., Vertes, A. A., and Yukawa, H. (2004b) Metabolic analysis of *Corynebacterium glutamicum* during lactate and succinate productions under oxygen deprivation conditions. *J. Mol. Microbiol. Biotechnol.* **7**, 182–196.

Janausch, I. G., Zientz, E., Tran, Q. H., Kröger, A., and Unden, G. (2002) C4-dicarboxylate carriers and sensors in bacteria. *Biochim. Biophys. Acta* **1553**, 39–56.

Jang, Y. S., Malaviya, A., Cho, C., Lee, J., and Lee, S. Y. (2012) Butanol production from renewable biomass by clostridia. *Bioresour. Technol.* **123**, 653–663.

Jeffries, T. W., and Jin, Y. S. (2004) Metabolic engineering for improved fermentation of pentoses by yeasts. *Appl. Microbiol. Biotechnol.* **63**, 495–509.

Jo, J. H., Seol, H. Y., Lee, Y. B., Kim, M. H., Hyun, H. H., and Lee, H. H. (2012) Disruption of genes for the enhanced biosynthesis of α-ketoglutarate in *Corynebacterium glutamicum*. *Can. J. Microbiol.* **58**, 278–286.

Kabus, A., Georgi, T., Wendisch, V. F., and Bott, M. (2007) Expression of the *Escherichia coli pntAB* genes encoding a membrane-bound transhydrogenase in *Corynebacterium glutamicum* improves L-lysine formation. *Appl. Microbiol. Biotechnol.* **75**, 47–53.

Kawaguchi, H., Vertes, A. A., Okino, S., Inui, M., and Yukawa, H. (2006) Engineering of a xylose metabolic pathway in *Corynebacterium glutamicum*. *Appl. Environm. Microbiol.* **72**, 3418–3428.

Kawaguchi, H., Sasaki, M., Vertes, A. A., Inui, M., and Yukawa, H. (2008) Engineering of an L-arabinose metabolic pathway in *Corynebacterium glutamicum*. *Appl. Microbiol. Biotechnol.* **77**, 1053–1062.

Kind, S., and Wittmann, C. (2011) Bio-based production of the platform chemical 1,5-diaminopentane. *Appl. Microbiol. Biotechnol.* **91**, 1287–1296.

Klaffl, S., Brocker, M., Kalinowski, J., Eikmanns, B. J., and Bott, M. (2013) Complex regulation of the phosphoenolpyruvate carboxykinase gene *pck* and characterization of its GntR-Type regulator IolR as a repressor of *myo*-inositol utilization genes in *Corynebacterium glutamicum*. *J. Bacteriol.* **195**, 4283–4296.

Krause, F. S., Blombach, B., and Eikmanns, B. J. (2010a) Metabolic engineering of *Corynebacterium glutamicum* for 2-ketoisovalerate production. *Appl. Environ. Microbiol.* **76**, 8053–8061.

Krause, F. S., Henrich, A., Blombach, B., Kramer, R., Eikmanns, B. J., and Seibold, G. M. (2010b) Increased glucose utilization in *Corynebacterium glutamicum* by use of maltose, and its application for the improvement of L-valine productivity. *Appl. Environm. Microbiol.* **76**, 370–374.

Krill, S. (2003) Verfahren zur Herstellung von 6-Methylheptanon. European patent 1366009 A1.

Lamsen, E. N., and Atsumi, S. (2012) Recent progress in synthetic biology for microbial production of C3-C10 alcohols. *Front. Microbiol.* **3**, 196.

van Leeuwen, B. N., van der Wulp, A. M., Duijnstee, I., van Maris, A. J., and Straathof, A. J. (2012) Fermentative production of isobutene. *Appl. Microbiol. Biotechnol.* **93**, 1377–1387.

Li, H., Cann, A. F., and Liao, J. C. (2010) Biofuels: biomolecular engineering fundamentals and advances. *Ann. Rev. Chem. Biomol. Eng.* **1**, 19–36.

Li, S., Wen, J., and Jia, X. (2011) Engineering *Bacillus subtilis* for isobutanol production by heterologous Ehrlich pathway construction and the biosynthetic 2-ketoisovalerate precursor pathway overexpression. *Appl. Microbiol. Biotechnol.* **91**, 577–589.

Li, Y., Chen, J., and Lun, S. Y. (2001) Biotechnological production of pyruvic acid. *Appl. Microbiol. Biotechnol.* **57**, 451–459.

Lindner, S. N., Seibold, G. M., Henrich, A., Krämer, R., and Wendisch, V. F. (2011) Phosphotransferase system-independent glucose utilization in *Corynebacterium glutamicum* by inositol permeases and glucokinases. *Appl. Environm. Microbiol.* **77**, 3571–3581.

Litsanov, B., Brocker, M., and Bott, M. (2012a) Toward homosuccinate fermentation: metabolic engineering of *Corynebacterium glutamicum* for anaerobic production of succinate from glucose and formate. *Appl. Environm. Microbiol.* **78**, 3325–3337.

Litsanov, B., Kabus, A., Brocker, M., and Bott, M. (2012b) Efficient aerobic succinate production from glucose in minimal medium with *Corynebacterium glutamicum*. *Microb. Biotechnol.* **5**, 116–128.

Litsanov, B., Brocker, M., and Bott, M. (2013) Glycerol as a substrate for aerobic succinate production in minimal medium with *Corynebacterium glutamicum*. *Microb. Biotechnol.* **6**, 189–195.

Litsanov, B., Brocker, M., Oldiges, M., and Bott, M. (2014) Succinic acid. In: Kondo A., and Bisaria V. (eds.) Bioprocessing of Renewable Resources to Commodity Bioproducts. John Wiley & Sons, Hoboken, NJ, USA, pp. 437–474.

McKinlay, J. B., Vieille, C., and Zeikus, J. G. (2007) Prospects for a bio-based succinate industry. *Appl. Microbiol. Biotechnol.* **76**, 727–740.

Matsushita, Y., Inomata, T., Hasegawa, T., and Fukushima, K. (2009) Solubilization and functionalization of sulfuric acid lignin generated during bioethanol production from woody biomass. *Bioresour. Technol.* **100**, 1024–1026.

Mimitsuka, T., Sawai, H., Hatsu, M., and Yamada, K. (2007) Metabolic engineering of *Corynebacterium glutamicum* for cadaverine fermentation. *Biosci. Biotechnol. Biochem.* **71**, 2130–2135.

Morbach, S., Junger, C., Sahm, H., and Eggeling, L. (2000) Attenuation control of *ilvBNC* in *Corynebacterium glutamicum*: evidence of leader peptide formation without the presence of a ribosome binding site. *J. Biosci. Bioeng.* **90**, 501–507.

Mussatto, S. I., Dragone, G., Guimaraes, P. M. R., Silva, J. P. A., Carneiro, L. M., Roberto, I. C., Vicente, A., Domingues, L., and Teixeira, J. A. (2010) Technological trends, global market, and challenges of bio-ethanol production. *Biotechnol. Adv.* **28**, 817–830.

Nakamura, C. E., and Whited, G. M. (2003) Metabolic engineering for the microbial production of 1,3-propanediol. *Curr. Opin. Biotechnol.* **14**, 454–459.

Nevoigt, E. (2008) Progress in metabolic engineering of *Saccharomyces cerevisiae*. *Microbiol. Mol. Biol. Rev.* **72**, 379–412.

Nielsen, J., Larsson, C., van Maris, A., and Pronk, J. (2013) Metabolic engineering of yeast for production of fuels and chemicals. *Curr. Opin. Biotechnol.* **24**, 398–404.

Nunan, D., Howatson, G., and van Someren, K. A. (2010) Exercise-induced muscle damage is not attenuated by beta-hydroxy-beta-methylbutyrate and alpha-ketoisocaproic acid supplementation. *J. Strength Cond. Res.* **24**, 531–537.

Ödman, P., Wellborn, W. B., and Bommarius, A. S. (2004) An enzymatic process to α-ketoglutarate from L-glutamate: the coupled system L-glutamate dehydrogenase/NADH oxidase. *Tetrahedron Asymmetry* **15**, 2933–2937.

Okano, K., Tanaka, T., Ogino, C., Fukuda, H., and Kondo, A. (2010) Biotechnological production of enantiomeric pure lactic acid from renewable resources: recent achievements, perspectives, and limits. *Appl. Microbiol. Biotechnol.* **85**, 413–423.

Okino, S., Inui, M., and Yukawa, H. (2005) Production of organic acids by *Corynebacterium glutamicum* under oxygen deprivation. *Appl. Microbiol. Biotechnol.* **68**, 475–480.

Okino, S., Noburyu, R., Suda, M., Jojima, T., Inui, M., and Yukawa, H. (2008a) An efficient succinic acid production process in a metabolically engineered *Corynebacterium glutamicum* strain. *Appl. Microbiol. Biotechnol.* **81**, 459–464.

Okino, S., Suda, M., Fujikura, K., Inui, M., and Yukawa, H. (2008b) Production of D-lactic acid by *Corynebacterium glutamicum* under oxygen deprivation. *Appl. Microbiol. Biotechnol.* **78**, 449–454.

Pang, X. A., Zhuang, X. L., Tang, Z. H., and Chen, X. S. (2010) Polylactic acid (PLA): research, development and industrialization. *Biotechnol. J.* **5**, 1125–1136.

Pátek, M., Krumbach, K., Eggeling, L., and Sahm, H. (1994) Leucine synthesis in *Corynebacterium glutamicum*: enzyme activities, structure of *leuA*, and effect of *leuA* inactivation on lysine synthesis. *Appl. Environ. Microbiol.* **60**, 133–140.

Reinscheid, D. J., Eikmanns, B. J., and Sahm, H. (1994) Characterization of the isocitrate lyase gene from *Corynebacterium glutamicum* and biochemical analysis of the enzyme. *J. Bacteriol.* **176**, 3474–3483.

Riedel, E., Nundel, M., and Hampl, H. (1996) alpha-Ketoglutarate application in hemodialysis patients improves amino acid metabolism. *Nephron* **74**, 261–265.

Rittmann, D., Lindner, S. N., and Wendisch, V. F. (2008) Engineering of a glycerol utilization pathway for amino acid production by *Corynebacterium glutamicum*. *Appl. Environm. Microbiol.* **74**, 6216–6222.

Roufs, J. B. (1996) Pyruvate: does it amp endurance and burn more fat? *Muscle Fitness* **57**, 195–197.

Saier, M. H., Tran, C. V., and Barabote, R. D. (2006) TCDB: the Transporter Classification Database for membrane transport protein analyses and information. *Nucleic Acids Res.* **34**, D181–D186.

Sakai, S., Tsuchida, Y., Nakamoto, H., Okino, S., Ichihashi, O., Kawaguchi, H., Watanabe, T., Inui, M., and Yukawa, H. (2007) Effect of lignocellulose-derived inhibitors on growth of and ethanol production by growth-arrested *Corynebacterium glutamicum* R. *Appl. Environm. Microbiol.* **73**, 6328–6328.

Sasaki, M., Jojima, T., Inui, M., and Yukawa, H. (2010) Xylitol production by recombinant *Corynebacterium glutamicum* under oxygen deprivation. *Appl. Microbiol. Biotechnol.* **86**, 1057–1066.

Sauer, M., Porro, D., Mattanovich, D., and Branduardi, P. (2008) Microbial production of organic acids: expanding the markets. *Trends Biotechnol.* **26**, 100–108.

Sauer, U., and Eikmanns, B. J. (2005) The PEP-pyruvate-oxaloacetate node as the switch point for carbon flux distribution in bacteria. *FEMS Microbiol. Rev.* **29**, 765–794.

Schiel-Bengelsdorf, B., Montoya, J., Linder, S., and Dürre, P. (2013) Butanol fermentation. *Environ. Technol.* **34**, 1691–1710.

Schneider, J., and Wendisch, V. F. (2011) Biotechnological production of polyamines by bacteria: recent achievements and future perspectives. *Appl. Microbiol. Biotechnol.* **91**, 17–30.

Schreiner, M. E., Fiur, D., Holatko, J., Patek, M., and Eikmanns, B. (2005) E1 enzyme of the pyruvate dehydrogenase complex in *Corynebacterium glutamicum*: molecular analysis of the gene and phylogenetic aspects. *J. Bacteriol.* **187**, 6005–6018.

Schröder, J., and Tauch, A. (2010) Transcriptional regulation of gene expression in *Corynebacterium glutamicum*: the role of global, master and local regulators in the modular and hierarchical gene regulatory network. *FEMS Microbiol. Rev.* **34**, 685–737.

Seliverstov, A. V., Putzer, H., Gelfand, M. S., and Lyubetsky, V. A. (2005) Comparative analysis of RNA regulatory elements of amino acid metabolism genes in Actinobacteria. *BMC Microbiol.* **5**, 54.

Smith, K. M., Cho, K. M., and Liao, J. C. (2010) Engineering *Corynebacterium glutamicum* for isobutanol production. *Appl. Microbiol. Biotechnol.* **87**, 1045–1055.

van Someren, K. A., Edwards, A. J., and Howatson, G. (2005) Supplementation with beta-hydroxy-beta-methylbutyrate (HMB) and alpha-ketoisocaproic acid (KIC) reduces signs and symptoms of exercise-induced muscle damage in man. *Int. J. Sport Nutr. Exerc. Metab.* **15**, 413–424.

Song, Y. Y., Matsumoto, K., Yamada, M., Gohda, A., Brigham, C. J., Sinskey, A. J., and Taguchi, S. (2012) Engineered *Corynebacterium glutamicum* as an endotoxin-free platform strain for lactate-based polyester production. *Appl. Microbiol. Biotechnol.* **93**, 1917–1925.

Song, Y., Matsumoto, K., Tanaka, T., Kondo, A., and Taguchi, S. (2013) Single-step production of polyhydroxybutyrate from starch by using α-amylase cell-surface displaying system of *Corynebacterium glutamicum*. *J. Biosci. Bioeng.* **115**, 12–14.

Stanko, R. T., Tietze, D. L., and Arch, J. E. (1992a) Body composition, energy utilization, and nitrogen metabolism with a 4.25-Mj/D low-energy diet supplemented with pyruvate. *Am. J. Clin. Nutr.* **56**, 630–635.

Stanko, R. T., Tietze, D. L., and Arch, J. E. (1992b) Body composition, energy utilization, and nitrogen metabolism with a severely restricted diet supplemented with dihydroxyacetone and pyruvate. *Am. J. Clin. Nutr.* **55**, 771–776.

Takahashi, C., Shirakawa, J., Tsuchidate, T., Okai, N., Hatada, K., Nakayama, H., Tateno, T., Ogino, C., and Kondo, A. (2012) Robust production of γ-amino butyric acid using recombinant *Corynebacterium glutamicum* expressing glutamate decarboxylase from *Escherichia coli*. *Enzyme Microb. Technol.* **51**, 171–176.

Takors, R., Bathe, B., Rieping, M., Hans, S., Kelle, R., and Huthmacher, K. (2007) Systems biology for industrial strains and fermentation processes – example: amino acids. *J. Biotechnol.* **129**, 181–190.

Teplan, V., Schuck, O., Knotek, A., Hajny, J., Horackova, M., Skibova, J., and Maly, J. (2001) Effects of low-protein diet supplemented with ketoacids and erythropoietin in chronic renal failure: a long-term metabolic study. *Ann. Transplant.* **6**, 47–53.

Teramoto, H., Inui, M., and Yukawa, H. (2011) Transcriptional regulators of multiple genes involved in carbon metabolism in *Corynebacterium glutamicum*. *J. Biotechnol.* **154**, 114–125.

Teschan, P. E., Beck, G. J., Dwyer, J. T., Greene, T., Klahr, S., Levy, A. S., Mitch, W. E., Snetselaar, L. G., Steinman, T. I., and Walser, M. (1998) Effect of a ketoacid-aminoacid-supplemented very low protein diet on the progression of advanced renal disease: a reanalysis of the MDRD feasibility study. *Clin. Nephrol.* **50**, 273–283.

Tjener, K., Stahnke, L. H., Andersen, L., and Martinussen, J. (2004) Addition of alpha-ketoglutarate enhances formation of volatiles by *Staphylococcus carnosus* during sausage fermentation. *Meat Sci.* **67**, 711–719.

Tsuge, Y., Yamamoto, S., Suda, M., Inui, M., and Yukawa, H. (2013) Reactions upstream of glycerate-1,3-bisphosphate drive *Corynebacterium glutamicum* D-lactate productivity under oxygen deprivation. *Appl. Microbiol. Biotechnol.* **97**, 6693–6703.

Vašicová, P., Pátek, M., Nešvera, J., Sahm, H., and Eikmanns, B. (1999) Analysis of the *Corynebacterium glutamicum dapA* promoter. *J. Bacteriol.* **181**, 6188–6191.

Vemuri, G. N., Eiteman, M. A., and Altman, E. (2002) Effects of growth mode and pyruvate carboxylase on succinic acid production by metabolically engineered strains of *Escherichia coli*. *Appl. Environ. Microbiol.* **68**, 1715–1727.

Vertes, A. A., Inui, M., and Yukawa, H. (2012) Postgenomic approaches to using corynebacteria as biocatalysts. *Annu. Rev. Microbiol.* **66**, 521–550.

Vogt, M., Haas, S., Klaffl, S., Polen, T., Eggeling, L., van Ooyen, J., and Bott, M. (2014) Pushing product formation to its limit: metabolic engineering of *Corynebacterium glutamicum* for L-leucine overproduction. *Metab. Eng.* **22**, 40–52.

Wang, C., Zhang, H., Cai, H., Zhou, Z., Chen, Y., Chen, Y., and Ouyang, P. (2014) Succinic acid production from corn cob hydrolysates by genetically engineered *Corynebacterium glutamicum*. *Appl. Biochem. Biotechnol.* **172**, 340–350.

Wen, M., Bond-Watts, B. B., and Chang, M. C. (2013) Production of advanced biofuels in engineered *E. coli*. *Curr. Opin. Chem. Biol.* **17**, 472–479.

Wendisch, V. F., Bott, M., and Eikmanns, B. J. (2006a) Metabolic engineering of *Escherichia coli* and *Corynebacterium glutamicum* for biotechnological production of organic acids and amino acids. *Curr. Opin. Microbiol.* **9**, 268–274.

Wendisch, V. F., Bott, M., Kalinowski, J., Oldiges, M., and Wiechert, W. (2006b) Emerging *Corynebacterium glutamicum* systems biology. *J. Biotechnol.* **124**, 74–92.

Wendisch, V. F., Meiswinkel, T. M., and Lindner, S. N. (2011) Use of glycerol in biotechnological applications. In: Montero G., and Stoytcheva M. (eds.) Biodiesel – Quality, Emissions and By-Products. InTech – Open Access Publisher, Rijeka, Croatia, pp. 305–340.

Werpy, T., and Peterson, G. (2004) Top Value Added Chemicals from Biomass. US Department of Energy, Oak Ridge, TN, USA.

Wieschalka, S., Blombach, B., and Eikmanns, B. J. (2012) Engineering *Corynebacterium glutamicum* for the production of pyruvate. *Appl. Microbiol. Biotechnol.* **94**, 449–459.

Wieschalka, S., Blombach, B., Bott, M., and Eikmanns, B. J. (2013) Bio-based production of organic acids with *Corynebacterium glutamicum*. *Microb. Biotechnol.* **6**, 87–102.

Yamamoto, S., Suda, M., Niimi, S., Inui, M., and Yukawa, H. (2013) Strain optimization for efficient isobutanol production using *Corynebacterium glutamicum* under oxygen deprivation. *Biotechnol. Bioeng.* **110**, 2938–2948.

Yukawa, H., and Inui, M. (eds.) (2013) *Corynebacterium glutamicum*: Biology and Biotechnology. Springer, Heidelberg, Germany.

Zanchi, N. E., Gerlinger-Romero, F., Guimaraes-Ferreira, L., de Siqueira, M. A., Felitti, V., Lira, F. S., Seelaender, M., and Lancha, A. H. (2011) HMB supplementation: clinical and athletic performance-related effects and mechanisms of action. *Amino Acids* **40**, 1015–1025.

Zeikus, J. G., Jain, M. K., and Elankovan, P. (1999) Biotechnology of succinic acid production and markets for derived industrial products. *Appl. Microbiol. Biotechnol.* **51**, 545–552.

Zelic, B., Gerharz, T., Bott, M., Vasic-Racki, B., Wandrey, C., and Takors, R. (2003) Fed-batch process for pyruvate production by recombinant *Escherichia coli* YYC202 strain. *Eng. Life Sci.* **3**, 299–305.

Zhou, S. D., Shanmugam, K. T., and Ingram, L. O. (2003) Functional replacement of the *Escherichia coli* D-(–)-lactate dehydrogenase gene (*ldhA*) with the L-(+)-lactate dehydrogenase gene (*ldhL*) from *Pediococcus acidilactici*. *Appl. Environ. Microbiol.* **69**, 2237–2244.

Zhu, N. Q., Xia, H. H., Wang, Z. W., Zhao, X. M., and Chen, T. (2013) Engineering of acetate recycling and citrate synthase to improve aerobic succinate production in *Corynebacterium glutamicum*. *Plos One* **8**, e60659.

Zhu, N., Xia, H., Yang, J., Zhao, X., and Chen, T. (2014) Improved succinate production in *Corynebacterium glutamicum* by engineering glyoxylate pathway and succinate export system. *Biotechnol. Lett.* **36**, 553–560.

Zhu, Y. H., Eiteman, M. A., Altman, R., and Altman, E. (2008) High glycolytic flux improves pyruvate production by a metabolically engineered *Escherichia coli* strain. *Appl. Environm. Microbiol.* **74**, 6649–6655.

Microbial Factory for the Production of Polyesters: A New Platform of *Corynebacterium glutamicum*

9

Yuyang Song, John Masani Nduko, Ken'ichiro Matsumoto and Seiichi Taguchi

Abstract

Corynebacterium glutamicum has been extensively employed for the industrial production of many value-added compounds such as food grade amino acids. Recently, many promising studies have focused on the engineering of this bacterial platform beyond the scope of amino acid production such as; for the production of microbial polyesters called polyhydroxyalkanoates (PHAs), which have the potential to replace petroleum-derived plastics. Thus far, *C. glutamicum* has been engineered by expressing PHA synthase and related monomer supplying enzymes to produce a variety of PHAs including poly(3-hydroxybutyrate) [P(3HB)] homopolymer and copolymers such as poly[3HB-*co*-3-hydroxyvalerate(3HV)]. In a current hot topic, the biosynthesis of poly(lactic acid)-like polyesters has been achieved in *C. glutamicum* by using an engineered PHA synthase (lactate-polymerizing enzyme: LPE). Furthermore, the arming *C. glutamicum* displaying α-amylase on its cell surface has provided an attractive platform for the production of the microbial polyesters from starch as a less expensive feedstock. These studies demonstrate the potential of this organism as a platform for microbial polymer factory.

Introduction

Corynebacterium glutamicum was first isolated in the 1950s from Japanese soil during a quest for natural L-glutamate producers (Kinoshita *et al.*, 1957). Since then, it has been thoroughly investigated and used as a generally-regarded-as-safe (GRAS) organism in fermentation industry for more than 50 years. The organism is a Gram-positive bacteria thus does not contain endotoxin, lipopolysaccharide (LPS) (Krämer, 1994) that is advantageous for food-grade amino acids production such as L-glutamate and L-lysine (Leuchtenberger *et al.*, 2005), as well as polyester production as mentioned in this chapter. After genetic engineering tools were developed, *C. glutamicum* genome was sequenced and made publicly available in 2003 (Kalinowski *et al.*, 2003). These milestones marked the beginning of extensive research efforts to metabolically engineer *C. glutamicum*, initially for L-glutamate and L-lysine production and later on for the production of a variety of products such as organic acids, alcohols or diamines. Therefore, there is much accumulated experiences concerning high-performance fermentation conditions which can be used to optimize the fermentation process (Krämer, 1994) that is also helpful for the production of microbial polyesters.

Bio-based polyesters are a class of materials produced from natural resources with potential as replacements for the petroleum-based plastics. Among the bio-based polyesters, polyhydroxyalkanoates (PHAs) and poly(lactic acid) (PLA) are typical examples, which can be applied to commodity uses. Moreover, these polymers possess biocompatibility and biodegradability (Doi, 1990; Jendrossek *et al.*, 1996), thus have been utilized in the medical field for drug delivery, making resolvable sutures, as material for medical implants and other related applications among others (Auras *et al.*, 2004; Nampoothiri *et al.*, 2010). Therefore, studies and development of the biological polyesters have gained unprecedented attention over the years in the world (Hazer and

Steinbuchel, 2007; Nomura and Taguchi, 2007; Chanprateep, 2010; Rehm, 2010; Haddouche *et al.*, 2011).

PHAs are produced by a wide variety of bacteria (Steinbuchel and Fuchtenbusch, 1998; Sudesh *et al.*, 2000). The biological function of these materials is that they serve as intracellular carbon reserve and energy storage compounds. These functions are similar to that of glycogen in mammalian systems and starch in plants (Inoue and Yoshie, 1992). However, when PHAs are extracted from the bacterial cells, they display material properties that are similar to some well-known plastics materials such as polypropylene and rubber (Steinbuchel and Fuchtenbusch, 1998). The monomeric composition of PHAs, which affects the polyester material property, basically depends on the substrate specificity of PHA synthase (or polymerase) (Steinbüchel *et al.*, 1992) and on the hydroxyacyl-CoAs (HA-CoAs) precursors supplied to the enzyme. The HA-CoAs supply is dependent on the metabolic pathways operating in the cell and on the external carbon source (Aldor and Keasling, 2003). The primary aims of the PHAs metabolic engineering include the management of different factors that determine physical properties, such as monomeric composition, molecular weight and copolymer microstructure, as well as optimizing yield. The most abundant PHA is poly (3-hydroxybutyrate) [P(3HB)], which bacteria synthesize from acetyl-CoA (Anderson and Dawes, 1990). Growing on glucose, the natural and recombinant bacteria can amass up to 90% of their dry weights as P(3HB) (Sudesh *et al.*, 2000), which make these microorganisms miniature bioplastic factories (Aldor and Keasling, 2003).

On the other hand, PLAs are synthesized in a fundamentally different production process, a chemo-process which involves the oligomerization of fermented lactic acid into lactide and the ring-opening polymerization of the lactide into PLA polyesters using metal catalysts (Auras *et al.*, 2004) (Fig. 9.1). PLAs have good mechanical properties, are highly transparent, have low environmental impact and thus, they are commonly used for packaging containers and stationery (Auras *et al.*, 2004; Tokiwa and Calabia, 2006). Additionally, PLAs serve as a component of nano-alloys in creating engineering plastics (Kricheldorf, 2001; Ozkoc and Kemaloglu, 2009). Although PLAs have good properties, their wide application in the medical and food uses is limited by the potential remnants of harmful heavy metal catalysts used during the chemo-process. Furthermore, the purification of lactic acid makes the

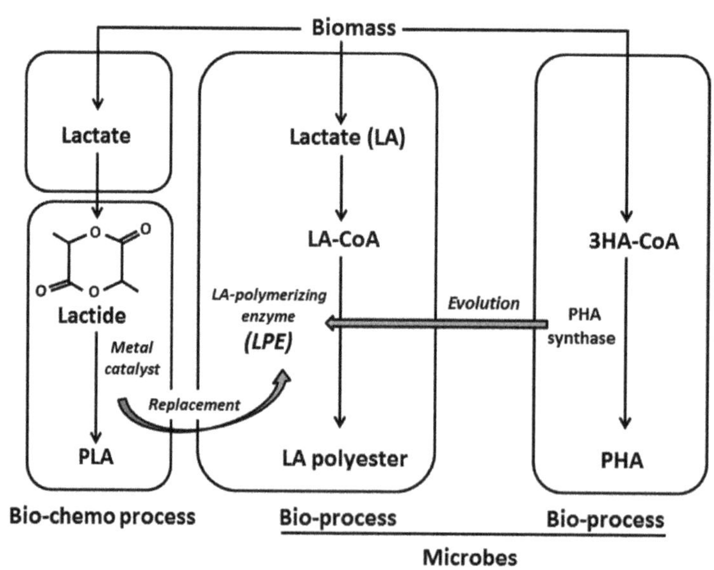

Figure 9.1 Process conversions for the production of lactate (LA)-based polyesters. The lactate-polymerizing enzyme (LPE) is the key to the process conversion.

process expensive; rendering this polyester more costly compared with the petroleum-derived polyesters (Corma *et al.*, 2007).

To solve these problems, the switch from the multistep chemo-process to a complete bioprocess for LA-based polyester production is thus preferable. The authors established a one-pot whole-cell biosynthesis system for LA-based polyester, P(LA-*co*-3HB) production in recombinant *Escherichia coli* using an engineered PHA synthase with LA-polymerizing activity (details are described below) (Taguchi and Doi, 2004; Taguchi *et al.*, 2008; Matsumoto and Taguchi, 2009). Based on this new bioprocess, LA-based polyesters joined as a member of microbial polyesters.

In both cases of productions of PHAs and LA-based polyesters, the Gram-negative bacterium *E. coli* has been widely used as a model system. It is a well-known fact that the Gram-negative bacteria contain LPS as a constituent of the outer membrane of their cell wall (Lee *et al.*, 1999; Furrer *et al.*, 2007; Valappil *et al.*, 2007). LPS is known to be co-purified with the PHA (Valappil *et al.*, 2007). The level of contamination of LPS can be reduced to some extent during P(3HB) preparation by chloroform, NaOH treatment (Lee *et al.*, 1999), or by a temperature-controlled method (Furrer *et al.*, 2007), but trace amounts of LPS still remain. In considering practical applications of LA-based polyester, especially for food-grade and biomedical demands, the use of endotoxin-free Gram-positive *C. glutamicum* which has GRAS status is preferable (Takors *et al.*, 2007).

This chapter will introduce the biosynthesis of P(3HB) in *C. glutamicum* and the enhancement of its production. Compared with recombinant *E. coli* and other typical Gram-negative PHA producers, *C. glutamicum* utilizes a narrower range of carbon sources that limits the variety of monomer constituents of PHAs. In the confines of the starting substance, the copolymer of 3HB and 3-hydroxyvalerate was successfully synthesized in this organism. Besides the safe nature of *C. glutamicum*, advantages of this bacterium for polyester synthesis became apparent when the strain synthesized PLA homopolymer-like polyester from glucose, and when glutamate was coproduced with PHA in the same batch. In addition, the cell-surface displaying technology conferred the ability to utilize starch on this strain.

Polyhydroxyalkanoate production in *Corynebacterium glutamicum*

Biosynthesis of poly(3-hydroxybutyrate) in *Corynebacterium glutamicum*

P(3HB) is the most commonly found PHA in natural PHA producers, which is synthesized in successive reactions involving supply of the monomer substrate, 3-hydroxybutyryl-CoA (3HB-CoA) and its polymerization by PHA synthase (PhaC). 3HB-CoA is synthesized via two steps (Fig. 9.2); β-ketothiolase (PhaA) condenses two acetyl-CoAs into acetoacetyl-CoA in the initial step, and the NADPH-dependent acetoacetyl-CoA reductase (PhaB) reduces acetoacetyl-CoA to 3-hydroxybutyryl-CoA (3HB-CoA) in the second step. The genome sequence of *C. glutamicum* indicated the absence of PHA synthase genes in this organism. Thus, to produce P(3HB) in non P(3HB) producing-organisms, the *phaCAB* genes were required.

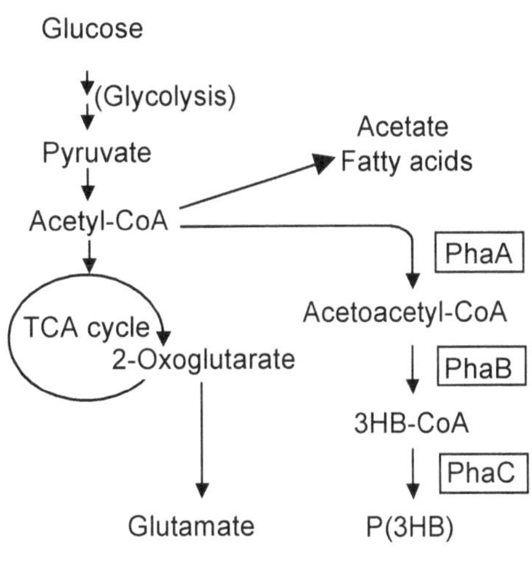

Figure 9.2 Metabolic linkage between glutamate synthetic pathway and P(3HB) synthetic pathway in *Corynebacterium glutamicum*. PhaA, β-ketothiolase; PhaB, NADPH-dependent acetoacetyl-CoA reductase; PhaC, PHA synthase. Boxed enzymes indicate exogenous enzymes.

Another required factor is the channelling of the carbon flux from TCA cycle to P(3HB) production. Fig. 9.2 shows the metabolic linkage between the glutamate and the P(3HB) synthetic pathways in *C. glutamicum,* both of which competitively utilize acetyl-CoA. It has been known that biotin limitation is key to promote glutamate formation. Thus conversely, the addition of excess biotin suppressed the glutamate production. Using this method, acetyl-CoA can be channelled into the P(3HB) synthesis pathway. In fact, the recombinant *C. glutamicum* harbouring the three genes encoding PhaA, PhaB and PhaC (Fig. 9.2) derived from *Ralstonia eutropha* produced 22.5 wt% P(3HB) from glucose (Jo *et al.*, 2006). The polymer had a number average molecular weight (*Mn*) and a polydispersity index of 2.1×10^5 and 1.63, respectively. These values were different from those of P(3HB) (*Mn* of 1.8×10^6 and polydispersity index of 1.8) synthesized in the recombinant *E. coli* expressing the same PHB biosynthetic genes from *R. eutropha.* The differences in these properties between P(3HB)s synthesized by both bacterial strains could be attributed to the differences between individual physiological features such as monomer supplying ability and the expression levels of P(3HB) biosynthetic genes.

Although the P(3HB) production in the recombinant *C. glutamicum* was attained by the artificial P(3HB) biosynthetic pathway deriving monomers from acetyl-CoA, the P(3HB) content in *C. glutamicum* was still lower than those in *E. coli* and *R. eutropha.* In retrospect, Jo *et al.* (2007) applied two experimental strategies to improve P(3HB) production in *C. glutamicum.* One was codon optimization of the N-terminal-coding region of the PHA synthase (PhaC) gene focusing on the codon usage preference for the translation system of *C. glutamicum.* The other was the replacement of wild-type *phaC* with a modified gene encoding a mutation of Gly4Asp (G4D), which enhanced the expression and production of PhaC and P(3HB) in *E. coli* (Normi *et al.*, 2005). The introduction of these engineered PHA synthase genes into *C. glutamicum* enhanced the expression of PhaC and the production of P(3HB). The recombinant *C. glutamicum* strain harbouring the codon optimized PhaC and G4D mutation of PhaC increased the P(3HB) production into 28.6 wt% and 29.9 wt%, respectively. Interestingly, these gene modifications also caused increases in the concentration of the translation products of the genes encoding monomer-supplying enzymes, PhaA and PhaB. This finding prompted them to carry out gene dosage of *PhaAB* using a double plasmid system, and the highest production (52.5 wt%) of P(3HB) was finally achieved by combining the gene dosage of *PhaAB* with codon optimization of PhaC. The molecular weight of P(3HB) was also increased by approximately 2-fold, as was the P(3HB) content. These results demonstrated that the strategies of codon optimization, point mutation and gene dosage of P(3HB) biosynthetic genes were effective in improving the production of P(3HB) and thus *C. glutamicum* could be a suitable host cell candidate for efficient P(3HB) production.

In a related study, Mastumoto *et al.* (2011) designed a new synthetic pathway involving the generation of acetoacetyl-CoA in order to produce P(3HB) in *C. glutamicum.* They used a newly isolated enzyme, designated acetoacetyl-CoA synthase (AACS), to catalyse a single condensation of acetyl-CoA and malonyl-CoA to give acetoacetyl-CoA in an irreversible reaction. Its enzymatic property contrasts with that of the PhaA enzyme, which exhibits reversible condensation reaction of two acetyl-CoA and prefers acetoacetyl-CoA thiolysis to acetoacetyl-CoA synthesis (Matsumoto *et al.*, 2011). The data showed that the AACS-catalysed generation of acetoacetyl-CoA was also functional in the biosynthesis of P(3HB) in *C. glutamicum.* No significant difference in P(3HB) production was observed between the recombinant constructs carrying the new and the conventional synthetic pathways. Thus AACS had sufficient ability, comparable to PhaA to produce P(3HB) in *C. glutamicum.* These result indicated that AACS is a monomer supplier for P(3HB) production using malonyl-CoA as a second substrate. In addition, the mode of reaction of AACS is irreversible, differing from that of PhaA. Therefore, the new synthetic pathway has a potential to increase monomer supply and P(3HB) production as compared with the conventional pathway.

To diversify the production system of *C. glutamicum* and maximize carbon source utilization, co-production of intracellular P(3HB) along with extracellular glutamate was attempted (Jo *et al.*, 2009). As reported by Jo *et al.*, *C. glutamicum* cells expressing *phaCAB* operon accumulated 36 wt% of PHB and secreted glutamate up to 18 g/l by a two-step cultivation technique. In addition, Liu *et al.* (2007) checked the impact of P(3HB) accumulation on glutamate production and the results suggested that the P(3HB) synthesis genes has a positive effect on glutamate production in *C. glutamicum*. These results suggest that *C. glutamicum* has a potential as host for simultaneous production of a number of bio-products, thereby decreasing the cost for polyester production.

Production of poly(3-hydroxybutyrate-*co*-3-hydroxyvalerate) copolymer

P(3HB) is an isotactic and a highly crystalline thermoplastic that is environmentally biodegradable (Kemnitzer *et al.*, 1993; Serafim *et al.*, 2004; Ivanova *et al.*, 2009). However, the closeness of the melting temperature (T_m) and the thermal degradation temperature along with the brittleness of P(3HB) limits its processability and the range of applications (Kemnitzer *et al.*, 1993; VanderHart *et al.*, 1995; Valentin *et al.*, 1999). One method to decipher these bottlenecks has been the copolymerization of 3HB with other hydroxyalkanoate (HA) monomers. In particular, copolymerization of 3HB with 3-hydroxyvalerate (3HV) (Yoshie *et al.*, 2004), 3-hydroxypropionate (3HP), 3-hydroxyhexanoate (3HHx) (Doi *et al.*, 1995), and 4-hydroxybutyrate (4HB) (Nakamura *et al.*, 1992) has been demonstrated to be effective for creating materials with improved physical and mechanical properties.

3HV is one of the attractive monomer units owing to its significant reduction of the crystallinity of P(3HB), resulting into lowered T_m and improved flexibility. The T_m of P(3HB-*co*-3HV) varies in the range of 65–170°C, dependent on 3HV fraction (Inoue, 1998). The lowered T_m facilitates moulding of the polyester during processing. In addition, elongation at break is increased along with 3HV faction, reaching up to 700% when 28 mol% 3HV was incorporated (Inoue and Yoshie, 1992). To incorporate 3HV into P(3HB) in the *C. glutamicum* platform, Matsumoto *et al.* (Matsumoto *et al.*, 2010) investigated the use of propionate, which is commonly used as the structurally related secondary precursor for the production of P(3HB-*co*-3HV) (Byrom, 1987; Rhie and Dennis, 1995). When fed into bacteria, propionate could be activated into propionyl-CoA by a propionyl-CoA synthetase involved in an intrinsic metabolic pathway (Fig. 9.3), and then condensed with acetyl-CoA to make 3-ketovaleryl-CoA through the action of PhaA. Subsequently, 3-ketovaleryl-CoA would be accepted as an alternate substrate by PhaB and ultimately be polymerized by PhaC along with 3HB to form P(3HB-*co*-3HV) copolymer (Fig. 9.3).

The requisite genes, *phaA*, *phaB* and *phaC* were introduced into *C. glutamicum* and the recombinant cells were cultivated on a medium supplemented with glucose and varying concentrations of propionate. Analysis of the cells confirmed the incorporation of 3HV to form P(3HB-*co*-3HV) with the amount of propionate added being directly proportional with the 3HV fraction in the copolymer. The P(3HB-*co*-3HV) obtained had similar molecular weights as P(3HB) produced in *C. glutamicum*, suggesting that 3HV incorporation had no significant effect on molecular weights of polymers produced in *C. glutamicum*.

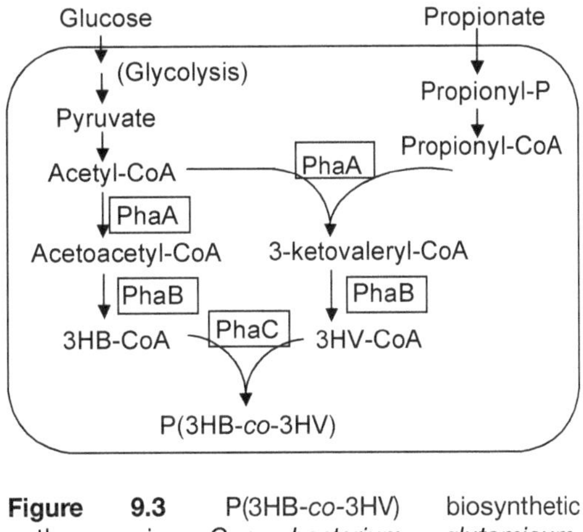

Figure 9.3 P(3HB-*co*-3HV) biosynthetic pathway in *Corynebacterium glutamicum*. PhaA: β-ketothiolase; PhaB: NADPH-dependent acetoacetyl-CoA reductase; PhaC: PHA synthase; 3HB-CoA: 3-hydroxybutyryl-CoA; propionyl-P: propionyl phosphate. Boxed enzymes indicate exogenous enzymes.

The study demonstrated that *C. glutamicum* is a versatile host for the production of a variety of polyesters. However, the transformation efficiencies of glucose and propionate into P(3HB-*co*-3HV) were 19% and 9.6%, respectively at the highest PHA production (6.0 g/l) condition. These efficiencies are low and calls for further efforts to improve polymer yields so as to reduce the cost of the carbon sources. The efficiency in the flask-scale fermentation could be improved by finely regulating fermentation using a bioreactor. In addition, co-production of the polymers with other products such as glutamate could be a good strategy to increase the cost effectiveness of the process.

Poly(lactic acid)-like polyester production in *Corynebacterium glutamicum*

As mentioned above, PLA is synthesized via multistep chemo-bioprocesses and the contamination by harmful heavy metal catalyst remains a problem. Thus, PLA production using the PHA biosynthetic pathway has attracted research interest. The key factor in this challenge is substrate specificity of the PHA synthase. Lactic acid shares the structural similarity with PHA monomers such as 3HB hence; the CoA ester of lactate, lactyl-CoA (LA-CoA) could potentially be polymerized in a similar manner by PHA synthase. However, although much effort has been done, the polymerization of LA-CoA using the native PHA synthases has been unsuccessful (Valentin and Steinbuchel, 1994; Yuan *et al.*, 2001). A breakthrough was achieved when significant LA-polymerizing activity was found in an engineered PHA synthase, which was the Ser325Thr/Gln481Lys mutant of PHA synthase from *Pseudomonas* sp. 61-3 [PhaC1$_{Ps}$(ST/QK)]. The engineered PHA synthase mutant have been created to acquire broad substrate specificity and enhanced activity for PHA production (Taguchi *et al.*, 2002; Taguchi and Doi, 2004; Takase *et al.*, 2004; Tsuge *et al.*, 2004; Nomura and Taguchi, 2007), but the mutations unexpectedly contributed to the recognition of LA-CoA as a substrate. Subsequently, the LA-polymerizing PHA synthase, designated as lactate-polymerizing enzyme (LPE), was expressed in the recombinant *E. coli* to synthesize P(LA-*co*-3HB) that was the first enzyme-driven microbial cell factory (MCF) for the synthesis of LA-based polyesters (Taguchi *et al.*, 2008). Although the prototype MCF produced P(LA-*co*-3HB) containing only small amount of LA units (6 mol%), the subsequent studies have demonstrated that the LA fraction in P(LA-*co*-3HB) can be increased up to 63 mol% by means of metabolic and enzyme engineering (Yamada *et al.*, 2009, 2010, 2011; Shozui *et al.*, 2010; Shozui *et al.*, 2011). However, the synthesis of P(LA-*co*-3HB) with LA fraction higher than this has been unsuccessful. If PhaA and PhaB were eliminated from this pathway, the polymer synthesis disappeared, namely PLA homopolymer is not synthesized using the MCF. Although the mechanism for this incapability of PLA biosynthesis has not been fully revealed, LPE presumably requires the presence of 3HB-CoA as a priming unit for incorporating LA units into the polymer chain that limited the range of LA fraction in the copolymer. Another problem was, as mentioned above, *E. coli* contains LPS, which is an unfavourable contaminant of food-contacting and medical use materials.

Thus, to upgrade this prototype system to the practical scale, especially for food-grade and biomedical demands, the Gram-positive *C. glutamicum* which has GRAS status, is endotoxin-free and extensively used for the industrial production of amino acids was considered as a good candidate for the production of LA-based polyester (Takors *et al.*, 2007). Because *C. glutamicum* produced P(3HB) and P(3HB-*co*-3HV) (Jo *et al.*, 2006, 2007, 2009; Matsumoto *et al.*, 2010), it also had potential to accumulate LA-based polyesters. The design of the metabolic pathway for the synthesis of LA-based polyester in *C. glutamicum* is shown in Fig. 9.4. This pathway consists of three elements: (i) generation of D-lactyl-CoA (LA-CoA) by D-lactate dehydrogenase (D-LDH) and propionyl-CoA transferase (PCT), (ii) 3HB-CoA generation, catalysed by PhaA and PhaB, and (iii) copolymerization of LA-CoA and 3HB-CoA, catalysed by LPE. A key factor in the construction of the metabolic pathway to produce LA-based polyesters in *C. glutamicum* was the stereochemistry of LA, because it has been demonstrated that LPE

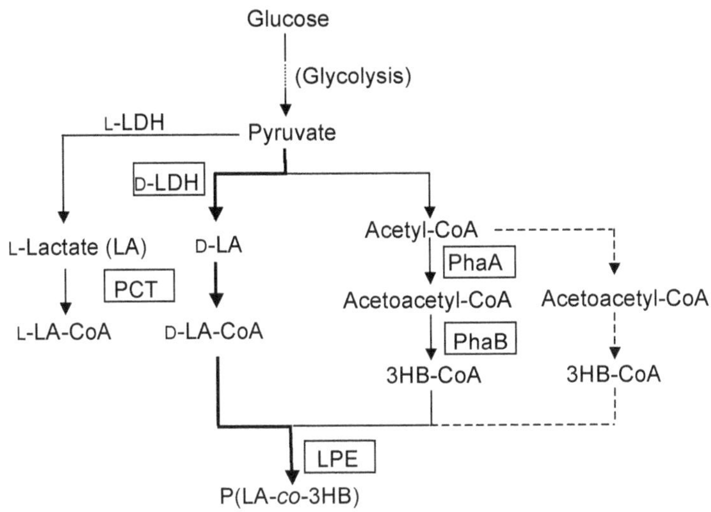

Figure 9.4 Metabolic pathways relevant to production of P(lactate-*co*-3-hydroxybutyrate) in engineered *Corynebacterium glutamicum*. L-LDH, L-lactate dehydrogenase; D-LDH, D-lactate dehydrogenase; PCT, propionyl-CoA transferase; PhaA, β-ketothiolase; PhaB, acetoacetyl-CoA reductase; LPE, lactate-polymerizing enzyme. Boxed enzymes indicate exogenous enzymes. Thick lines indicate reinforced pathways. Dashed lines indicate proposed pathways; a small amount of 3HB-CoA was supplied by an intrinsic, but uncharacterized, pathway.

has strict stereospecificity towards D-LA-CoA (Taguchi *et al.*, 2008; Tajima *et al.*, 2009; Yamada *et al.*, 2009). However, the *C. glutamicum* strain is known to produce mainly L-LA, as reported in several previous studies (Inui *et al.*, 2004; Okino *et al.*, 2008; Toyoda *et al.*, 2009), whereas *E. coli* can produce D-LA (Bunch *et al.*, 1997; Chang *et al.*, 1999). Therefore, *C. glutamicum* was remodelled as a D-LA producer by introduction of D-LDH from *E. coli* (Song *et al.*, 2012).

In fact, the recombinant *C. glutamicum* expressing the five relevant genes produced P(LA-*co*-3HB). Surprisingly, the P(LA-*co*-3HB) produced contained extremely high LA fractions (96.8 mol%). This result was in striking contrast to the results obtained by the *E. coli* system using the same set of genes and same carbon source (Yamada *et al.*, 2009, 2010; Shozui *et al.*, 2010).

The low 3HB fraction in the copolymers synthesized in *C. glutamicum* suggested that the 3HB supplying pathway in this strain was very weak, and prompted the authors to evaluate the essentiality of the 3HB-supplying pathway. Indeed, the recombinant *C. glutamicum* expressing D-LDH, PCT and LPE synthesized P(LA-*co*-3HB) with LA fraction of 99.6 mol%. This result was a sharp contrast to the *E. coli* result in that PhaAB elimination led to the total loss of polymer synthesis (Taguchi *et al.*, 2008). However, it should be noted that the small amount of 3HB units were still found in the copolymer, indicating that there is an endogenous 3HB-CoA supplying pathway in *C. glutamicum* and this pathway might enable this strain to synthesize the PLA-like polymer without PhaAB. Furthermore, the finding of this pathway provided a useful insight into how to synthesize PLA-like polyesters in MCFs, i.e. a weak 3HB-CoA supplying pathway is needed for synthesizing PLA-like and LA-enriched polymers.

Poly(3-hydroxybutyrate) production from starch based on *Corynebacterium glutamicum* with cell surface displaying of α-amylase

The aforementioned studies have demonstrated the potential of *C. glutamicum* as a platform for polyester production. However, the economic considerations of production cost suggest that the carbon source, mainly glucose used for polyester production has a high impact on the production cost of P(3HB) and PHA, affecting their commercialization potential. There has been a considerable interest in using inexpensive substrates as an alternative to glucose in PHA production. One such substrate is starch, which is a renewable carbon source available in large quantities. Nevertheless, prior to fermentation, starch need to be hydrolysed into glucose by a two-step process, liquefaction and saccharification, which need physicochemical and enzymatic treatment (Rusendi and Sheppard, 1995). For example, Chen treated starch by an enzymatic reactive extrusion process for polyester production (Chen *et al.*, 2006); hydrolysed starch was used as substrate for P(3HB) production in *Halomonas boliviensis* (Quillaguamán *et al.*, 2005). Similarly, hydrolysed corn starch was also utilized for P(3HB) synthesis using recombinant *E. coli* (Fonseca *et al.*, 2008). The treatment before these fermentation renders the use of starch less economically viable. One strategy to decrease the costs of using starch for microbial polyester production is the direct utilization of starch instead of using enzyme-treated starch.

To meet this goal, the cell surface display technology is a powerful platform, which allows peptides and/or proteins fusing with the anchoring motifs to be localized on the external surface of the cell membrane of the host bacteria. The display of glycosidases enables microbes to hydrolyse and simultaneously utilize polysaccharides, such as starch if α-amylase is displayed, as a carbon source. Additionally, the displayed enzymes were reported to maintain their activity during long time cultivation (Yamakawa *et al.*, 2012) that could be an advantage over the enzyme-secreting system. As an example, the α-amylase displaying *C. glutamicum* was demonstrated to produce L-lysine from starch (Tateno *et al.*, 2007).

Song *et al.* (2013) attempted the application of the α-amylase displaying *C. glutamicum* to the production of P(3HB) from starch (Fig. 9.5). The challenge in this attempt was to achieve a certain utilization rate of starch sufficient for the polymer synthesis, because PHA production is induced by the excess amount of carbon source. To examine this, the recombinant *C. glutamicum* strain dually expressing α-amylase and the P(3HB) biosynthetic enzymes was grown on starch. The cells exhibited a significant α-amylase activity and P(3HB) accumulation. Furthermore, the productivity of P(3HB) from starch was even slightly higher than that from glucose. Previous research also indicated such benefit of using starch than that of glucose in *C. glutamicum* for L-lysine

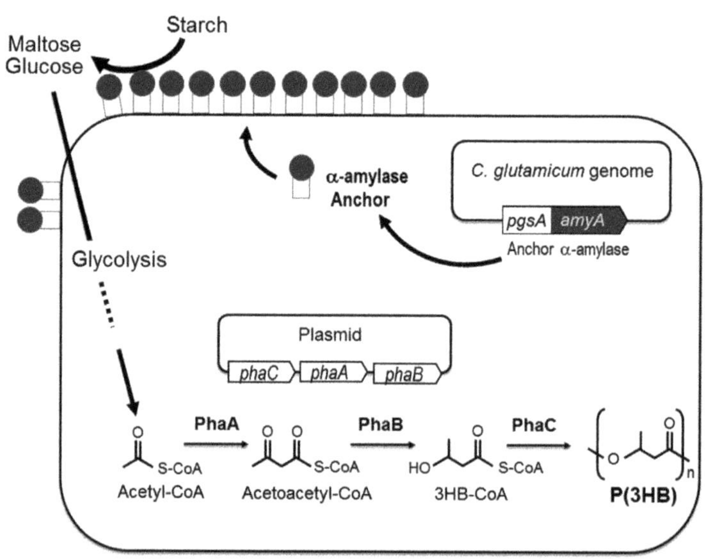

Figure 9.5 Combination of cell surface displayed amylase and P(3HB) biosynthetic pathway in *Corynebacterium glutamicum*. The α-amylase gene is integrated into genome DNA while P(3HB) biosynthetic genes are in plasmid. The cells produce P(3HB) from starch.

production (Tateno *et al.*, 2007). This phenomenon was interpreted that starch was utilized slowly compared with glucose, and therefore, the carbon flux was channelled towards the production in the later stage of cultivation rather than cell growth. This result suggested that starch served as a good substrate for P(3HB) production, and the α-amylase displaying *C. glutamicum* is useful for the reduction of the cost of microbial polyester productions.

Conclusion

C. glutamicum is a useful platform for the production of various endotoxin-free microbial polyesters. In particular, the capability of synthesizing PLA-like polyester is reported only in this organism. The biological PLA has an advantage over chemical PLA, which potentially contains harmful heavy metal catalyst. At present, overcoming the low productivity of LA-based polyesters is the main task towards practical polyester production using *C. glutamicum*. In addition, cell-surface displaying system has been shown to be a powerful tool to enable this organism to utilize less expensive starch as a feedstock. The cell engineering technology will be combined with the aforementioned metabolic and enzyme engineering approaches to upgrade the *C. glutamicum* polyester factory.

References

Aldor, I. S., and Keasling, J. D. (2003) Process design for microbial plastic factories: metabolic engineering of polyhydroxyalkanoates. *Curr. Opin. Biotech.* **14**, 475–483.

Anderson, A. J., and Dawes, E. A. (1990) Occurrence, metabolism, metabolic mole, and industrial uses of bacterial polyhydroxyalkanoates. *Microbiol. Rev.* **54**, 450–472.

Auras, R., Harte, B., and Selke, S. (2004) An overview of polylactides as packaging materials. *Macromol. Biosci.* **4**, 835–864.

Bunch, P. K., Mat-Jan, F., Lee, N., and Clark, D. P. (1997) The *ldhA* gene encoding the fermentative lactate dehydrogenase of *Escherichia coli*. *Microbiology* **143**, 187–195.

Byrom, D. (1987) Polymer synthesis by microorganisms: technology and economics. *Trends Biotechnol.* **5**, 246–250.

Chang, D. E., Jung, H. C., Rhee, J. S., and Pan, J. G. (1999) Homofermentative production of D- or L-lactate in metabolically engineered *Escherichia coli* RR1. *Appl. Environ. Microbiol.* **65**, 1384–1389.

Chanprateep, S. (2010) Current trends in biodegradable polyhydroxyalkanoates. *J. Biosci. Bioeng.* **110**, 621–632.

Chen, C. W., Don, T.-M., and Yen, H.-F. (2006) Enzymatic extruded starch as a carbon source for the production of poly(3-hydroxybutyrate-*co*-3-hydroxyvalerate) by *Haloferax mediterranei*. *Process Biochem.* **41**, 2289–2296.

Corma, A., Iborra, S., and Velty, A. (2007) Chemical routes for the transformation of biomass into chemicals. *Chem. Rev.* **107**, 2411–2502.

Doi, Y. (1990) Microbial Polyesters. VCH Publisher, New York, NY, USA.

Doi, Y., Kitamura, S., and Abe, H. (1995) Microbial synthesis and characterization of poly(3-hydroxybutyrate-*co*-3-hydroxyhexanoate). *Macromolecules* **28**, 4822–4828.

Fonseca, G. G., de Arruda-Caulkins, J. C., and Vasconcellos Antonio, R. (2008) Production and characterization of poly-(3-hydroxybutyrate) from recombinant *Escherichia coli* grown on cheap renewable carbon substrates. *Waste Manage. Res.* **26**, 546–552.

Furrer, P., Zinn, M., and Panke, S. (2007) Efficient recovery of low endotoxin medium-chain-length poly([*R*]-3-hydroxyalkanoate) from bacterial biomass. *J. Microbiol. Methods* **69**, 206–213.

Haddouche, R., Poirier, Y., Delessert, S., Sabirova, J., Pagot, Y., Neuveglise, C., and Nicaud, J.-M. (2011) Engineering polyhydroxyalkanoate content and monomer composition in the oleaginous yeast *Yarrowia lipolytica* by modifying the β-oxidation multifunctional protein. *Appl. Microbiol. Biotechnol.* **91**, 1327–1340.

Hazer, B., and Steinbuchel, A. (2007) Increased diversification of polyhydroxyalkanoates by modification reactions for industrial and medical applications. *Appl. Microbiol. Biotechnol.* **74**, 1–12.

Inoue, Y. (1998) Solid-state structure and properties of bacterial copolyesters. *J. Mol. Struct.* **441**, 119–127.

Inoue, Y., and Yoshie, N. (1992) Structure and physical properties of bacterially synthesized polyesters. *Prog. Polym. Sci.* **17**, 571–610.

Inui, M., Murakami, S., Okino, S., Kawaguchi, H., Vertes, A. A., and Yukawa, H. (2004) Metabolic analysis of *Corynebacterium glutamicum* during lactate and succinate productions under oxygen deprivation conditions. *J. Mol. Microbiol. Biotechnol.* **7**, 182–196.

Ivanova, G., Serafim, L. S., Lemos, P. C., Ramos, A. M., Reis, M. A. M., and Cabrita, E. J. (2009) Influence of feeding strategies of mixed microbial cultures on the chemical composition and microstructure of copolyesters P(3HB-*co*-3HV) analyzed by NMR and statistical analysis. *Magn. Res. Chem.* **47**, 497–504.

Jendrossek, D., Schirmer, A., and Schlegel, H. G. (1996) Biodegradation of polyhydroxyalkanoic acids. *Appl. Microbiol. Biotechnol.* **46**, 451–463.

Jo, S. J., Maeda, M., Ooi, T., and Taguchi, S. (2006) Production system for biodegradable polyester polyhydroxybutyrate by *Corynebacterium glutamicum*. *J. Biosci. Bioeng.* **102**, 233–236.

Jo, S. J., Matsumoto, K., Leong, C. R., Ooi, T., and Taguchi, S. (2007) Improvement of poly(3-hydroxybutyrate) [P(3HB)] production in *Corynebacterium glutamicum* by codon optimization, point mutation and gene dosage of P(3HB) biosynthetic genes. *J. Biosci. Bioeng.* **104**, 457–463.

Jo, S. J., Leong, C. R., Matsumoto, K., and Taguchi, S. (2009) Dual production of poly (3-hydroxybutyrate) and glutamate using variable biotin concentrations in *Corynebacterium glutamicum*. *J. Biosci. Bioeng.* **107**, 409–411.

Kalinowski, J., Bathe, B., Bartels, D., Bischoff, N., Bott, M., Burkovski, A., Dusch, N., Eggeling, L., Eikmanns, B. J., Gaigalat, L., *et al.* (2003) The complete *Corynebacterium glutamicum* ATCC 13032 genome sequence and its impact on the production of L-aspartate-derived amino acids and vitamins. *J. Biotechnol.* **104**, 5–25.

Kemnitzer, J. E., McCarthy, S. P., and Gross, R. A. (1993) Preparation of predominantly syndiotactic poly(β-hydroxybutyrate) by the tributyltin methoxide catalyzed ring-opening polymerization of racemic β-butyrolactone. *Macromolecules* **26**, 1221–1229.

Kinoshita, S., Udaka, S., and Shimono, M. (1957) Studies on the amino acid fermentation – Part I. Production of L-glutamic acid by various microorganisms. *J. Gen. Appl. Microbiol.* **3**, 193–205.

Krämer, R. (1994) Secretion of amino acids by bacteria: physiology and mechanism. *FEMS Microbiol. Rev.* **13**, 75–93.

Kricheldorf, H. R. (2001) Syntheses and application of polylactides. *Chemosphere* **43**, 49–54.

Lee, S. Y., Choi, J. I., Han, K., and Song, J. Y. (1999) Removal of endotoxin during purification of poly(3-hydroxybutyrate) from Gram-negative bacteria. *Appl. Environ. Microbiol.* **65**, 2762–2764.

Leuchtenberger, W., Huthmacher, K., and Drauz, K. (2005) Biotechnological production of amino acids and derivatives: current status and prospects. *Appl. Microbiol. Biotechnol.* **69**, 1–8.

Liu, Q., Ouyang, S.-P., Kim, J., and Chen, G.-Q. (2007) The impact of PHB accumulation on L-glutamate production by recombinant *Corynebacterium glutamicum*. *J. Biotechnol.* **132**, 273–279.

Matsumoto, K., and Taguchi, S. (2009) Enzymatic and whole-cell synthesis of lactate-containing polyesters: toward the complete biological production of polylactate. *Appl. Microbiol. Biotechnol.* **85**, 921–932.

Matsumoto, K., Kitagawa, K., Jo, S. J., Song, Y., and Taguchi, S. (2010) Production of poly(3-hydroxybutyrate-*co*-3-hydroxyvalerate) in recombinant *Corynebacterium glutamicum* using propionate as a precursor. *J. Biotechnol.* **152**, 144–146.

Matsumoto, K., Yamada, M., Leong, C. R., Jo, S. J., Kuzuyama, T., and Taguchi, S. (2011) A new pathway for poly(3-hydroxybutyrate) production in *Escherichia coli* and *Corynebacterium glutamicum* by functional expression of a new acetoacetyl-coenzyme A synthase. *Biosci. Biotechnol. Biochem.* **75**, 364–366.

Nakamura, S., Doi, Y., and Scandola, M. (1992) Microbial synthesis and characterization of poly(3-hydroxybutyrate-*co*-4-hydroxybutyrate). *Macromolecules* **25**, 4237–4241.

Nampoothiri, M. K., Nair, N. R., and John, R. P. (2010) An overview of the recent developments in polylactide (PLA) research. *Biores. Technol.* **101**, 8493–8501.

Nomura, C., and Taguchi, S. (2007) PHA synthase engineering toward superbiocatalysts for custom-made biopolymers. *Appl. Microbiol. Biotechnol.* **73**, 969–979.

Normi, Y. M., Hiraishi, T., Taguchi, S., Abe, H., Sudesh, K., Najimudin, N., and Doi, Y. (2005) Characterization and properties of G4X mutants of *Ralstonia eutropha* PHA synthase for poly(3-hydroxybutyrate) biosynthesis in *Escherichia coli*. *Macromol. Biosci.* **5**, 197–206.

Okino, S., Suda, M., Fujikura, K., Inui, M., and Yukawa, H. (2008) Production of D-lactic acid by *Corynebacterium glutamicum* under oxygen deprivation. *Appl. Microbiol. Biotechnol.* **78**, 449–454.

Ozkoc, G., and Kemaloglu, S. (2009) Morphology, biodegradability, mechanical, and thermal properties of nanocomposite films based on PLA and plasticized PLA. *J. Appl. Polymer Sci.* **114**, 2481–2487.

Quillaguamán, J., Hashim, S., Bento, F., Mattiasson, B., and Hatti-Kaul, R. (2005) Poly(β-hydroxybutyrate) production by a moderate halophile, *Halomonas boliviensis* LC1 using starch hydrolysate as substrate. *J. Appl. Microbiol.* **99**, 151–157.

Rehm, B. H. A. (2010) Bacterial polymers: biosynthesis, modifications and applications. *Nat. Rev. Microbiol.* **8**, 578–592.

Rhie, H. G., and Dennis, D. (1995) The function of *ackA* and *pta* genes is necessary for poly(3-hydroxybutyrate-*co*-3-hydroxyvalerate) synthesis in recombinant pha+ *Escherichia coli*. *Can. J. Microbiol.* **41**, 200–206.

Rusendi, D., and Sheppard, J. D. (1995) Hydrolysis of potato processing waste for the production of poly-β-hydroxybutyrate. *Bioresour. Technol.* **54**, 191–196.

Serafim, L. S., Lemos, P. C., Oliveira, R., and Reis, M. A. M. (2004) Optimization of polyhydroxybutyrate production by mixed cultures submitted to aerobic dynamic feeding conditions. *Biotechnol. Bioeng.* **87**, 145–160.

Shozui, F., Matsumoto, K., Nakai, T., Yamada, M., and Taguchi, S. (2010) Biosynthesis of novel terpolymers poly(lactate-*co*-3-hydroxybutyrate-*co*-3-hydroxyvalerate)s in lactate-overproducing mutant *Escherichia coli* JW0885 by feeding propionate as a precursor of 3-hydroxyvalerate. *Appl. Microbiol. Biotechnol.* **85**, 949–954.

Shozui, F., Matsumoto, K.i., Motohashi, R., Sun, J., Satoh, T., Kakuchi, T., and Taguchi, S. (2011) Biosynthesis of a lactate (LA)-based polyester with a 96 mol% LA fraction and its application to stereocomplex formation. *Polymer Degr. Stabil.* **96**, 499–504.

Song, Y., Matsumoto, K., Yamada, M., Gohda, A., Brigham, C., Sinskey, A., and Taguchi, S. (2012) Engineered *Corynebacterium glutamicum* as an endotoxin-free platform strain for lactate-based polyester production. *Appl. Microbiol. Biotechnol.* **93**, 1917–1925.

Song, Y., Matsumoto, K.i., Tanaka, T., Kondo, A., and Taguchi, S. (2013) Single-step production of polyhydroxybutyrate from starch by using α-amylase cell-surface displaying system of *Corynebacterium glutamicum*. *J. Biosci. Bioeng.* **115**, 12–14.

Steinbuchel, A., and Fuchtenbusch, B. (1998) Bacterial and other biological systems for polyester production. *Trends Biotechnol.* **16**, 419–427.

Steinbüchel, A., Hustede, E., Liebergell, M., Pieper, U., Timm, A., and Valentin, H. (1992) Molecular basis for biosynthesis and accumulation of polyhydroxyalkanoic acids in bacteria. *FEMS Microbiol. Rev.* **103**, 217–230.

Sudesh, K., Abe, H., and Doi, Y. (2000) Synthesis, structure and properties of polyhydroxyalkanoates: biological polyesters. *Prog. Polym. Sci.* **25**, 1503–1555.

Taguchi, S., and Doi, Y. (2004) Evolution of polyhydroxyalkanoate (PHA) production system by "enzyme evolution": successful case studies of directed evolution. *Macromol. Biosci.* **4**, 145–156.

Taguchi, S., Nakamura, H., Hiraishi, T., Yamato, I., and Doi, Y. (2002) *In vitro* evolution of a polyhydroxybutyrate synthase by intragenic suppression-type mutagenesis. *J. Biochem.* **131**, 801–806.

Taguchi, S., Yamada, M., Matsumoto, K., Tajima, K., Satoh, Y., Munekata, M., Ohno, K., Kohda, K., Shimamura, T., Kambe, H., *et al.* (2008) A microbial factory for lactate-based polyesters using a lactate-polymerizing enzyme. *Proc. Natl. Acad. Sci. U.S.A.* **105**, 17323–17327.

Tajima, K., Satoh, Y., Satoh, T., Itoh, R., Han, X. R., Taguchi, S., Kakuchi, T., and Munekata, M. (2009) Chemo-enzymatic synthesis of poly(lactate-*co*-(3-hydroxybutyrate)) by a lactate-polymerizing enzyme. *Macromolecules* **42**, 1985–1989.

Takase, K., Matsumoto, K.i., Taguchi, S., and Doi, Y. (2004) Alteration of substrate chain-length specificity of type II synthase for polyhydroxyalkanoate biosynthesis by *in vitro* evolution: *in vivo* and *in vitro* enzyme assays. *Biomacromolecules* **5**, 480–485.

Takors, R., Bathe, B., Rieping, M., Hans, S., Kelle, R., and Huthmacher, K. (2007) Systems biology for industrial strains and fermentation processes – example: amino acids. *J. Biotechnol.* **129**, 181–190.

Tateno, T., Fukuda, H., and Kondo, A. (2007) Production of L-Lysine from starch by *Corynebacterium glutamicum* displaying α-amylase on its cell surface. *Appl. Microbiol. Biotechnol.* **74**, 1213–1220.

Tokiwa, Y., and Calabia, B. (2006) Biodegradability and biodegradation of poly(lactide). *Appl. Microbiol. Biotechnol.* **72**, 244–251.

Toyoda, K., Teramoto, H., Inui, M., and Yukawa, H. (2009) The *ldhA* gene, encoding fermentative L-lactate dehydrogenase of *Corynebacterium glutamicum*, is under the control of positive feedback regulation mediated by LldR. *J. Bacteriol.* **191**, 4251–4258.

Tsuge, T., Saito, Y., Narike, M., Muneta, K., Normi, Y. M., Kikkawa, Y., Hiraishi, T., and Doi, Y. (2004) Mutation effects of a conserved alanine (Ala510) in type I polyhydroxyalkanoate synthase from *Ralstonia eutropha* on polyester biosynthesis. *Macromol. Biosci.* **4**, 963–970.

Valappil, S. P., Boccaccini, A. R., Bucke, C., and Roy, I. (2007) Polyhydroxyalkanoates in Gram-positive bacteria: insights from the genera *Bacillus* and *Streptomyces*. *Antonie Van Leeuwenhoek* **91**, 1–17.

Valentin, H. E., and Steinbuchel, A. (1994) Application of enzymatically synthesized short-chain-length hydroxy fatty acid coenzyme A thioesters for assay of polyhydroxyalkanoic acid synthases. *Appl. Microbiol. Biotechnol.* **40**, 699–709.

Valentin, H. E., Berger, P. A., Gruys, K. J., Filomena de Andrade Rodrigues, M., Steinbuchel, A., Tran, M., and Asrar, J. (1999) Biosynthesis and characterization of poly(3-hydroxy-4-pentenoic acid). *Macromolecules* **32**, 7389–7395.

VanderHart, D. L., Orts, W. J., and Marchessault, R. H. (1995) ^{13}C NMR determination of the degree of cocrystallization in random copolymers of poly(β-hydroxybutyrate-co-β-hydroxyvalerate). *Macromolecules* **28**, 6394–6400.

Yamada, M., Matsumoto, K., Nakai, T., and Taguchi, S. (2009) Microbial production of lactate-enriched poly[(*R*)-lactate-*co*-(*R*)-3-hydroxybutyrate] with novel thermal properties. *Biomacromolecules* **10**, 677–681.

Yamada, M., Matsumoto, K., Shimizu, K., Uramoto, S., Nakai, T., Shozui, F., and Taguchi, S. (2010) Adjustable mutations in lactate (LA)-polymerizing enzyme for the microbial production of LA-based polyesters with tailor-made monomer composition. *Biomacromolecules* **11**, 815–819.

Yamada, M., Matsumoto, K., Uramoto, S., Motohashi, R., Abe, H., and Taguchi, S. (2011) Lactate fraction dependent mechanical properties of semitransparent poly(lactate-*co*-3-hydroxybutyrate)s produced by control of lactyl-CoA monomer fluxes in recombinant *Escherichia coli*. *J. Biotechnol.* **154**, 255–260.

Yamakawa, S., Yamada, R., Tanaka, T., Ogino, C., and Kondo, A. (2012) Repeated fermentation from raw starch using *Saccharomyces cerevisiae* displaying both glucoamylase and α-amylase. *Enzyme Microb. Technol.* **50**, 343–347.

Yoshie, N., Saito, M., and Inoue, Y. (2004) Effect of chemical compositional distribution on solid-state structures and properties of poly(3-hydroxybutyrate-*co*-3-hydroxyvalerate). *Polymer* **45**, 1903–1911.

Yuan, W., Jia, Y., Tian, J., Snell, K. D., Muh, U., Sinskey, A. J., Lambalot, R. H., Walsh, C. T., and Stubbe, J. (2001) Class I and III polyhydroxyalkanoate synthases from *Ralstonia eutropha* and *Allochromatium vinosum*: characterization and substrate specificity studies. *Arch. Biochem. Biophys.* **394**, 87–98.

Biotechnological Application of *Corynebacterium glutamicum* Under Oxygen Deprivation

10

Toru Jojima, Masayuki Inui and Hideaki Yukawa

Abstract

Traditional bioprocesses using *Corynebacterium glutamicum* as biocatalysts, including amino acid fermentation process, are operated under aerobic conditions. An emerging bioprocess in which the microorganism is utilized under oxygen deprivation, so-called growth-arrested bioprocess, is from these traditional fermentation processes. This novel bioprocess is expected to play a major role in the production of biofuels, commodity and fine chemicals, besides the aforementioned amino acids. In this chapter, we describe the physiology of *C. glutamicum* under oxygen-deprived conditions with respect to recent advances in development of the microorganism as a mainstay biocatalysts used in growth-arrested bioprocess, with special emphasis on potential applications relevant to future lignocellulosic biorefineries.

Introduction

Corynebacterium glutamicum has traditionally been widely used for industrial production of amino acids, especially L-glutamate and L-lysine. These are produced via fermentation processes operated under aerobic conditions where vigorous oxygen supply is required to maintain an optimal supply of oxygen during the production. It follows that aeration and agitation therefore are crucial control factors affecting production yields realized with such processes (Ikeda, 2003). Recently, bioprocesses based on *C. glutamicum* but utilizing cells under oxygen-deprived conditions have started to emerge. They differ from conventional processes in their tremendous potential to produce higher yields of not only amino acids, but also biofuels, commodity and fine chemicals. These bioprocesses utilize the inherent ability of *C. glutamicum* to remain metabolically active under environments of low oxygen concentrations; the cells metabolize sugars to produce reduced metabolites under such conditions that do not favour cellular growth. The concept of a bioprocess utilizing such growth-arrested cells of *C. glutamicum* was proposed in the late 1980s by Yukawa and co-workers (Terasawa *et al.*, 1989), but the first demonstration was not until 2004 (Inui *et al.*, 2004a,b). In its essence, aerobically cultivated cells of *C. glutamicum* are compactly packed into a simple reactor and fed with a simple medium composed of essential mineral salts and a carbon source, usually a sugar. As the conversion of the sugar to product occurs without aeration, cells do not actively grow, and hence led to the name 'growth-arrested bioprocess'. The detailed history of the development of this growth-arrested bioprocess is described elsewhere (Jojima *et al.*, 2013). The focus of this chapter is on recent advances in the development of biocatalysts for growth-arrested bioprocess, with special emphasis on its foreseeable applications in future lignocellulosic biorefineries where a range of value-added chemicals and fuels will be produced from lignocellulosic biomass. Numerous efforts have been made to adapt *C. glutamicum* to lignocellulosic biorefineries bearing in mind the increasing urgency to develop technologies that mitigate the harmful effects of greenhouse gas emissions. Technologically speaking, the growth-arrested bioprocess possesses ideal characteristics for actualizing a lignocellulosic biorefinery future, details of which are expounded upon here. But, like all emerging technologies, there exist considerable obstacles that must be overcome before the bioprocess can be considered adequately developed and commercially viable. Therefore, challenges that lie ahead of industrial application of the growth-arrested bioprocess will be outlined.

Basics of the bioprocess: utilization of *Corynebacterium glutamicum* under oxygen deprivation

Oxygen-deprived cells of *C. glutamicum* can consume a variety of sugars, such as glucose, fructose, mannose and sucrose to produce lactate, succinate and acetate, as the major products while cellular proliferation remains suppressed during the bioconversion reaction. Such uninterrupted metabolic activity under oxygen deprivation has been recorded for 360 hours (Okino *et al.*, 2005). Bicarbonate accelerates sugar consumption and increases production of succinate relative to lactate (Fig. 10.1) (Inui *et al.*, 2004b; Okino *et al.*, 2005). Pyruvate also accelerates sugar consumption, but does so without leading to changed ratios of metabolites. Since additions of both bicarbonate and pyruvate is known to decrease intracellular NADH/NAD$^+$ ratio, it follows that intracellular redox balance (NADH/NAD$^+$ ratio) is a crucial factor controlling sugar consumption in *C. glutamicum* cells under oxygen deprivation. It is thought that a more favourable redox balance releases inhibition of glyceraldehyde 3-phosphate dehydrogenase (GAPDH) otherwise present under high NADH/NAD$^+$ ratio (Dominguez *et al.*, 1993). Cofactor engineering studies of the valine production pathway of the microorganism lend further support for this release of GAPDH activity (see the section 'Metabolic engineering *Corynebacterium glutamicum* for production of value-added chemicals via the growth-arrested bioprocess', below).

The sugar metabolism pathways of *C. glutamicum* under oxygen deprivation (Fig. 10.2), point at the reductive arm of the TCA cycle as the major contributor to succinate production. Conversely, phosphoenolpyruvate carboxylase (EC 4.1.1.31; PEPC) activity is largely responsible for TCA cycle anaplerosis under oxygen deprivation (Inui *et al.*, 2004b); in contrast, in aerobic conditions, pyruvate carboxylase (EC 6.4.1.1) activity plays the dominant role (Park *et al.*, 1997; Petersen *et al.*, 2000).

Because growth of *C. glutamicum* ceases upon oxygen deprivation, regulatory machinery activated within cells enable them to adapt to the prevailing low oxygen environment. Using transcriptome analysis, differential expression of genes under oxygen deprivation has been demonstrated; the expression of a total of 161

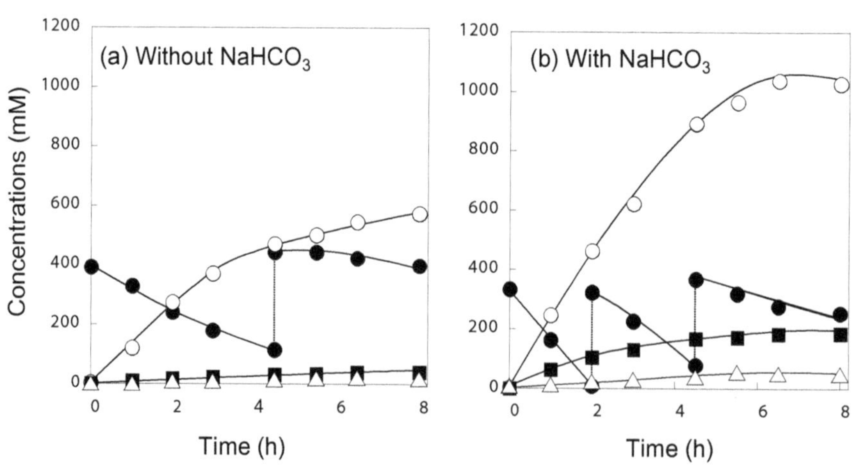

Figure 10.1 Metabolic profiles of *Corynebacterium glutamicum* in the absence (a) and presence (b) of sodium bicarbonate under oxygen deprivation. Glucose (filled circle), lactic acid (empty circle), succinic acid (square) and acetic acid (triangle) are represented.

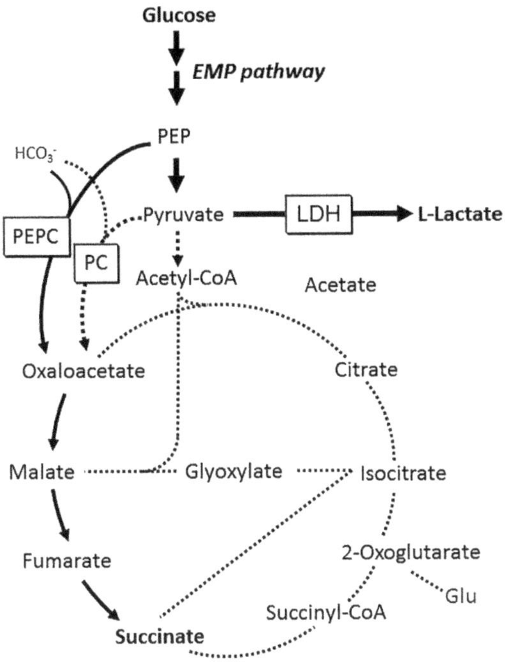

Figure 10.2 Sugar metabolic pathways of *Corynebacterium glutamicum* under oxygen deprivation. Active pathways under the conditions are indicated by solid lines. Enzyme names are blocked. LDH; lactate dehydrogenase, PC; pyruvate decarboxylase, PEPC; phosphoenolpyruvate carboxylase.

genes is up-regulated in excess of 2-fold, whereas that of a further 221 genes is down-regulated in excess of 2-fold compared with their expression levels under aerobic conditions (Inui *et al.*, 2007). These 383 genes include genes encoding glycolytic and TCA cycle enzymes; *gapA, pgk, tpi, ppc, ldhA* and *mdh* are up-regulated while *gltA, acn, icd* and *sucABCD* are down-regulated. The changes in sugar metabolism in response to availability of oxygen are clearly regulated at transcriptional level in *C. glutamicum* in which elaborate endogenous regulation mechanisms to adapt low-oxygen environments must function. Indeed, it was recently confirmed that *C. glutamicum*, long considered an aerobe, does grow anaerobically by nitrate respiration (Nishimura *et al.*, 2007; Takeno *et al.*, 2007). The ability of *C. glutamicum* to adapt its metabolic machinery to survive oxygen deprivation appears to be a survival strategy against environmental changes, considering its natural soil habitat.

Based on the physiological response of *C. glutamicum* to oxygen deprivation, the growth-arrested bioprocess that separates a cell growth phase from a production phase has been developed at RITE. Decoupling growth from production permits maximal utilization of the benefits associated with each of the phases in this growth-arrested bioprocess. The fact is that productivity of conventional fermentation processes is strongly dependent upon the growth rate of the production strain, but optimal conditions for microbial growth may not necessarily be optimal for production of target compounds. Moreover, modern genetic engineering or traditional mutation strategies for improving productivity often impose detrimental effects on growth of production strains, but this is much less problematic in growth-arrested bioprocesses. It is also possible to produce compounds that inhibit growth of microorganisms using the growth-arrested bioprocess, something that would be extremely problematic in conventional bioprocesses. Therefore, engineered strains of C. glutamicum able to produce up to 7.8% (w/v) of ethanol under oxygen deprivation are possible (Sakai *et al.*, 2007) despite microbial growth being strongly inhibited at less than 4% (w/v). Such merits accordingly indicate great promise for the growth-arrested bioprocesses in the production of a wide variety of compounds.

Metabolic engineering *Corynebacterium glutamicum* for production of value-added chemicals via the growth-arrested bioprocess

Wild-type *C. glutamicum* can produce organic acids from sugars under oxygen deprivation, but metabolic engineering extends this capability further. General strategies employed vary from engineering totally new biosynthetic pathways, enhancement of select product pathways to inactivation of competing biosynthetic pathways (Table 10.1). Detailed discussion of these efforts have been documented elsewhere (Blombach and Eikmanns, 2011; Zahoor *et al.*, 2012; Jojima *et al.*, 2013). The focus of this section is to highlight the major challenges that must be surmounted in

Table 10.1 Production of value-added compounds via growth-arrested bioprocess with *Corynebacterium glutamicum*

Strain	Product	Yield (g/g)	Productivity (g/l/h)	Reference
Gly3/pCRD914	Alanine	0.85	4.9	Yamamoto *et al.* (2012)
Iso7	Isobutanol	0.20	0.33	Blombach *et al.* (2011)
C. glutamicum ΔldhA -pCRB204	D-Lactic acid	0.87	4	Okino *et al.* (2008)
	Ethanol	0.27	30[a]	Inui *et al.* (2004a)
BOL-3/pAN6-*gap*	Succinic acid	1.1[b]	2.5	Litsanov *et al.* (2012)
C. glutamicum ΔldhA -pCRA717	Succinic acid	0.9	12	Okino *et al.* (2008)
Val-9	Valine	0.57	6.2	Hasegawa *et al.* (2013)
CtXR7	Xylitol	0.92[c]	7.9	Sasaki *et al.* (2009)

[a]Maximum productivity.
[b]Formate was used as co-substrate.
[c]Our calculation from data in the reference.

order to establish high-performance mainstay *C. glutamicum* strains for the growth-arrested bioprocess.

Importance of redox balance

As discussed above, the reduction of NAD^+ to NADH is an important driver of *C. glutamicum* cellular metabolism. It follows that re-oxidation of NADH formed through glycolysis is essential for efficient sugar catabolism. But oxygen, the most preferred oxidizing compound, is unavailable in the growth-arrested bioprocess. Different metabolites must therefore be consumed to generate the required reducing power. Accordingly, an appropriate balance between oxidative and reductive reactions must be designed into the metabolic pathways in order for high productivity and high yield of the bioprocess under oxygen deprivation to be realized. Additional bicarbonate or pyruvate has been shown to enhance the rate of glucose consumption of wild-type *C. glutamicum*, leading to decreased intracellular $NADH/NAD^+$ ratio. Alternatively, mutant strains with inactivated lactate dehydrogenase (LDH) or PEPC-encoding genes exhibit slower glucose consumption than wild-type strain, retaining higher intracellular $NADH/NAD^+$ ratios than the wild-type strain (Inui *et al.*, 2004b). These observations demonstrate that intracellular redox balance strongly influences productivity of the growth-arrested bioprocess.

The design of an efficient L-valine-producing strain highlights the importance of redox balance in oxygen-deprived production (Hasegawa *et al.*, 2012, 2013). In L-valine biosynthetic pathway by *C. glutamicum*, synthesis of 1 mole of L-valine from glucose generates 2 moles of NADH via glycolysis but consumes 2 moles of NADPH, one at acetohydroxyacid isomeroreductase (AHAIR)-catalysed reaction step and the other at the step where regeneration of L-glutamate as an amino-group donor for the transaminase reaction occurs. Because no transhydrogenase activity is known in *C. glutamicum*, conversion of NADH to NADPH, if possible, is presumed to be inefficient in the growth-arrested bioprocess. By modifying cofactor requirements of the L-valine synthetic pathway so that the pathway relies solely on NADH, this inefficiency is eliminated. This is effected through a change of cofactor dependence of AHAIR to NADH and introduction of leucine dehydrogenase (LeuDH) in place of transaminase B and leads to increased glucose consumption and decreased the $NADH/NAD^+$ ratio under oxygen deprivation. There is clear correlation between glucose consumption rates and intracellular $NADH/NAD^+$ ratios in these mutant strains (Fig. 10.3). These examples demonstrate the importance of redox balance as a means to achieve high productivity under oxygen deprivation.

In the same way decreased intracellular $NADH/NAD^+$ ratio promotes high productivity in the growth-arrested bioprocess, curtailed reducing power that suppresses succinate yield in oxygen-deprived *C. glutamicum* (Litsanov *et*

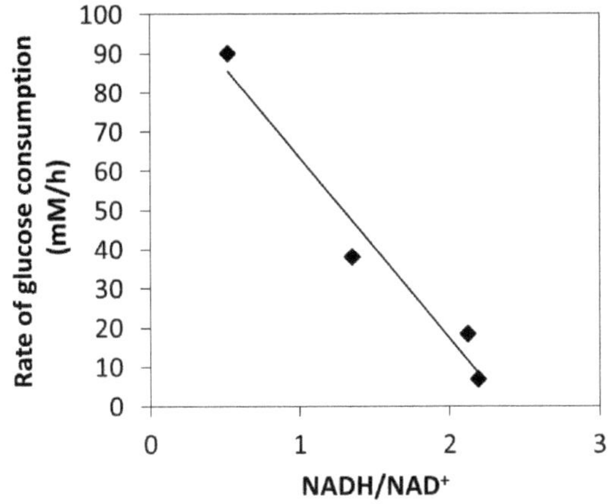

Strain	*ilvBN*	*ilvC*	*ilvC*^{TM}	*ilvD*	*ilvE*	*leudh*
1	+	+		+	+	
2	+	+		+		+
3	+		+		+	
4	+		+	+		+

Figure 10.3 Correlation of glucose consumption rate and intracellular NADH/NAD$^+$ ratio in L-valine-producing *Corynebacterium glutamicum* strains. Numbers indicate recombinant strains where genes shown in the table are overexpressed. The genes for NADH-preferring mutant AHAIR and leucine dehydrogenase are coded by *ilvC*^{TM} and *leudh*, respectively. Data are adapted from Hasegawa *et al*. (2012).

al., 2012). Succinate is produced via the reductive arm of the TCA cycle in *C. glutamicum* under oxygen deprivation where 4 moles of NADH is consumed through the synthesis of 2 moles of succinate from 1 mole of glucose whereas 2 moles of NADH is generated via glycolysis. A solution to this is through engineering the gene encoding formate dehydrogenase into the strain. By so doing, formate can be utilized as co-substrate to fuel reducing equivalents and CO_2 for carboxylation of PEP and pyruvate to form oxaloacetate, resulting in co-utilization of glucose and formate and a 20% increased production of succinate.

Modifications in the glycolytic pathway

To improve the productivity of any given production strain, overexpression of enzyme(s) specific for biosynthesis of target metabolite is a typical strategy. It is in this context that overexpression of glycolytic enzymes appears attractive at a glance, as it is expected to accelerate glucose consumption of the production strain. However, this strategy has met very limited success so far. For instance, a *Saccharomyces cerevisiae* strain overexpressing five glycolytic enzymes (GAPDH, phosphoglycerate mutase, phosphoglycerate kinase, enolase and pyruvate kinase) in tandem with two fermentative enzymes (pyruvate decarboxylase and alcohol dehydrogenase) achieves no increase in ethanol productivity in batch cultivation (Peter Smits *et al*., 2000). Instead all that is realized is accelerated CO_2 formation. Fortunately, recent similar studies using *C. glutamicum* under oxygen deprivation have resulted in significant improvement in glucose consumption. Improved alanine production in a metabolically engineered *C. glutamicum* strain overexpressing GAPDH is anchored in the improved glucose consumption of the strain (Jojima *et al*., 2010). The reason that this approach is effective in this case is that GAPDH activity is the rate-limiting enzyme of oxygen-deprived cells of *C. glutamicum*. The GAPDH-overexpressing strain doubles glucose consumption and alanine production rates when compared with a control strain under oxygen deprivation. Interestingly, the GAPDH overexpression has no effect on the glucose consumption the rate of either strain during aerobic cultivation (Jojima *et al.*, 2010).

In the same vein, chromosomal integration of four genes encoding GAPDH, glucose 6-phosphate isomerase (PGI), 6-phosphofructokinase (PFK) and pyruvate kinase (PYK) results in an alanine-producing recombinant strain that simultaneously overexpresses GAPDH, PFK and PYK and exhibits 13% increased glucose consumption under oxygen deprivation *C. glutamicum* (Yamamoto *et al.*, 2012). Additional integration of the gene PGI further improves glucose consumption. Unexpectedly, alanine yields of both strains are also improved by the overexpression of the glycolytic enzymes, probably by the sheer fact that improvement in alanine productivity is higher than corresponding improvement in production of assorted by-products; genes encoding the by-product formation pathways are controlled via their native promoters whereas that for alanine dehydrogenase responsible for the conversion of pyruvate to alanine is controlled by the stronger *tac* promoter. Consequently enzymes responsible for by-product formation are probably expressed in insufficient quantities to match the rate of enhanced glycolytic flux. Furthermore, change in extracellular metabolite ratios leads to reduced intracellular $NADH/NAD^+$ ratio, which could be also responsible, at least partially, for the enhanced glucose consumption. Again, the enhanced glucose consumption of these recombinant strains is observed only under oxygen-deprived conditions (Yamamoto *et al.*, 2012). The implication here is that glycolytic flux of alanine-producing *C. glutamicum* is primarily controlled only by the EMP pathway under oxygen deprivation, while other metabolic pathways such as the pentose phosphate pathway, the TCA cycle and other metabolic pathways functioning under aerobic conditions may also contribute to the control on glucose consumption under aerobic (cell growing) conditions. Although detailed mechanisms for control of glycolytic flux in oxygen-deprived cells of *C. glutamicum* are not clear, it is obvious that overexpression of glycolytic enzymes seems to be useful in metabolic engineering for the growth-arrested bioprocess with *C. glutamicum*.

Application of the growth-arrested bioprocess to lignocellulosic biorefinery

Concerns about global climate change and energy security necessitate a variety of efforts to develop an integrated bioprocess, so-called biorefinery, to produce commodity chemicals and fuels from biomass. These bioproducts must be produced in a sustainable manner, meaning that non-food-based biomass resources including lignocellulosic biomass are expected to be leading candidate feedstocks of large-scale biorefineries. Growth-arrested bioprocesses have the potential to bring to fruition such lignocellulosic biorefineries, as will be described below.

Enhanced efficiency of sugar substrate utilization

Lignocellulosic biomass primarily consists of cellulose, hemicellulose and lignin. Typically, the glucose homopolymer cellulose accounts for 35–50% of lignocellulosic biomass and is more readily susceptible to biological degradation than the 15–25% that comprises hemicellulose, a heteropolymer composed of hexoses and pentoses. For efficient substrate utilization, it is necessary that any biocatalyst used in any lignocellulosic biorefinery process is able to produce value-added chemicals at a high yield and with high productivity from both pentose and hexose sugars. Unfortunately, no known microorganism able to naturally degrade these different sugars with any semblance of acceptable production efficiency has been isolated.

Typically, therefore, genes encoding absent metabolic pathways must be introduced into the genome of a microorganism to extend the substrate spectrum. This strategy has been employed to develop recombinant *C. glutamicum* strains that utilize xylose (Kawaguchi *et al.*, 2006), arabinose (Kawaguchi *et al.*, 2008), galactose (Barrett *et al.*, 2004), cellobiose (Kotrba *et al.*, 2003) and glycerol (Rittmann *et al.*, 2008). These efforts have met only moderate success in extending the substrate spectrum of *C. glutamicum*, given that the recombinant strains so developed retain the characteristic preferential utilization of glucose when present in a mixture of 'non-native' sugars that are cannot be consumed by the wild-type strain. Of particular concern is the fact that the rate of xylose consumption by recombinant *C. glutamicum* is lower than that of glucose under oxygen deprived conditions even after extensive efforts to optimize the expression levels of genes encoding xylose metabolic enzymes (Sasaki *et al.*, 2008). Clearly, laboratory optimization of gene expression presents big obstacles. However, the discovery that *C. glutamicum* ATCC31831 does utilize arabinose exceptionally well in the presence of glucose provided important clues that have facilitated the solution of the problem of slow consumption of xylose and arabinose in mixed sugar substrates (Kawaguchi *et al.*, 2009). The gene cluster for arabinose metabolism in strain ATCC31831 contains a gene encoding the sugar transporter (*araE*) in addition to an operon encoding the metabolic enzymes (*araBAD*). Transformation of a pentose-utilizing strain of *C. glutamicum* R with *araE* accelerates the rates of both arabinose and xylose consumption under aerobic as well as oxygen-deprived conditions. In a mixture containing glucose and either xylose or arabinose, the recombinant strain transformed with *araE* simultaneously consumes the pentoses and glucose at equivalent rates (Sasaki *et al.*, 2009), indicating the effectiveness of the pentose sugar transporter engineering on improved of sugar utilization.

Biocatalyst tolerance to inhibitory compounds in lignocellulose hydrolysates

Prior to enzymatic hydrolysis of lignocellulosic biomass to produce fermentable sugars, the biomass must necessarily be subjected to physical and/or thermochemical pre-treatment to improve accessibility of constituent sugars to saccharifying enzymes. This pre-treatment process is crucial in order to minimize the dose of the very expensive saccharifying enzymes. Unfortunately the pre-treatment process produces undesirable by-products that inhibit consequent saccharification and fermentation steps (Klinke *et al.*, 2004). These inhibitory compounds are extensively known to reduce process productivity and yield of, in particular, ethanol production. Thus, it is highly desirable that biocatalysts used in the lignocellulosic biorefineries exhibit as much tolerance to these inhibitory compounds as possible. Similar to most bacteria, aerobic growth of *C. glutamicum* is inhibited by furans (furfural and hydroxymethylfurfural) and acetic acid, the typical inhibitory compounds present in lignocellulose hydrolysate (Sakai *et al.*, 2007; Rumbold *et al.*, 2009). This suggests that productivity of any process utilizing *C. glutamicum* will be adversely affected if the microorganism is used under aerobic conditions to convert hydrolysate. Fortunately, the sugar metabolism of oxygen-deprived cells of *C. glutamicum* shows high tolerance to individual inhibitory compounds or mixtures thereof (Sakai *et al.*, 2007). This is most probably attributable to the prevailing growth-arrested conditions. This has permitted use recombinant *C. glutamicum* to efficiently produce ethanol from a corn stover hydrolysate without necessitating any detoxification of the substrate (our unpublished results), demonstrating the merits of the growth-arrested bioprocess as the basis of lignocellulosic biorefineries.

Future trends

During most of the last decade, considerable successes have been realized in development of the growth-arrested bioprocess and providing proof-of-concept of its applicability in lignocellulosic biorefineries. Enough data are available to move to the next stage of actually industrializing the process. The challenge going forward is to develop of cost-effective methods for preparation of biocatalyst (i.e. cells). Since oxygen-deprived cells of *C. glutamicum* retain metabolic activity for prolonged durations, either repeated batch processes with cell-recycling or continuous production can be easily contemplated. Either of these production modes is possible with minimal costs of cell preparation per batch. Therefore, investigating the mechanisms underlying prolonged maintenance of metabolic activity under oxygen deprivation should be important in order to provide further insights into extending the life of the biocatalyst.

The challenge of better understanding of control mechanisms of glycolytic flux under oxygen deprivation remains. The fact that the productivity of the bioprocess remains lower than that of chemical processes hinders potential industrial application of the bioprocess. Further deciphering of the molecular mechanism of accelerated glucose consumption by overexpression of glycolytic enzymes should open a new avenue to improve productivity of the growth-arrested bioprocess.

Acknowledgement

We thank Crispinus A. Omumasaba (RITE) for critical reading of the manuscript.

References

Barrett, E., Stanton, C., Zelder, O., Fitzgerald, G., and Ross, R. P. (2004) Heterologous expression of lactose- and galactose-utilizing pathways from lactic acid bacteria in *Corynebacterium glutamicum* for production of lysine in whey. *Appl. Environ. Microbiol.* **70**, 2861–2866.

Blombach, B., and Eikmanns, B. J. (2011) Current knowledge on isobutanol production with *Escherichia coli*, *Bacillus subtilis* and *Corynebacterium glutamicum*. *Bioeng. Bugs* **2**, 346–350.

Dominguez, H., Nezondet, C., Lindley, N. D., and Cocaign, M. (1993) Modified carbon flux during oxygen limited growth of *Corynebacterium glutamicum* and the consequences for amino acid overproduction. *Biotech. Lett.* **15**, 449–454.

Hasegawa, S., Uematsu, K., Natsuma, Y., Suda, M., Hiraga, K., Jojima, T., Inui, M., and Yukawa, H. (2012) Improvement of the redox balance increases L-valine production by *Corynebacterium glutamicum* under oxygen deprivation. *Appl. Environ. Microbiol.* **78**, 865–875.

Hasegawa, S., Suda, M., Uematsu, K., Natsuma, Y., Hiraga, K., Jojima, T., Inui, M., and Yukawa, H. (2013) Engineering of *Corynebacterium glutamicum* for high yield L-valine production under oxygen deprivation conditions. *Appl. Environ. Microbiol.* **79**, 1250–1257.

Ikeda, M. (2003) Amino acid production processes. *Adv. Biochem. Eng. Biotechnol.* **79**, 1–35.

Inui, M., Kawaguchi, H., Murakami, S., Vertès, A. A., and Yukawa, H. (2004a) Metabolic engineering of *Corynebacterium glutamicum* for fuel ethanol production under oxygen-deprivation conditions. *J. Mol. Microbiol. Biotechnol.* **8**, 243–254.

Inui, M., Murakami, S., Okino, S., Kawaguchi, H., Vertès, A. A., and Yukawa, H. (2004b) Metabolic analysis of *Corynebacterium glutamicum* during lactate and succinate productions under oxygen deprivation conditions. *J. Mol. Microbiol. Biotechnol.* 7, 182–196.

Inui, M., Suda, M., Okino, S., Nonaka, H., Puskas, L. G., Vertès, A. A., and Yukawa, H. (2007) Transcriptional profiling of *Corynebacterium glutamicum* metabolism during organic acid production under oxygen deprivation conditions. *Microbiology* **153**, 2491–2504.

Jojima, T., Fujii, M., Mori, E., Inui, M., and Yukawa, H. (2010) Engineering of sugar metabolism of *Corynebacterium glutamicum* for production of amino acid L-alanine under oxygen deprivation. *Appl. Microbiol. Biotechnol.* **87**, 159–165.

Jojima, T., Inui, M., and Yukawa, H. (2013) Biorefinery applications of *Corynebacterium glutamicum*. In: Yukawa, H., and Inui, M. (eds.) *Corynebacterium glutamicum*. Springer, Berlin Heidelberg, Germany, pp. 149–172.

Kawaguchi, H., Vertès, A. A., Okino, S., Inui, M., and Yukawa, H. (2006) Engineering of a xylose metabolic pathway in *Corynebacterium glutamicum*. *Appl. Environ. Microbiol.* **72**, 3418–3428.

Kawaguchi, H., Sasaki, M., Vertès, A. A., Inui, M., and Yukawa, H. (2008) Engineering of an L-arabinose metabolic pathway in *Corynebacterium glutamicum*. *Appl. Microbiol. Biotechnol.* **77**, 1053–1062.

Kawaguchi, H., Sasaki, M., Vertès, A. A., Inui, M., and Yukawa, H. (2009) Identification and functional analysis of the gene cluster for L-arabinose utilization in *Corynebacterium glutamicum*. *Appl. Environ. Microbiol.* **75**, 3419–3429.

Klinke, H. B., Thomsen, A. B., and Ahring, B. K. (2004) Inhibition of ethanol-producing yeast and bacteria by degradation products produced during pre-treatment of biomass. *Appl. Microbiol. Biotechnol.* **66**, 10–26.

Kotrba, P., Inui, M., and Yukawa, H. (2003) A single V317A or V317M substitution in Enzyme II of a newly identified beta-glucoside phosphotransferase and utilization system of *Corynebacterium glutamicum* R extends its specificity towards cellobiose. *Microbiology* **149**, 1569–1580.

Litsanov, B., Kabus, A., Brocker, M., and Bott, M. (2012) Efficient aerobic succinate production from glucose in minimal medium with *Corynebacterium glutamicum*. *Microb. Biotechnol.* **5**, 116–128.

Nishimura, T., Vertès, A. A., Shinoda, Y., Inui, M., and Yukawa, H. (2007) Anaerobic growth of *Corynebacterium glutamicum* using nitrate as a terminal electron acceptor. *Appl. Microbiol. Biotechnol.* **75**, 889–897.

Okino, S., Inui, M., and Yukawa, H. (2005) Production of organic acids by *Corynebacterium glutamicum* under oxygen deprivation. *Appl. Microbiol. Biotechnol.* **68**, 475–480.

Park, S., Shaw-Reid, C., Sinskey, A., and Stephanopoulos, G. (1997) Elucidation of anaplerotic pathways in *Corynebacterium glutamicum* via ^{13}C-NMR spectroscopy and GC-MS. *Appl. Microbiol. Biotechnol.* **47**, 430–440.

Peter Smits, H., Hauf, J., Müller, S., Hobley, T. J., Zimmermann, F. K., Hahn-Hägerdal, B., Nielsen, J., and Olsson, L. (2000) Simultaneous overexpression of enzymes of the lower part of glycolysis can enhance the fermentative capacity of *Saccharomyces cerevisiae*. *Yeast* **16**, 1325–1334.

Petersen, S., de Graaf, A. A., Eggeling, L., Möllney, M., Wiechert, W., and Sahm, H. (2000) *In vivo* quantification of parallel and bidirectional fluxes in the anaplerosis of *Corynebacterium glutamicum*. *J. Biol. Chem.* **275**, 35932–35941.

Rittmann, D., Lindner, S. N., and Wendisch, V. F. (2008) Engineering of a glycerol utilization pathway for amino acid production by *Corynebacterium glutamicum*. *Appl. Environ. Microbiol.* **74**, 6216–6222.

Rumbold, K., van Buijsen, H. J., Overkamp, K. M., van Groenestijn, J. W., Punt, P. J., and van der Werf, M. J. (2009) Microbial production host selection for converting second-generation feedstocks into bioproducts. *Microb. Cell Fact.* **8**, 64.

Sakai, S., Tsuchida, Y., Nakamoto, H., Okino, S., Ichihashi, O., Kawaguchi, H., Watanabe, T., Inui, M., and Yukawa, H. (2007) Effect of lignocellulose-derived inhibitors on growth of and ethanol production by growth-arrested *Corynebacterium glutamicum* R. *Appl. Environ. Microbiol.* **73**, 2349–2353.

Sasaki, M., Jojima, T., Inui, M., and Yukawa, H. (2008) Simultaneous utilization of D-cellobiose, D-glucose, and D-xylose by recombinant *Corynebacterium glutamicum* under oxygen-deprived conditions. *Appl. Microbiol. Biotechnol.* **81**, 691–699.

Sasaki, M., Jojima, T., Kawaguchi, H., Inui, M., and Yukawa, H. (2009) Engineering of pentose transport in *Corynebacterium glutamicum* to improve simultaneous utilization of mixed sugars. *Appl. Microbiol. Biotechnol.* **85**, 105–115.

Takeno, S., Ohnishi, J., Komatsu, T., Masaki, T., Sen, K., and Ikeda, M. (2007) Anaerobic growth and potential for amino acid production by nitrate respiration in *Corynebacterium glutamicum*. *Appl. Microbiol. Biotechnol.* **75**, 1173–1182.

Terasawa, M., Kakinuma, N., Shikata, K., and Yukawa, H. (1989) New process for L-isoleucine production. *Process Biochem.* **24**, 60–61.

Yamamoto, S., Gunji, W., Suzuki, H., Toda, H., Suda, M., Jojima, T., Inui, M., and Yukawa, H. (2012) Overexpression of genes encoding glycolytic enzymes in *Corynebacterium glutamicum* enhances glucose metabolism and alanine production under oxygen deprivation conditions. *Appl. Environ. Microbiol.* **78**, 4447–4457.

Zahoor, A., Lindner, S. N., and Wendisch, V. F. (2012) Metabolic engineering of *Corynebacterium glutamicum* aimed at alternative carbon sources and new products. *Comp. Struct. Biotechnol. J.* **3**, e201210004.

Corynebacterium glutamicum as a Platform Organism for the Secretory Production of Heterologous Proteins

11

Roland Freudl

Abstract

Corynebacterium glutamicum is widely used as an industrial workhorse for the production of amino acids and other low molecular weight compounds. However, an increasing amount of evidence indicates that this microorganism likewise might possess a great potential as an alternative platform organism for the secretory production of heterologous proteins. Besides giving an overview on the major protein secretion routes present in *C. glutamicum*, i.e. the general secretion (Sec) system and the twin-arginine translocation (Tat) system, this review summarizes the attempts that have been made so far to use *C. glutamicum* as a host system for the expression and subsequent secretion of foreign proteins. In addition, this review also addresses recent progress with respect to producer strain improvement and process optimization.

Introduction

The secretory production of biotechnologically relevant recombinant proteins possesses several important advantages over intracellular expression strategies. For example, the secretion of some proteins into the growth medium of an expression host can prevent their accumulation in the cytoplasm as insoluble inclusion bodies. Furthermore, secretion can reduce or alleviate toxic effects that some proteins exert on the production host. In addition, some proteins require the formation of disulfide bonds in order to reach their active state, a critical step in the folding process that is prevented in the reducing environment of the cytoplasm. In contrast, export of the respective proteins out the cytoplasm into an oxidizing compartment such as the periplasm of Gram-negative bacteria or the cell wall area of Gram-positive bacteria allows the formation of disulfide bonds which, in most organisms, is assisted by disulfide oxidoreductases that are present in the respective extracytosolic locations. But most importantly, the secretion of recombinant proteins into the culture medium of the production host greatly facilitates the downstream processing and therefore can significantly reduce the production costs of a desired target protein (Quax, 1997).

Gram-positive bacteria are considered especially interesting host organisms for the secretory production of proteins since they usually lack an outer membrane and proteins that are exported across the cytoplasmic membrane are directly released into the external environment. For this reason, various organisms of this group of bacteria (mainly various *Bacillus* species) are used since many years in the secretory production of technical enzymes such as proteases, lipases, or amylases, resulting in yields of more than 20 g/l culture medium (Harwood and Cranenburgh, 2008). However, such high yields have been obtained almost exclusively with naturally secreted homologous enzymes that originate either from the production host itself or from one of its close relatives. The situation very often changes drastically in a negative way when the secretion of heterologous proteins is attempted in the very same expression hosts, resulting either in disappointingly low yields or in no secretion of the target protein at all. For this, several bottlenecks in the secretory pathway have been observed that, alone or in combination, can drastically reduce the amounts of the desired target protein in the culture supernatant. Frequently observed problems are (1) an inefficient translocation of the target protein across the cytoplasmic membrane, (2) an inefficient or wrong folding of the target protein on the *trans*-side of the cytoplasmic membrane or (3) the degradation of the target protein by cell wall-associated and/or secreted host proteases (Li *et al.*, 2004). For example, *Bacillus subtilis* possesses at least seven secreted and five cell wall-associated proteases that can attack the desired heterologous target proteins as soon as they emerge on the *trans*-side of the cytoplasmic membrane (Westers *et al.*, 2004). Since the demand for recombinant proteins is steadily increasing, there is an urgent need to explore alternative host systems with respect to their ability to express and secrete heterologous proteins of biotechnological interest that cannot efficiently be produced by standard host organisms such as the commonly used *B. subtilis*.

So far, *C. glutamicum* has mainly been used as an industrial workhorse for the production of amino acids and other low-molecular weight compound (Leuchtenberger *et al.*, 2005; Becker and Wittmann, 2012). However, as summarized in this review, an increasing number of reports indicate that *C. glutamicum* also possesses a great potential as an alternative host for the secretory expression of foreign proteins.

Protein export systems of *Corynebacterium glutamicum*

The first step during secretion of a desired protein into the growth medium is its transport across the cytoplasmic membrane. In *C. glutamicum*, two main export pathways (the general secretion or Sec pathway and the twin-arginine translocation or Tat pathway) exist for the transport of proteins across the plasma membrane that fundamentally differ with respect to the folding status of the substrate proteins during the actual translocation event.

The general secretion (Sec) protein export pathway

The major route for exporting proteins across the bacterial cytoplasmic membrane is the general secretion (Sec) pathway which transports its substrates in a more or less unfolded conformation. This pathway has been extensively studied in the Gram-negative model bacterium *Escherichia coli* and, to a somewhat lesser extent, also in *B. subtilis*, the latter being considered as the model organism for the Gram-positive group of bacteria (Yuan *et al.*, 2010; Chatzi *et al.*, 2013; Freudl, 2013).

Most Sec substrates that are translocated across the plasma membrane are initially synthesized as precursors with an amino-terminal signal peptide, which is cleaved off from the mature protein during or shortly after membrane translocation. Sec signal peptides possess a tripartite overall structure comprising a positively charged amino-terminal region (n-domain), a hydrophobic core (h-domain) and a polar carboxyl-terminal region (c-domain) that contains the signal peptidase recognition site (Rusch and Kendall, 2007). Signal peptides of Sec-dependent lipoproteins possess a so-called lipobox motif (L^{-3}-$(A/S)^{-2}$-$(A/G)^{-1}\downarrow C^{+1}$) including a highly conserved cysteine residue to which the lipid moiety is attached (von Heijne, 1989). A survey of genome of the *C. glutamicum* R strain using the signal peptide prediction program SignalP 3.0 (Bendtsen *et al.*, 2004) revealed the presence of 405 putative signal peptides in this organism (Watanabe *et al.*, 2009). Subsequently, each signal peptide was fused to a signal peptide-less α-amylase from *Geobacillus stearothermophilus* and the resulting hybrid precursor proteins were expressed in *C. glutamicum* R. Secretion of the α-amylase reporter protein by the respective recombinant strains was assayed *in situ* on iodine-stained starch-containing agar plates. Of the 405 signal peptides tested, 108 showed significant export of the amylase reporter in this plate assay. Further classification of these 108 signal peptides showed that 98 (including 8 signal peptides containing a putative lipobox motif) very likely mediated amylase export via the Sec pathway, since amylase secretion driven by these signal peptides was unaffected in a *C. glutamicum* R Δ*tatC* mutant strain (Watanabe *et al.*, 2009 and see below). However, it should be kept in mind that the export pathway selectivity (i.e. Sec versus Tat) of a given signal peptide is significantly influenced by the fused mature protein (Tullman-Ercek *et al.*, 2007) and, for this reason, a detailed analysis of the unaltered bona fide full-length precursors would be required to allow a final signal peptide classification in *C. glutamicum*.

Sec-dependent protein translocation in bacteria can occur in either a co- or a post-translational way. In the co-translational mode, Sec substrates possessing highly hydrophobic signal peptides are recognized already during their synthesis at the ribosome by a signal recognition particle (SRP) consisting of the Ffh protein and a small cytoplasmic RNA (scRNA) that tightly binds to these hydrophobic signals. The SRP-ribosome-nascent chain (RNC) complex subsequently docks to a Sec translocon-bound SRP receptor (FtsY). In the next step, the RNC is then further transferred to the translocation pore (Saraogi and Shan, 2011). Like in all other bacteria, the components of the SRP protein targeting pathway are present also in *C. glutamicum*. In contrast to the situation in *E. coli* and other Proteobacteria, Gram-positive bacteria including *C. glutamicum* generally lack the targeting chaperone SecB that, in the post-translational mode, interacts with newly synthesized precursor proteins, maintains their unfolded, translocation-competent state and targets them to the SecA component of the Sec translocase (Bechtluft *et al.*, 2010). So far, it is not clear which proteins functionally substitute for SecB in Gram-positive bacteria and it has been proposed that general chaperones GroEL–GroES/DnaK–DnaJ–GrpE/trigger factor (Lyon *et al.*, 1998; Wu *et al.*, 1998) or a soluble form of SecA (Herbort *et al.*, 1999) might be involved in maintaining the export competence of fully synthesized Sec-dependent precursor proteins.

Subsequently to the targeting step, the Sec-dependent precursor proteins are handed over to the Sec translocase in the membrane which consists of the peripheral membrane protein SecA, the membrane-integral core components SecY, SecE, SecG, and the membrane-integral accessory components SecD and SecF. SecYEG constitutes the actual translocation channel that provides a path for proteins that are translocated to the *trans*-side of the membrane. In addition, SecYEG also mediates the insertion of integral membrane proteins containing hydrophobic transmembrane segments that are moved into the lipid bilayer through a lateral gate of the translocation pore (Dalbey *et al.*, 2011). The energy for the actual translocation in the co-translational pathway is provided by further elongation of the substrate at the ribosome (du Plessis *et al.*, 2011). In the post-translational mode of export, translocation is driven by the motor protein SecA. The translocation ATPase SecA interacts with the SecYEG pore and, after delivery of a respective substrate protein, the substrate is pushed by SecA through the SecYEG pore in a stepwise and ATP-dependent manner (Lycklama a Nijeholt and Driessen, 2012). Furthermore, a complex of the accessory factors SecD and SecF has recently been shown to exert a proton-motive force-dependent pulling force on the substrate from the extracytosolic side of the membrane (Tsukazaki *et al.*, 2011). During or shortly after membrane translocation, the signal peptide is cleaved off from the Sec precursors by a type I signal peptidase (LepB), allowing the release of the respective mature proteins on the *trans*-side of the membrane (Auclair *et al.*, 2012). In the case of lipoproteins, signal peptide processing is performed by type-II signal peptidase (LspA) (Hayashi and Wu, 1990). Although the mechanism of Sec-dependent protein export as outlined above has been elucidated almost exclusively in *E. coli*, the presence of all relevant components (except SecB) in *C. glutamicum* (Fig. 11.1) strongly suggests that both Sec translocases function in a very similar way.

Besides the highly conserved Sec export pathway, an accessory Sec pathway exists in *C. glutamicum*. Like some other (mostly pathogenic) Gram-positive bacteria such as *Mycobacterium* species, *C. glutamicum* possesses a second SecA protein, termed SecA2 (Caspers and Freudl, 2008; Feltcher and Braunstein, 2012). The *C. glutamicum* SecA2 is involved in the export of most likely only a limited subset of substrates, the nature of which is unknown so far. Nevertheless, the finding that both SecA proteins are essential in *C. glutamicum* indicates that at least one protein, whose extracytoplasmic location is required for viability, cannot be translocated across the plasma membrane by the SecA1-driven housekeeping Sec system (Caspers and Freudl, 2008).

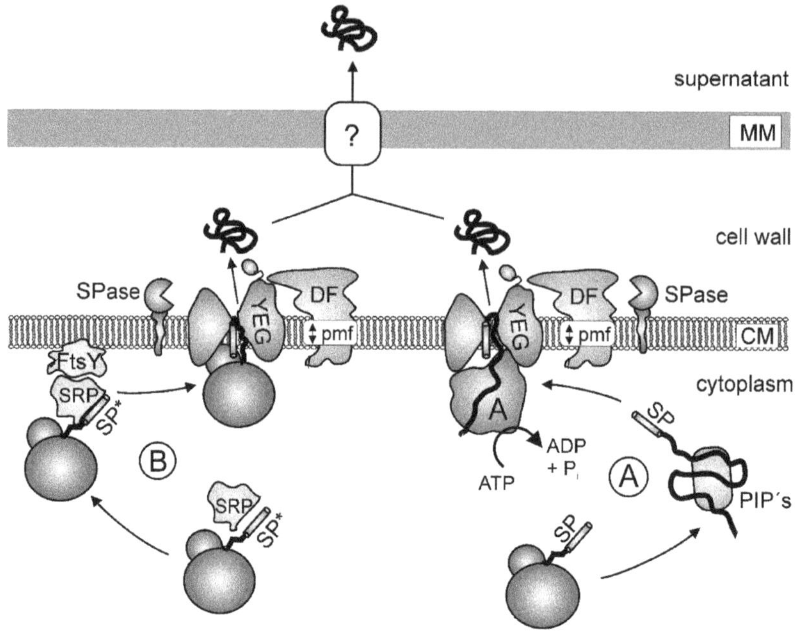

Figure 11.1 The general secretion (Sec) protein export pathway of *Corynebacterium glutamicum*. (A) In the post-translational mode, Sec-dependent precursor proteins are kept in an export-competent state by post-translationally interacting proteins (PIPs) such as the general chaperones GroELS/DnaK-DnaJ-GrpE/trigger factor or the soluble form of SecA. The signal peptide (SP) is recognized by the SecA protein which pushes the protein through the SecYEG protein conducting channel in a stepwise and ATP-dependent manner. In addition, SecDF exerts a proton motive force (pmf)-dependent pulling force on the substrate from the *trans*-side of the cytoplasmic membrane (CM). During or shortly after translocation, the signal peptide is removed by signal peptidase (SPase) and the mature protein is released on the *trans*-side of the CM. (B) In the co-translational mode, Sec substrates possessing highly hydrophobic signal peptides (SP*) are recognized at the ribosome by the signal recognition particle (SRP). Subsequently, the ribosome-nascent chain (RNC)–SRP complex docks to the SRP-receptor FtsY and the RNC is then further transferred to the SecYEG translocation pore such that ribosomal exit site is in close proximity to SecYEG. The energy for translocation in the co-translational export mode is provided by further elongation of the substrate at the ribosome. So far, it is completely unclear how proteins, after their Sec-dependent translocation across the CM, are transported across the outer mycomembrane (MM) of *C. glutamicum* and, so far, no components of a dedicated mycomembrane protein translocation system (if existent at all) have been identified.

The twin-arginine translocation (Tat) export pathway

In many (but not all) bacteria, another major protein translocation pathway exists that, in contrast to the Sec pathway, has the remarkable ability to export completely folded protein across the cytoplasmic membrane. This pathway has been designated twin-arginine translocation (Tat) pathway due to the fact that a characteristic amino acid motif, including two consecutive arginine residues, can be identified in the signal peptides of the respective precursor proteins (Berks, 1996). Many of the Tat substrates are proteins that have to recruit a cofactor in the cytoplasm and as a prerequisite have to acquire a folded status prior to export. In addition, also cofactor-less proteins are exported via the Tat route, presumably because their rapid folding kinetics precludes transport via the Sec pathway (Palmer and Berks, 2012). Both Sec and Tat signal peptides possess a similar tripartite structure consisting of an amino-terminal n-domain, a central hydrophobic h-domain, and a carboxyl-terminal c-domain. However, in Tat signal peptides a conserved consensus motif (S/T^{-1}-**R**-**R**-X^{+1}-F^{+2}-L^{+3}K^{+4}) is located at the boundary between the n- and h-domains. The amino acid residues of the Tat consensus motif are required for the productive recognition and binding of Tat precursor proteins by the Tat translocase and their importance for the membrane translocation of Tat-dependent precursors has been analysed in numerous mutagenesis studies in *E. coli* and *B. subtilis* (e.g. Mendel *et al.*, 2008; Lausberg *et al.*, 2012). A computer analysis of the genome of *C. glutamicum* using the Tat signal peptide prediction program TATFIND (version 1.2) identified 15 putative Tat substrates in this organism (Schaerlaekens *et al.*, 2004). In addition, several signal peptides that do not possess a recognizable Tat consensus motif have been claimed to confer Tat-dependent export of an α-amylase reporter protein in *C. glutamicum* (Watanabe *et al.*, 2009), a finding which is somewhat at odds with the present view of precursor recognition by the Tat translocase and requires further investigations.

In *E. coli* and other Gram-negative bacteria, a functional unit of a Tat translocase consists of three integral membrane proteins, i.e. TatA, TatB, and TatC. Another Tat protein, TatE, is a paralogue of TatA that is present in some, but not all Gram-negative bacteria and, at least in *E. coli*, the *tatE* gene is expressed at a level 50–200 times lower than *tatA* (Jack *et al.*, 2001). Therefore, *tatE* is commonly regarded as a cryptic gene duplication of *tatA*. A complex of TatB and TatC (the TatBC receptor complex) recognizes and binds the signal peptides of Tat substrates by virtue of their Tat consensus motifs (Alami *et al.*, 2003; Lausberg *et al.*, 2012). Besides acting as the primary substrate receptor, TatC also possesses an insertase activity that positions the signal peptide and possibly also some amino acids of the early mature region of Tat precursor protein into a membrane-embedded, advanced stage binding pocket that jointly is formed by TatB and TatC (Fröbel *et al.*, 2012; Lausberg *et al.*, 2012). Next, homo-oligomeric complexes of TatA are recruited to the TatBC–substrate complex in a proton motive force-dependent manner (Mori and Cline, 2002). Most current models propose that differently sized TatA complexes form protein conducting channels that are fitted to the sizes and shapes of the different Tat substrates and through which the folded proteins are transferred across the membrane (Gohlke *et al.*, 2005). An alternative model proposes that the association of TatA multimers with the substrate-loaded TatBC complex results in a punctual weakening of the membrane, thereby allowing substrate translocation without the need of a translocation pore (Brüser and Sanders, 2003). Following this still poorly understood translocation step, the Tat signal peptides are cleaved off from their corresponding mature regions by signal peptidase and the mature proteins are released on the *trans*-side of the membrane. A deviation of this scenario is observed in Gram-positive bacteria with low-GC content genomes (such as *B. subtilis*). In these organisms, no TatB is present and Tat-dependent protein translocation is mediated by so-called TatAC-type minimal translocases in which the TatB function is carried out by a bifunctional TatA subunit (Blaudeck *et al.*, 2005; Barnett *et al.*, 2008).

Like *E. coli*, *C. glutamicum* and other Gram-positive bacteria with high-GC content genomes possess a TatABC(E)-type translocase (Fig. 11.2) (Kikuchi *et al.*, 2006). Inactivation of the *C. glutamicum* Tat translocase by deleting *tatA* or *tatC* prevented export of a reporter protein consisting of the signal peptide of the *E. coli* Tat substrate trimethylamine *N*-oxide reductase (TorA) fused to green fluorescence protein (GFP) and caused a severe growth defect of the corresponding mutant strains. Although not yet experimentally analysed so far in *C. glutamicum*, recent results from *Streptomyces coelicolor* (Keller *et al.*, 2012) strongly suggest that a major reason for the poor growth of *C. glutamicum tat* mutants might be the mislocalization of the membrane-integral Rieske protein QcrA. QcrA is an important component of the bc_1 aa_3 supercomplex of the aerobic respiratory chain (Niebisch and Bott, 2003) and, based on the situation in *S. coelicolor* (Keller *et al.*, 2012), very likely also in *C. glutamicum* depends on a functional Tat system for the membrane translocation of its iron–sulfur cluster domain and thereby its correct integration into the cytoplasmic membrane. Whereas the chromosomally encoded TatE was not sufficient to compensate for the lack of TatA in a *C. glutamicum ΔtatA* mutant strain, the overexpression of plasmid-encoded TatE in the *ΔtatA* mutant partially relieved the export defect of TorA-GFP showing that, also in *C. glutamicum*, TatA and TatE are functionally redundant and, furthermore, that *tatE* most likely is a cryptic gene as it is in *E.*

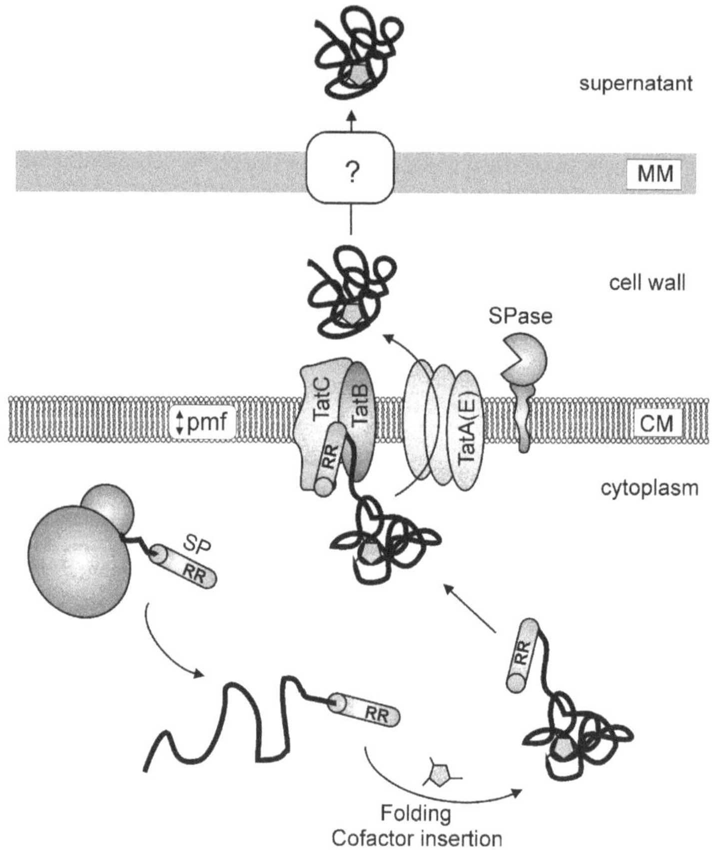

Figure 11.2 The twin-arginine translocation (Tat) protein export pathway of *Corynebacterium glutamicum*. After folding and, if required (as shown here), cofactor insertion, preproteins containing a signal peptide (SP) with a twin-arginine motif (RR) are recognized by a receptor complex consisting of TatC and TatB. Subsequently, homo-oligomeric complexes of TatA(E) are recruited to the substrate-loaded receptor complex in a proton motive force (pmf)-dependent manner, followed by the translocation of the substrate across the cytoplasmic membrane (CM). Most current models propose that TatA(E) multimers might form a substrate-fitted protein conducting channel. In an alternative view, the TatA(E) multimers might lead to a punctual weakening of the membrane and thereby allow a TatC-driven translocation of the substrate without the need for a translocation pore. Following to substrate translocation, the signal peptide is removed by signal peptidase (SPase) and the mature protein is released on the *trans*-side of the membrane. So far, it is completely unclear how proteins, after their Tat-dependent translocation across the CM, are transported across the outer mycomembrane (MM) of *C. glutamicum* and, so far, no components of a dedicated mycomembrane protein translocation system (if existent at all) have been identified.

coli. Surprisingly, the deletion of *tatB* had only a moderate negative effect on growth and allowed a significant level of Tat-dependent export (i.e. 9% compared with a Tat wild-type strain) of the TorA-GFP reporter, indicating that TatB might not be absolutely essential in *C. glutamicum* in a mechanistic sense, but rather contributes to the efficiency of Tat transport in this organism (Kikuchi *et al.*, 2006).

Secretory production of heterologous proteins by *Corynebacterium glutamicum*

C. glutamicum possesses only very low extracellular proteolytic activities that potentially can diminish the yields of secreted heterologous proteins, an important feature that sets this microorganism apart from many other commonly used secretory expression hosts, such as *B. subtilis*. Besides this, other features exist that make *C. glutamicum* a very interesting alternative choice when an industrial secretory production of heterologous proteins is envisaged, such as its GRAS (generally regarded as safe) status, its long established tradition in industry, and the wealth of knowledge that has been generated over the years with respect to its physiology, its molecular cell biology, its systems biology, and very importantly, its handling in large scale industrial fermentation processes. Surprisingly, however, only a very limited number of reports exist in the literature describing the use of *C. glutamicum* as a host organism for the expression and secretion of heterologous protein (Table 11.1 and below). Although the earliest report on the successful secretion of a foreign protein dates back as far as 1986, only recently some more systematic approaches have been undertaken to explore and expand the potential of *C. glutamicum* also as a protein production platform organism.

Export via the Sec pathway

In 1986, Smith and co-workers were the first that reported on the expression of a heterologous secretory protein in *C. glutamicum*. Introduction of a plasmid encoding an α-amylase from *Bacillus amyloliquefaciens* into *C. glutamicum* resulted in the formation of clearing zones around the colonies of the recombinant strain on iodine-stained, starch-containing agar plates. Although not analysed in any further depth, these results already strongly suggested that *C. glutamicum* is able to mediate the secretion of a foreign protein into its growth medium (Smith *et al.*, 1986). Next, Paradis and co-workers described the successful secretion an exoglucanase (Cex) and an endoglucanase (CenA) from the cellulolytic Gram-positive bacterium *Cellulomonas fimi* into the supernatant of *C. glutamicum*. Upon expression of the respective genes in *C. glutamicum*, the corresponding enzyme activities were almost exclusively found in the culture supernatant (Paradis *et al.*, 1987). Expression of an α-amylase from *Streptomyces griseus* in *C. glutamicum* likewise resulted in the secretion of the corresponding enzyme into the supernatant and *N*-terminal sequencing of the first nine amino acid residues of the secreted enzyme revealed that processing of the signal peptide had occurred exactly at the same position as is observed in the native host *S. griseus* (Cadenas *et al.*, 1992). Later on, further α-amylases such as those derived from *Streptococcus bovis* (Tateno *et al.*, 2007) and *G. stearothermophilus* (Watanabe *et al.*, 2009) have been secreted via the Sec export pathway into the culture supernatant of *C. glutamicum*. Also, in the case of a nuclease from *Staphylococcus aureus*, a successful secretion of active enzyme into the supernatant was observed and, also for the nuclease, it could be demonstrated that the processing of the pre-nuclease signal peptide by the *C. glutamicum* signal peptidase occurred exactly at the same position as in the native host *S. aureus*, confirming that substrate recognition also by the *C. glutamicum* signal peptidase occurs at the Ala-X-Ala consensus motif that is present in Sec-dependent precursor proteins (Liebl *et al.*, 1992). Subsequently, it was shown that also heterologous proteases can be expressed and secreted in *C. glutamicum*. Expression of subtilisin (AprE) from *B. subtilis* harbouring its native signal peptide in *C. glutamicum* resulted in the release of active protease into the supernatant. The same result was obtained for a basic protease (BprV) from the Gram-negative bacterium *Dichelobacter nodosus*, provided that the AprE signal peptide was hooked up in front of the mature BprV protein. In contrast, BprV was not secreted by its native signal peptide, a result that led the authors to speculate that signal peptides derived from Gram-negative bacteria might not be functional in *C. glutamicum* (Billman-Jacobe *et al.*, 1995). Another target protein that was shown to be efficiently secreted by *C. glutamicum* was the fibronectin-binding protein 85A (antigen 85A) from *M. tuberculosis*. The heterologous secretion of biologically active antigen 85A could be triggered either by the native antigen 85A signal peptide or, alternatively and even to a somewhat greater extent, by the signal peptide of the *C. glutamicum* surface layer protein PS2 (CspB) (Salim *et al.*, 1997). In 2001, Adham and co-workers reported on the successful secretion by *C. glutamicum* of plasmid-encoded xylanase (Xys1) and cellulase (Cel1) from *Streptomyces halstedii* (Adham *et al.*, 2001a). In an extension of this work, the *xys1* gene was also integrated into the genome in a locus located downstream of the cell division gene *ftsZ*, resulting in a stable *C. glutamicum* strain with xylanolytic properties (Adham *et

Table 11.1 Secretory production of heterologous proteins with *Corynebacterium glutamicum*

Protein	Source organism	Signal peptide	Host strain	Maximum yield	Comments	Reference
Sec-dependent export						
α-Amylase	*Bacillus amyloliquefaciens*	Native	ATCC 13869	n.d.		Smith *et al*. (1986)
Exoglucanase Cex	*Cellulomonas fimi*	Native	ATCC 21798	0.3 U/ml		Paradis *et al*. (1987)
Endoglucanase CenA	*Cellulomonas fimi*	Native	ATCC 21798	35 U/ml		Paradis *et al*. (1987)
α-Amylase	*Streptomyces griseus*	Native	ATCC 21798 derivative	4 nanokatals/10^8 cells		Cadenas *et al*. (1992)
Nuclease	*Staphylococcus aureus*	Native	AS019/R163	20 mg/l		Liebl *et al*. (1992)
Subtilisin AprE	*Bacillus subtilis*	Native	AS019	0.5 mg/l		Billman-Jacobe *et al*. (1995)
Basic protease BprV	*Dichelobacter nodosus*	AprE $^{B.s.}$	AS019	2.5 mg/l		Billman-Jacobe *et al*. (1995)
Antigen 85A	*Mycobacterium tuberculosis*	Native/CspB $^{C.g.}$	CGL1017	n.d.		Salim *et al*. (1997)
Xylanase Xys1	*Streptomyces halstedii*	Native	ATCC 13869	20 U/ml		Adham *et al*. (2001a)
Cellulase Cel1	*Streptomyces halstedii*	Native	ATCC 13869	5 U/ml		Adham *et al*. (2001a)
SAM-P45 protease	*Streptomyces albogriseolus*	CspA $^{C.a.}$	ATCC 13869	78 U/l		Kikuchi *et al*. (2003)
Transglutaminase TG	*Streptomyces mobaraensis*	CspA $^{C.g.}$/CspB $^{C.g.}$; CspA $^{C.a.}$	ATCC 13869	235 mg/l pro-TG		Kikuchi *et al*. (2003)
Transglutaminase TG	*Streptomyces mobaraensis*	CspA $^{C.a.}$	ATCC 13869	142 mg/l mature TG	SAM-P45 co-secretion	Date *et al*. (2003)
Transglutaminase TG	*Streptomyces mobaraensis*	CspA $^{C.a.}$	ATCC 13869	881 mg/l pro-TG	Hybrid prodomain	Date *et al*. (2004)
X_{22} Xylanase	*Aspergillus nidulans*	native/Xys1 $^{S.h.}$	R31	6 U/ml		Diaz *et al*. (2004)
Epidermal growth factor	Human	CspA $^{C.a.}$	YDK010	156 mg/l per day		Date *et al*. (2006)
α-Amylase	*Streptococcus bovis*	CspB $^{C.g.}$	ATCC 13032	6300 U/l		Tateno *et al*. (2007)
α-Amylase	*Geobacillus stearothermophilus*	98 different *C. glutamicum* Sec signal peptides	R	0.1 U/ml	Sec signal peptide library $^{C.g.}$	Watanabe *et al*. (2009)

Cutinase	*Fusarium solani pisi*	Six different *Bacillus subtilis* Sec signal peptides	ATCC 13032	13 kU/l		Rohe *et al*. (2012)
Endoxylanase XynA-Ba	*Bacillus* species	PorB $^{C.g.}$	ATCC 13032	n.d.		An *et al*. (2013)
Endoxylanase XynA-St	*Streptomyces coelicolor*	PorB $^{C.g.}$	ATCC 13032	615 mg/l		An *et al*. (2013)
Endoxylanase XynA-St	*Streptomyces coelicolor*	PorB $^{C.g.}$	ATCC 13032	746 mg/l	Synthetic promoter	Yim *et al*. (2013)
M18 scFv	Murine	PorB $^{C.g.}$	ATCC 13032	13 mg/l		Yim *et al*. (2013)
M18 scFv	Murine	PorB $^{C.g.}$	ATCC 13032	68 mg/l	Codon optimization, synthetic promoter	Yim *et al*. (2014)
Tat-dependent export						
GFP	*Aequorea victoria*	TorA $^{E.c.}$/IMD $^{A.g.}$	YDK010	n.d.		Kikuchi *et al*. (2006)
Isomaltodextranase IMD	*Arthrobacter globiformis*	Native	YDK010	100 mg/l		Kikuchi *et al*. (2006)
Transglutaminase TG	*Streptomyces mobaraensis*	IMD $^{S.m.}$	YDK010	100 mg/l pro-TG		Kikuchi *et al*. (2006)
Transglutaminase TG	*Streptomyces mobaraensis*	TorA $^{E.c.}$/IMD $^{S.m.}$	ATCC 13869	846 mg/l pro-TG	TatABC overproduction	Kikuchi *et al*. (2009)
GFP	*Aequorea victoria*	TorA $^{E.c.}$/PhoD $^{C.g.}$/PhoD $^{B.s.}$	ATCC 13032	n.d.	Comparison C.g., B.s., S.c.	Meissner *et al*. (2007)
Protein glutaminase PG	*Chryseobacterium proteolyticum*	TorA $^{E.c.}$/IMD $^{S.m.}$	YDK010	183 mg/l pro-PG		Kikuchi *et al*. (2008)
Protein glutaminase PG	*Chryseobacterium proteolyticum*	TorA $^{E.c.}$	ATCC 13869	1237 mg/l pro-PG	TatABC overproduction	Kikuchi *et al*. (2009)
α-Amylase	*Geobacillus stearothermophilus*	10 different *C. glutamicum* Tat signal peptides	R	0.3 U/ml	Tat signal peptide library $^{C.g.}$	Watanabe *et al*. (2009)
Sorbitol-xylitol oxidase SoXy	*Streptomyces coelicolor*	TorA $^{E.c.}$	ATCC 13032	n.d.	FAD cofactor	Scheele *et al*. (2013)
GFP	*Aequorea victoria*	CGR0949 $^{C.g.}$	R	2.8 g/l	*SigB* deletion mutant	Watanabe *et al*. (2013)

n.d., not determined; B.s., *Bacillus subtilis*; C.g., *Corynebacterium glutamicum*; C.a., *Corynebacterium ammoniagenes*; S.h., *Streptomyces halstedii*; E.c., *Escherichia coli*; A.g., *Arthrobacter globiformis*; S.m., *Streptomyces mobaraensis*; S.c., *Staphylococcus carnosus*.

al., 2001b). Likewise, further xylanases (endoxylanase XynA-Ba from *Bacillus* sp. and XynA-St from *S. coelicolor*, respectively) were secreted in an active form into the *C. glutamicum* supernatant by using the signal peptide of the *C. glutamicum* porin PorB (An *et al.*, 2013). By selecting the best performing promoter out of a set of strong synthetic promoters for *xynA-St* gene expression, a XynA-St yield of 746 mg/l could be obtained after a 35.6-hour fed-batch cultivation process (Yim *et al.*, 2013). Transglutaminase of *Streptomyces mobaraensis*, an enzyme with potential applications in the food industry, is synthesized as a pre-pro-enzyme in its natural host. Activation of the enzyme requires the removal of the propeptide which inhibits enzyme activity. Expression of the native pre-pro-form of the transglutaminase in *C. glutamicum* did not result in secretion of pro-transglutaminase into the supernatant, indicating that the native signal peptide does not confer export of the pre-pro-protein across the plasma membrane. In contrast, secretion of enzymatically inactive pro-transglutaminase could be achieved by using different corynebacterial signal peptides (i.e. derived from CspA and CspB from *C. glutamicum* and CspA from *Corynebacterium ammoniagenes*). Interestingly, the highest yields were observed when the *C. ammoniagenes* CspA signal peptide was used (Kikuchi *et al.*, 2003). SAM-P45, a secreted subtilisin-like serine protease from *Streptomyces albogriseolus* is likewise synthesized as a pre-pro-enzyme, whereby cleavage of the propeptide is performed in an autocatalytic reaction. Co-secretion of pro-SAM-P45 (with the aid of the *C. ammoniagenes* CspA signal peptide) and pro-transglutaminase resulted in pro-peptide cleavage from SAM-P45 (via autocatalysis) and, subsequently, also from pro-transglutaminase (via the generated SAM-45 activity), thereby resulting in the formation of active mature transglutaminase in the *C. glutamicum* supernatant (Kikuchi *et al.*, 2003; Date *et al.*, 2003). By using a chimeric propeptide derived from the propeptides of the transglutaminases from *S. mobaraensis* and *Streptomyces cinnamoneus* instead of the native propeptide, the yield of secreted hybrid pro-transglutaminase could be further increased by 25%, resulting in more than 880 mg of secreted protein per litre supernatant after 40 hours of cultivation under the examined growth conditions. Although the exact reasons for the latter improvement are unknown, the authors speculated that the chimeric propeptide might result in a comparably slower folding kinetics of the precursor protein in the cytoplasm and, as a consequence, a more effective translocation by the *C. glutamicum* Sec translocase (Date *et al.*, 2004). With the X_{22} xylanase (XynA) from *Aspergillus nidulans*, the first enzyme from a eukaryotic organism was expressed and successfully secreted in *C. glutamicum* using either its own signal peptide or the signal peptide from the Xys1 xylanase of *S. halstedii*. In both cases, similar enzymatic activities (6 units/l) could be recovered from the supernatant after 96 hours of growth of the respective recombinant strains, indicating that the respective signal peptides from *Streptomyces* and *Aspergillus* had a very similar efficiency with respect to XynA secretion in *C. glutamicum* (Diaz *et al.*, 2004). The potential of *C. glutamicum* for the production of heterologous eukaryotic proteins was further demonstrated by the successful secretion of a human protein. The human epidermal growth factor (hEGF) is a 53 amino acid polypeptide that contains six cysteine residues that form three disulfide bonds. Using the *C. glutamicum cspB* promoter and the CspA signal peptide from *C. ammoniagenes* to drive expression and Sec-dependent export across the plasma membrane, respectively, 156 mg/l/day of hEGF could be obtained in the culture supernatant of the respective recombinant bacteria. The finding that the secreted hEGF was biologically active furthermore indicated that the hEGF secreted by *C. glutamicum* was correctly folded and, importantly, had formed its disulfide bonds in a correct manner, indicating that *C. glutamicum* might be promising host organism not only for heterologous enzymes, but also for the production of pharmaceutically important human proteins (Date *et al.*, 2006). Finally, by using the PorB signal peptide for Sec-dependent membrane translocation, the secretion (resulting in a yield of 13 mg/l) of an active murine single-chain variable antibody fragment (M18-scFv) that is specific for the anthrax toxin PA has recently been described (Yim *et al.*, 2013). Subsequently, by optimizing the codon usage of the M18-scFv-encoding gene towards the codon preferences of *C. glutamicum* and by using a strong synthetic promoter for gene expression, the concentration of M18-scFv in the culture supernatant of the respective recombinant *C. glutamicum* production strain could be further increased to 68 mg/l (Yim *et al.*, 2014).

Export via the Tat pathway

In 2006, Kikuchi and co-workers were the first to describe the Tat-dependent export of a heterologous protein by *C. glutamicum*. Expression of a hybrid precursor consisting of the Tat-specific signal peptide of the *E. coli* Tat substrate trimethylamine *N*-oxide reductase (TorA) fused to green fluorescent protein (GFP) resulted in the secretion of GFP into the supernatant of a *C. glutamicum* Tat wild-type strain. In contrast, accumulation of GFP in the cytosol and no GFP secretion was observed when the TorA-GFP precursor was expressed in a *C. glutamicum ΔtatC* mutant strain. These results clearly showed that (1) *C. glutamicum* possesses an active Tat system and (2) that the Tat export route can be used for the secretory production of a heterologous protein (Kikuchi *et al.*, 2006). In the same study, the precursor of a secreted isomaltodextranase (IMD) of *Arthrobacter globiformis* which possesses a signal peptide with a twin-arginine motif was expressed in *C. glutamicum*. Secretion of mature IMD protein into the culture supernatant was observed in a *C. glutamicum* wild-type, but not in a *ΔtatC* mutant strain. In addition, Tat-dependent secretion of pro-transglutaminase from *S. mobaraensis* was observed when the pro-transglutaminase was fused to the IMD signal peptide (Kikuchi *et al.*, 2006). In a comparative analysis between three different Gram-positive bacteria (*B. subtilis, Staphylococcus carnosus,* and *C. glutamicum*), Meissner and co-workers demonstrated that *C. glutamicum* was superior to the two other species with respect to secretion of the GFP model protein. Although in all three bacteria GFP could be translocated across the cytoplasmic membrane in a Tat-dependent manner with the help of the TorA signal peptide, striking differences were observed with respect to the final localization and the folding status of the exported GFP protein. In *S. carnosus,* the exported GFP was not released into the supernatant but rather was trapped entirely in the cell wall. In *B. subtilis*, the GFP was secreted into the culture medium; however, in an inactive (i.e. non-fluorescent) conformation. In marked contrast, *C. glutamicum* effectively secreted GFP in an active and therefore correctly folded form. These results indicated that (1) a comparative evaluation of different Gram-positive host bacteria is an important step on the way to an efficient Tat-mediated secretory protein production process and (2) that *C. glutamicum* is in fact a promising alternative host organism that, at least in the case of GFP, outperforms even the well-established *B. subtilis* expression system (Meissner *et al.*, 2007). Protein glutaminase (PG) is an enzyme that is secreted by the Gram-negative bacterium *Chryseobacterium proteolyticum* and possesses potential applications in the food industry. Using either the TorA or the IMD signal peptide, efficient Tat-dependent secretion of PG into the supernatant of *C. glutamicum* could be achieved. Interestingly, expression of PG fused to two different Sec-type signal peptides (i.e. the native PG signal peptide and the *C. ammoniagenes* CspA signal peptide) did not result in PG secretion, indicating that PG folds too rapidly in *C. glutamicum* to be handled by the Sec pathway (Kikuchi *et al.*, 2008). Some biotechnologically interesting enzymes (such as carbohydrate oxidases or ω-transaminases) are excluded from Sec-dependent secretory production since they depend on a tightly or covalently bound cofactor for their activity and, for this reason, require that their folding and cofactor insertion has to take place in the cytosol. Since *C. glutamicum* has been shown to be an excellent host for the Tat-dependent production of cofactor-less proteins (such as GFP; see above), it was subsequently analysed whether it is likewise possible to secrete cofactor-containing proteins via the same export route. And, in fact, the expression of a hybrid precursor consisting of the TorA signal peptide fused to the FAD cofactor-containing cytosolic enzyme sorbitol-xylitol oxidase (SoXy) from *S. coelicolor* resulted in Tat-dependent secretion of enzymatically active SoXy protein into the culture supernatant of *C. glutamicum*, showing for the first time for any recombinant host organism that, also for cofactor-containing proteins, a secretory production is a feasible and promising alternative to conventional intracellular expression strategies (Scheele *et al.*, 2013).

Improving secretory protein production by *Corynebacterium glutamicum*

The choice of the best possible signal peptide is one of the most critical steps on the way to an efficient secretory production process for a desired target protein. In *B. subtilis*, the comparative analysis of Sec-dependent secretion of two different model target proteins (a cutinase from the fungus *Fusarium solani pisi* and an esterase of metagenomic origin) by using more than 150 different signal peptides clearly showed that the amounts of the secreted target proteins in the supernatant were strongly affected by the nature of the signal peptide that was used for membrane translocation. Strikingly, it also became clear that the best signal peptide for a given target protein is not necessarily a good signal peptide for another unrelated target protein and vice versa, indicating that for every individual target protein, an optimally fitted signal peptide has to be identified (Brockmeier *et al.*, 2006). In a similar approach, 405 predicted signal peptides from *C. glutamicum* R were tested with respect to their ability to secrete an α-amylase from *G. stearothermophilus* into the *C. glutamicum* R supernatant. Out of the entire pool, 108 signal peptides (further classified into 90 Sec-type, 10 Tat-type and 8 putative lipoprotein-type signal peptides) showed detectable export of the α-amylase reporter and, also in this study, drastic differences in the amounts of secreted α-amylase were observed for different signal peptides. Notably, however, 11 of these 108 signal peptides resulted in 50- to 150-fold increased amounts of secreted α-amylase when compared with one of the most studied *C. glutamicum* signal peptides (i.e. the PS2 (CspB)-derived signal peptide) that had empirically been chosen in other previous secretion studies. Surprisingly, the highest yield was observed when the normally Sec-dependent α-amylase was redirected into the Tat pathway by the CgR0949 signal peptide, showing that switching the protein export pathway can be a very effective additional option for further process improvement (Watanabe *et al.*, 2009). Although a much lower number of signal peptide–target protein combinations has been analysed, an influence of the nature of the signal peptide on the yield of a given target protein has likewise been observed in various other studies on the secretion of heterologous proteins via the Sec- (e.g. Kikuchi *et al.*, 2003; Rohe *et al.*, 2012) or via the Tat-pathway (e.g. Kikuchi *et al.*, 2006; Meissner *et al.*, 2007) of *C. glutamicum*.

Since the number of export sites is limited both in the Sec and in the Tat pathway, increasing the amounts of critical components of the respective secretion machineries is another possibility for improving *C. glutamicum* as a protein production host. Co-overproduction of various combinations of *C. glutamicum* Sec proteins, i.e. SecD + SecF, SecE + SecD + SecF, and SecY + SecD + SecF with an α-amylase from *S. griseus* resulted in 1.5-, 2.3-, and 2.5-fold increased amounts of the secreted amylase in the culture supernatants of the respective engineered *C. glutamicum* strains (Berens *et al.*, 2006). Also, for the Tat-pathway of *C. glutamicum*, the export capacity could be significantly increased by increasing the amounts of translocase components (Kikuchi *et al.*, 2009). Co-overproduction in *C. glutamicum* of a precursor consisting of the TorA signal peptide fused to the pro-form of the *C. proteolyticum* protein glutaminase (PG) with TatC alone or TatC in combination with TatA, but not of TatA or TatB alone resulted in a more than 3-fold increase of the amount of pro-PG in the culture supernatant, indicating that the amount of TatC represents the primary limiting bottleneck that determines the number of functional Tat translocases that can be formed in a given cell. Furthermore, compared with TatC or TatA + TatC overproduction, the concomitant overproduction of TatA + TatB + TatC resulted in a further 3-fold increase of the amount of secreted pro-PG, showing that, next to TatC, the amount of TatB is a second-layer bottleneck that limits Tat-dependent export in *C. glutamicum*. In addition, also the Tat-dependent secretion of a second substrate, pro-transglutaminase from *S. mobaraensis* possessing either the IMD or the TorA signal peptide was significantly increased when TatABC were co-overproduced. Interestingly, compared with the relatively small effect on the IMD signal peptide-bearing precursor, the positive effect was much more pronounced for the precursor that was equipped with the TorA signal peptide, indicating that screening for an optimal signal peptide is an important step also in the case when an expression strain is used that has been engineered for a higher number of Tat translocases (Kikuchi *et al.*, 2009).

Starting with a *C. glutamicum* R strain that expresses a precursor consisting of the CgR0949 Tat signal peptide fused to GFP, a random transposon mutagenesis has recently been described that allowed for the isolation of mutant strains that showed an up to 5-fold increased accumulation of GFP in the culture supernatant (Watanabe *et al.*, 2013). In the best producer (termed *gem1* for GFP-enhanced mutant 1), it was subsequently shown that inactivation of the gene for the group 2 sigma factor SigB was responsible for the hyperproduction phenotype. SigB is a non-essential sigma factor that has a role in the response to several environmental stresses and in the modulation of global gene expression during transition from exponential growth to stationary phase (Larisch *et al.*, 2007). Interestingly, the beneficial effect of *sigB* disruption could be observed also for another Tat-dependently secreted protein (*G. stearothermophilus* α-amylase fused to the CgR0949 signal peptide). Even more striking, an improved production of the α-amylase was also observed when the α-amylase was secreted via the Sec pathway by using a Sec signal peptide, indicating that the loss of SigB might have a global beneficial effect on protein productivity, at least for some of the signal peptide–target protein combinations that were analysed in the respective study. Although it has been speculated that, among other possible reasons, a reduced proteolytic activity as a consequence of the *sigB* deletion might be responsible for the observed higher yields of target proteins in the supernatant (Watanabe *et al.*, 2013), the exact reason for the improved protein productivity of a *sigB* deletion mutant is unclear so far and has to await further studies.

Using ^{13}C metabolic flux analysis of a *C. glutamicum* strain that secretes the transglutaminase from *S. mobaraensis* via the Tat pathway into the supernatant, Umakoshi and co-workers noticed an inverse correlation between the NADH/NAD^+ ratio and transglutaminase production. Since NADH is oxidized to NAD^+ in the lactate producing reaction catalysed by lactate dehydrogenase, an increased lactate production was attempted to lower the NADH/NAD^+ ratio by raising the pH of the culture from 6.7 to 7.2. The authors hypothesized that an increase of the extracellular pH might result in an increase in the amount of the dissociated form of lactate which cannot easily re-enter the cells and that, by this means, lactate production and concomitantly transglutaminase secretion could possibly be stimulated. And, in fact, lactate production and transglutaminase yields were 2-fold and 1.4-fold higher at pH 7.2 than at pH 6.7, respectively, showing that systems biology approaches might represent another promising option with respect to the optimization of protein production processes in *C. glutamicum* (Umakoshi *et al.*, 2011).

Finally, a highly automated and parallelized workflow has recently been described for the efficient and rapid optimization of the Sec-dependent secretory production of a cutinase from *F. solani pisi* with *C. glutamicum*. A microtitre-based cultivation system (BioLector) that allows 48 parallel cultivations was fully integrated in a robotic liquid-handling platform and this microbioreactor platform was used for the rapid optimization of important process parameters such as medium composition, the concentration of the inductor IPTG that was used to induce expression of the cutinase-encoding gene from a *tac* promoter, and the induction time. Besides this, the described system was also successfully applied to reliably identify the most efficient signal peptide out of three differently efficient Sec signal peptides that have been used to drive Sec-dependent secretion of the cutinase, suggesting that such an automated cultivation platform might be an effective way to screen large signal peptide libraries for the best signal peptide for a desired target protein. Importantly, all biological and bioprocess parameters that were optimized at microtitre scale showed perfect scalable results to 1-l and 20-l stirred tank bioreactors (Rohe *et al.*, 2012).

Concluding remarks

The results obtained so far clearly support the view that *C. glutamicum* possesses a considerable potential as an alternative host system for the secretory expression of heterologous proteins. Proteins derived from various pro- and eukaryotic sources have been successfully expressed and subsequently routed to the extracellular growth medium of *C. glutamicum* via either the Sec or the Tat protein export pathway. Tools for improving the secretory production of desired heterologous target proteins such as a *C. glutamicum* signal peptide library or strains possessing increased amounts of critical components of either the Sec or the Tat secretion machinery have been developed. Yields of 881 mg/l for the *S. mobaraensis* transglutaminase (Date *et al.*, 2004) and even 2.8 g/l for GFP (Watanabe *et al.*, 2013) have been described that, for heterologous proteins, are quite remarkable and only about one order of magnitude lower than the highly optimized yields that are obtained for host-derived, naturally secreted technical enzymes (such as proteases, amylases, or lipases) with *Bacillus* species as secretory production organisms. Despite these already very promising results, further improvements of *C. glutamicum* as a protein production platform organism can be expected in the future for example from the combined application of various systems biology techniques (i.e. transcriptomics, proteomics, metabolomics; all of which are already well established for *C. glutamicum*) to conditions of expression and secretion of different heterologous proteins. Such global analyses will very likely pave the way to second generation optimization strategies that, alone or in combination with already existing strategies, might result in even higher product yields than those reported so far.

Given the fact that, like other members of the Corynebacterianeae, *C. glutamicum* possesses an outer membrane composed of mycolic acids and other complex glycolipids (i.e. the so-called mycomembrane) that acts as an efficient permeability barrier for hydrophilic molecules (Bayan *et al.*, 2003; Hoffmann *et al.*, 2008; Zuber *et al.*, 2008), it is somewhat surprising that even heterologous proteins are efficiently secreted into the culture supernatant and no significant trapping in the intermembrane space is observed (Scheele *et al.*, 2013). Therefore, one of the most intriguing unsolved questions from a microbiologist's point of view with respect to the mechanism of protein secretion in this class of diderm Gram-positive microorganisms is how proteins are translocated across the mycomembrane into the external medium. So far, this last step in the *C. glutamicum* secretion pathway has remained completely enigmatic and, up to now, no components of a dedicated mycomembrane protein translocation system (if existent at all) have been identified. Clearly, the elucidation of the mechanism of protein transport across the *C. glutamicum* outer membrane will not only be very exciting for fundamental microbiology reasons, but also will be of considerable biotechnological interest.

Acknowledgement

The financial support from Bio.NRW, a funding initiative of the local government of North Rhine-Westphalia, is gratefully acknowledged (ExpressO; grant No w0805wb003c).

References

Adham, S. A. I., Honrubia, P., Diaz, M., Fernadez-Abalos, J., Santamaria, R. I., and Gil, J. A. (2001a) Expression of the genes coding for the xylanase Xys1 and the cellulase Cel1 from the straw-decomposing *Streptomyces halstedii* JM8 cloned into the amino-acid producer *Brevibacterium lactofermentum* ATCC13869. *Arch. Microbiol.* **177**, 91–97.

Adham, S. A. I., Campelo, A. B., Ramos, A., and Gil, J. A. (2001b) Construction of a xylanase-producing strain of *Brevibacterium lactofermentum* by stable integration of an engineered *xysA* gene from *Streptomyces halstedii* JM8. *Appl. Environ. Microbiol.* **67**, 5425–5430.

Alami, M., Lüke, I., Deitermann, S., Eisner, G., Koch, H.-G., Brunner, J., and Müller, M. (2003) Differential interaction between a twin-arginine signal peptide and its translocase in *Escherichia coli. Mol. Cell* **12**, 937–946.

An, S. J., Yim, S. S., and Jeong, K. J. (2013) Development of a secretion system for the production of heterologous proteins in *Corynebacterium glutamicum* using the porin B signal peptide. *Protein Expr. Purif.* **89**, 251–257.

Auclair, S. M., Bhanu, M. K., and Kendall, D. A. (2012) Signal peptidase I: cleaving the way to mature proteins. *Protein Sci.* **21**, 13–25.

Barnett, J. P., Eijlander, R. T., Kuipers, O. P., and Robinson, C. (2008) A minimal Tat system from a Gram-positive organism. A bifunctional TatA subunit participates in discrete TatAC and TatA complexes. *J. Biol. Chem.* **283**, 2534–2542.

Bayan, N., Houssin, C., Chami, M., and Leblon, G. (2003) Mycomembrane and S-layer: two important structures of *Corynebacterium glutamicum* cell envelope with promising biotechnology applications. *J. Biotechnol.* **104**, 55–67.

Bechtluft, P., Nouwen, N., Tans, S. J., and Driessen, A. J. M. (2010) SecB – A chaperone dedicated to protein translocation. *Mol. BioSyst.* **6**, 620–627.

Becker, J., and Wittmann, C. (2012) Bio-based production of chemicals, materials and fuels – *Corynebacterium glutamicum* as versatile cell factory. *Curr. Opin. Biotechnol.* **23**, 631–640.

Bendtsen, J. D., Nielsen, H., von Heijne, G., and Brunak, S. (2004) Improved prediction of signal peptides: SignalP 3.0. *J. Mol. Biol.* **340**, 783–795.

Berens, S., Kalinowski, J., and Pühler, A. (2006) *Corynebacterium glutamicum* strain with enhanced secretion activity. United States Patent US 6,982,162 B2.

Berks, B. C. (1996) A common export pathway for proteins binding complex redox cofactors? *Mol. Microbiol.* **22**, 393–404.

Billman-Jacobe, H., Wang, L. F., Kortt, A., Stewart, D., and Radford, A. (1995) Expression and secretion of heterologous proteases by *Corynebacterium glutamicum*. *Appl. Environ. Microbiol.* **61**, 1610–1613.

Blaudeck, N., Kreutzenbeck, P., Müller, M., Sprenger, G. A., and Freudl, R. (2005) Isolation and characterization of bifunctional *Escherichia coli* TatA mutant proteins that allow efficient Tat-dependent protein translocation in the absence of TatB. *J. Biol. Chem.* **280**, 3426–3432.

Brockmeier, U., Caspers, M., Freudl, R., Jockwer, A., Noll, T., and Eggert, T. (2006) Systematic screening of all signal peptides from *Bacillus subtilis*: a powerful strategy in optimizing heterologous protein secretion in Gram-positive bacteria. *J. Mol. Biol.* **362**, 393–402.

Brüser, T., and Sanders, C. (2003) An alternative model of the twin arginine translocation system. *Microbiol. Res.* **158**, 7–17.

Cadenas, R. F., Gil, J. A., and Martin, J. F. (1992) Expression of *Streptomyces* genes encoding extracellular enzymes in *Brevibacterium lactofermentum*: secretion proceeds by removal of the same leader peptide as in *Streptomyces lividans*. *Appl. Microbiol. Biotechnol.* **38**, 362–369.

Caspers, M., and Freudl, R. (2008) *Corynebacterium glutamicum* possesses two *secA* homologous genes that are essential for viability. *Arch. Microbiol.* **189**, 605–610.

Chatzi, K. E., Sardis, M. F., Karamanou, S., and Economou, A. (2013) Breaking on through to the other side: protein export through the bacterial Sec system. *Biochem. J.* **449**, 25–37.

Dalbey, R. E., Wang, P., and Kuhn, A. (2011) Assembly of bacterial inner membrane proteins. *Annu. Rev. Biochem.* **80**, 161–187.

Date, M., Yokoyama, K.-i., Umezawa, Y., Matsui, H., and Kikuchi, Y. (2003) Production of native-type *Streptoverticillium mobaraense* transglutaminase in *Corynebacterium glutamicum*. *Appl. Environ. Microbiol.* **69**, 3011–3014.

Date, M., Yokoyama, K.-i., Umezawa, Y., Matsui, H., and Kikuchi, Y. (2004) High level expression of *Streptomyces mobaraensis* transglutaminase in *Corynebacterium glutamicum* using a chimeric pro-region from *Streptomyces cinnamoneus* transglutaminase. *J. Biotechnol.* **110**, 219–226.

Date, M., Itaya, H., Matsui, H., and Kikuchi, Y. (2006) Secretion of human epidermal growth factor by *Corynebacterium glutamicum*. *Lett. Appl. Microbiol.* **42**, 66–70.

Diaz, M., Adham, S. A. I., Ramon, D., Gil, J. A., and Santamaria, R. I. (2004) *Streptomyces lividans* and *Brevibacterium lactofermentum* as heterologous hosts for the production of X_{22} xylanase from *Aspergillus nidulans*. *Appl. Microbiol. Biotechnol.* **65**, 401–406.

Feltcher, M. E., and Braunstein, M. (2012) Emerging themes in SecA2-mediated protein export. *Nat. Rev. Microbiol.* **10**, 779–789.

Freudl, R. (2013) Leaving home ain't easy – protein export systems in Gram-positive bacteria. *Res. Microbiol.* **164**, 664–674.

Fröbel, J., Rose, P., Lausberg, F., Blümmel, A.-S., Freudl, R., and Müller, M. (2012) Transmembrane insertion of twin-arginine signal peptides is driven by TatC and regulated by TatB. *Nat. Commun.* **3**, 1311.

Gohlke, U., Pullan, L., McDevitt, C. A., Porcelli, I., de Leeuw, E., Palmer, T., Saibil, H. R., and Berks, B. C. (2005) The TatA component of the twin-arginine protein transport system forms channel complexes of variable diameter. *Proc. Natl. Acad. Sci. U.S.A.* **102**, 10482–10486.

Harwood, C. R., and Cranenburgh, R. (2008) *Bacillus* protein secretion: an unfolding story. *Trends Microbiol.* **16**, 73–79.

Hayashi, S., and Wu, H. C. (1990) Lipoproteins in bacteria. *J. Bioenerg. Biomembr.* **22**, 451–471.

von Heijne, G. (1989) The structure of signal peptides from bacterial lipoproteins. *Protein Eng.* **2**, 531–534.

Herbort, M., Klein, M., Manting, E. H., Driessen, A. J. M., and Freudl, R. (1999) Temporal expression of the *Bacillus subtilis secA* gene, encoding a central component of the preprotein translocase. *J. Bacteriol.* **181**, 493–500.

Hoffmann, C., Leis, A., Niederweis, M., Plitzko, J. M., and Engelhardt, H. (2008) Disclosure of the mycobacterial outer membrane: cryo-electron tomography and vitreous sections reveal the lipid bilayer structure. *Proc. Natl. Acad. Sci. U.S.A.* **105**, 3963–3967.

Jack, R. L., Sargent, F., Berks, B. C., Sawers, G., and Palmer, T. (2001) Constitutive expression of *Escherichia coli tat* genes indicates an important role for the twin-arginine translocase during aerobic and anaerobic growth. *J. Bacteriol.* **183**, 1801–1804.

Keller, R., de Keyzer, J., Driessen, A. J. M., and Palmer, T. (2012) Co-operation between different targeting pathways during integration of a membrane protein. *J. Cell Biol.* **199**, 303–315.

Kikuchi, Y., Date, M., Yokoyama, K.-i. Umezawa, Y., and Matsui, H. (2003) Secretion of active-form *Streptoverticillum mobaraense* transglutaminase by *Corynebacterium glutamicum*: processing of the pro-transglutaminase by a cosecreted subtilisin-like protease from *Streptomyces albogriseolus*. *Appl. Environ. Microbiol.* **69**, 358–366.

Kikuchi, Y., Date, M., Itaya, H., Matsui, K., and Wu, L.-F. (2006) Functional analysis of the twin-arginine translocation pathway in *Corynebacterium glutamicum* ATCC 13869. *Appl. Environ. Microbiol.* **72**, 7183–7192.

Kikuchi, Y., Itaya, H., Date, M., Matsui, K., and Wu, L.-F. (2008) Production of *Chryseobacterium proteolyticum* protein-glutaminase using the twin-arginine translocation pathway in *Corynebacterium glutamicum*. *Appl. Microbiol. Biotechnol.* **78**, 67–74.

Kikuchi, Y., Itaya, H., Date, M., Matsui, K., and Wu, L.-F. (2009) TatABC overexpression improves *Corynebacterium glutamicum* Tat-dependent protein secretion. *Appl. Environ. Microbiol.* **75**, 603–607.

Larisch, C., Nakunst, D., Hüser, A. T., Tauch, A., and Kalinowski, J. (2007) The alternative sigma factor SigB of *Corynebacterium glutamicum* modulates global gene expression during transition from exponential growth to stationary phase. *BMC Genomics* **8**, 4.

Lausberg, F., Fleckenstein, S., Kreutzenbeck, P., Fröbel, J., Rose, P., Müller, M., and Freudl, R. (2012) Genetic evidence for a tight cooperation of TatB and TatC during productive recognition of twin-arginine (Tat) signal peptides in *Escherichia coli*. *PLoS ONE* **7**, e39867.

Leuchtenberger, W., Huthmacher, K., and Drau, K. (2005) Biotechnological production of amino acids and derivatives: current status and prospects. *Appl. Microbiol. Biotechnol.* **69**, 1–8.

Li, W., Zhou, X., and Lu, P. (2004) Bottlenecks in the expression and secretion of heterologous proteins in *Bacillus subtilis*. *Res. Microbiol.* **155**, 605–610.

Liebl, W., Sinskey, A. J., and Schleifer, K.-H. (1992) Expression, secretion, and processing of staphylococcal nuclease by *Corynebacterium glutamicum*. *J. Bacteriol.* **174**, 1854–1861.

Lycklama a Nijeholt, J. A., and Driessen, A. J. M. (2012) The bacterial Sec-translocase: structure and mechanism. *Philos. Trans. R. Soc. Lond. B Biol. Sci.* **367**, 1016–1028.

Lyon, W. R., Gibson, C. M., and Caparon, M. G. (1998) A role for trigger factor and an Rgg-like regulator in the transcription, secretion and processing of the cysteine proteinase of *Streptococcus pyogenes*. *EMBO J.* **17**, 6263–6275.

Mendel, S., McCarthy, A., Barnett, J. P., Eijlander, R. T., Nenninger, A., Kuipers, O. P., and Robinson, C. (2008) The *Escherichia coli* TatABC system and a *Bacillus subtilis* TatAC-type system recognise three distinct targeting determinants in twin-arginine signal peptides. *J. Mol. Biol.* **375**, 661–672.

Meissner, D., Vollstedt, A., van Dijl, J. M., and Freudl, R. (2007) Comparative analysis of twin-arginine (Tat)-dependent protein secretion of a heterologous model protein (GFP) in three different Gram-positive bacteria. *Appl. Microbiol. Biotechnol.* **76**, 633–642.

Mori, H., and Cline, K. (2002) A twin arginine signal peptide and the pH gradient trigger reversible assembly of the thylakoid ΔpH/Tat translocase. *J. Cell Biol.* **157**, 205–210.

Niebisch, A., and Bott, M. (2003) Purification of a cytochrome bc_1-aa_3 supercomplex with quinol oxidase activity from *Corynebacterium glutamicum*. Identification of a fourth subunit of cytochrome aa_3 oxidase and mutational analysis of diheme cytochrome c_1. *J. Biol. Chem.* **278**, 4339–4346.

Palmer, T., and Berks, B. C. (2012) The twin-arginine translocation (Tat) protein export pathway. *Nat. Rev. Microbiol.* **10**, 483–496.

Paradis, F. W., Warren, R. A. J., Kilburn, D. G., and Miller, Jr. R. C. (1987) The expression of *Cellulomonas fimi* cellulase genes in *Brevibacterium lactofermentum*. *Gene* **61**, 199–206.

du Plessis, D. J., Nouwen, N., and Driessen, A. J. M. (2011) The Sec translocase. *Biochim. Biophys. Acta* **1808**, 851–865.

Quax, W. J. (1997) Merits of secretion of heterologous proteins from industrial microorganisms. *Folia Microbiol.* **42**, 99–103.

Rohe, P., Venkanna, D., Kleine, B., Freudl, R., and Oldiges, M. (2012) An automated workflow for enhancing microbial bioprocess optimization on a novel microbioreactor platform. *Microb. Cell Fact.* **11**, 144.

Rusch, S. L., and Kendall, D. A. (2007) Interactions that drive Sec-dependent bacterial protein transport. *Biochemistry* **46**, 9665–9673.

Salim, K., Haedens, V., Content, J., Leblon, G., and Huygen, K. (1997) Heterologous expression of the *Mycobacterium tuberculosis* gene encoding antigen 85A in *Corynebacterium glutamicum*. *Appl. Environ. Microbiol.* **63**, 4392–4400.

Saraogi, I., and Shan, S.-O. (2011) Molecular mechanism of co-translational protein targeting by the signal recognition particle. *Traffic* **12**, 535–542.

Schaerlaekens, K., Van Mellaert, L., Lammertyn, E., Geukens, N., and Anné, J. (2004) The importance of the Tat-dependent protein secretion pathway in *Streptomyces* as revealed by phenotypic changes in *tat* deletion mutants and genome analysis. *Microbiology (UK)* **150**, 21–31.

Scheele, S., Oertel, D., Bongaerts, J., Evers, S., Hellmuth, H., Maurer, K.-H., Bott, M., and Freudl, R. (2013) Secretory production of an FAD cofactor-containing cytosolic enzyme (sorbitol-xylitol oxidase from *Streptomyces coelicolor*) using the twin-arginine translocation (Tat) pathway of *Corynebacterium glutamicum*. *Microb. Biotechnol.* **6**, 202–206.

Smith, M. D., Flickinger, J. L., Lineberger, D. W., and Schmidt, B. (1986) Protoplast transformation in coryneform bacteria and introduction of an alpha-amylase gene from *Bacillus amyloliquefaciens* into *Brevibacterium lactofermentum*. *Appl. Environ. Microbiol.* **51**, 634–639.

Tateno, H., Fukuda, H., and Kondo, A. (2007) Direct production of L-lysine from raw corn starch by *Corynebacterium glutamicum* secreting *Streptococcus bovis* α-amylase using *cspB* promoter and signal sequence. *Appl. Microbiol. Biotechnol.* **77**, 533–541.

Tsukazaki, T., Mori, H., Echizen, Y., Ishitani, R., Fukai, S., Tanaka, T., Perederina, A., Vassylyev, D. G., Kohno, T., Maturana, A. D., Ito, K., and Nureki, O. (2011) Structure and function of a membrane component SecDF that enhances protein export. *Nature* **474**, 235–238.

Tullman-Ercek, D., DeLisa, M. P., Kawaraski, Y., Iranpour, P., Ribnicky, B., Palmer, T., and Georgiou, G. (2007) Export pathway selectivity of *Escherichia coli* twin arginine translocation signal peptides. *J. Biol. Chem.* **282**, 8309–8316.

Umakoshi, M., Hirasawa, T., Furusawa, C., Takenaka, Y., Kikuchi, Y., and Shimizu, H. (2011) Improving protein secretion of a transglutaminase-secreting *Corynebacterium glutamicum* recombinant strain on the basis of ^{13}C metabolic flux analysis. *J. Biosci. Bioeng.* **112**, 595–601.

Watanabe, K., Tsuchida, Y., Okibe, N., Teramoto, H., Suzuki, N., Inui, M., and Yukawa, H. (2009) Scanning the *Corynebacterium glutamicum* R genome for high-efficiency secretion signal sequences. *Microbiology* **155**, 741–750.

Watanabe, K., Teramoto, H., Suzuki, N., Inui, M., and Yukawa, H. (2013) Influence of SigB inactivation on *Corynebacterium glutamicum* protein secretion. *Appl. Microbiol. Biotechnol.* **97**, 4917–4926.

Westers, L., Westers, H., and Quax, W. J. (2004) *Bacillus subtilis* as cell factory for pharmaceutical proteins: a biotechnological approach to optimize the host organism. *Biochim. Biophys. Acta* **1694**, 299–310.

Wu, S.-C., Ye, R., Wu, X.-C., Ng, S.-C., and Wong, S.-L. (1998) Enhanced secretory production of a single-chain antibody fragment from *Bacillus subtilis* by coproduction of molecular chaperones. *J. Bacteriol.* **180**, 2830–2835.

Yim, S. S., An, S. J., Kang, M., Lee, J., and Jeong, K. J. (2013) Isolation of fully synthetic promoters for high-level gene expression in *Corynebacterium glutamicum*. *Biotechnol. Bioeng.* **110**, 2959–2969.

Yim, S. S., An, S. J., Choi, J. W., Ryu, A. J., and Jeong, K. J. (2014) High-level secretory production of recombinant single-chain variable fragment (scFv) in *Corynebacterium glutamicum*. *Appl. Microbiol. Biotechnol.* **98**, 273–284.

Yuan, J., Zweers, J. C., van Dijl, J. M., and Dalbey, R. E. (2010) Protein transport across and into cell membranes in bacteria and archaea. *Cell. Mol. Life Sci.* **67**, 179–199.

Zuber, B., Chami, M., Houssin, C., Dubochet, J., Griffiths, G., and Daffé, M. (2008) Direct visualization of the outer membrane of mycobacteria and corynebacteria in their native state. *J. Bacteriol.* **190**, 5672–5680.

Genetically Encoded Biosensors for Strain Development and Single-cell Analysis of *Corynebacterium glutamicum*

12

Nurije Mustafi, Michael Bott and Julia Frunzke

Abstract

Genetically encoded single-cell biosensors have long been used for the detection of pollutants and stress stimuli in the field of microengineering and bioremediation. In industrial microbiology their potential for the detection of various small molecules in single cells and their implementation in strain development and high-throughput (HT) screenings has rarely been exploited so far. However, several recent studies revealed suchlike sensor systems, which transform the production of inconspicuous small molecules into an optical readout (e.g. fluorescence), as a powerful tool for single-cell quantification of metabolites. Recent studies reported on the construction of transcription factor-based biosensors for the intracellular detection of amino acid production in *Corynebacterium glutamicum*. In this chapter we will introduce the general design and characterization of sensor performance features. An overview of all currently available *C. glutamicum* biosensors is provided, including several sensors for the detection of amino acids and stress stimuli in single cells. Finally, we introduce the emerging field of applications for metabolite biosensors in flow cytometry-based HT screening approaches and live cell imaging of metabolite production using microfluidic lab-on-a-chip devices. First results, outlined in this chapter, already highlight the great potential of biosensor-driven strain development and single-cell analysis and promise the revolution of traditional approaches towards a 'bright' future of industrial microbiology.

Introduction: biosensors

Corynebacterium glutamicum, isolated in 1957 due to its capability to secrete L-glutamate, represents nowadays one of the most important platform organisms in industrial biotechnology (Eggeling and Bott, 2005; Wendisch, 2007; Burkovski, 2008). During the last decades, metabolic engineering of *C. glutamicum* revealed the great potential of this species not only for amino acid production, but also for the production of organic acids, polymer precursors and biofuels (Blombach and Eikmanns, 2011; Becker and Wittmann, 2012; Wieschalka *et al.*, 2012). Traditional ways of strain development include random mutagenesis followed by diverse strategies for selection and screening for the desired traits, or rational approaches involving the targeted alteration of specific genetic elements. For the establishment of a successful screening approach, a key issue is the detection and quantification of the particular metabolite of interest. A readily accessible phenotype, linked to product formation, significantly impacts success and throughput of the respective procedure (Dietrich *et al.*, 2010; Schallmey *et al.*, 2014). This is, however, given only for a minor fraction of small molecules such as, for instance, some antibiotics or carotenoids (An *et al.*, 1991; Azuma *et al.*, 1992; Albrecht *et al.*, 2000). In these cases product formation can be directly observed by the production of a chromophore. The majority of biotechnologically relevant metabolites are, however, inconspicuous small molecules, which are classically assayed by low-throughput techniques such as liquid or gas chromatography or NMR providing average data for the microbial population. Measurement of metabolite production at single-cell resolution is still hardly feasible (Heinemann and Zenobi, 2011).

In this context, the application of biosensors converting metabolite production into an optical readout has the potential to revolutionize the status quo of strain development and analysis in the field of biotechnology. During the last 10–20 years a variety of biosensors based on cellular regulatory elements has been developed. Various classifications of biosensors are described and resumed in several reviews (Dietrich *et al.*, 2010; van der Meer, 2010; Zhang and Keasling, 2011; Michener *et al.*, 2012; Schallmey *et al.*, 2014). In the following, four different categories of biosensor designs and architecture are presented, namely whole-cell biosensors, biosensors including single enzymes or multistep enzymatic pathways, Förster (fluorescence) resonance energy transfer (FRET)-based sensors and sensors based on genetic regulatory circuits.

Whole-cell biosensors based on natural or engineered auxotrophic strains for specific molecules represent an important class of biosensors and provide an easy accessible phenotype: cellular growth (Fig. 12.1A) (Burkholder, 1951; Payne *et al.*, 1977). Several studies addressed this idea and developed whole-cell biosensors by introducing constitutively expressed reporter genes (e.g. *gfp*) into auxotrophic bacterial strains (Erickson *et al.*, 2000; Chalova *et al.*, 2007; Kim *et al.*, 2010; Bertels *et al.*, 2012). This approach enables a direct correlation of reporter output (e.g. fluorescence) to the concentration of the molecule of interest in the environment of the sensor strain. Another class of biosensors is based on *in vivo* single- or multistep enzymatic pathways which transform the presence of a colourless molecule of interest into a detectable optical readout of the cell (Fig. 12.1B). In most cases, this is linked to the synthesis of natural chromophores as end or intermediate product of the enzymatic pathway (de la Pena Mattozzi *et al.*, 2006; Bernhardt *et al.*, 2007; Santos and Stephanopoulos, 2007). Here, in contrast to other biosensor constructs, no classical reporter markers (such as fluorescence

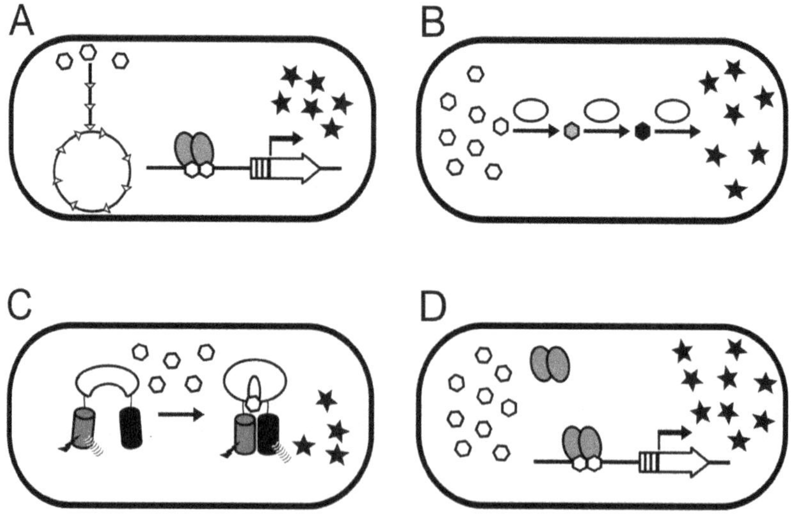

Figure 12.1 Categories of biosensor design. (A) Cellular growth of auxotrophic whole-cell biosensors reports on the bio-availability of the target molecule (⬡). Additionally, reporter gene expression (★) can be set under control of a constitutively regulated promoter. (B) In biosensors based on single- or multistep enzyme pathways, detection of the target molecule (⬡) and reporter output (natural chromophore, (⬢) is provided through enzymatic activities. (C) In FRET-based sensors, binding of the target molecule (⬡) conducts conformational changes of the sensor domain which enables a fluorescence readout (★) by energy transfer from the donor to the acceptor chromophore. (D) In sensors based on genetic regulatory circuits, the target molecule (⬡) is detected by a sensor molecule (RNA, DNA or protein) which in turn activates transcription from its target promoter fused to a reporter gene (adapted from Dietrich *et al.*, 2010).

proteins) are required to access a phenotype for small molecule detection. Despite the successful application of the described biosensors in various studies, several problems remain unsolved. One major bottleneck is the limited versatility of these biosensor constructs. In fact, a rather small fraction of biological, biotechnologically, or pharmaceutically interesting molecules are natural chromophores or can be linked to the synthesis of one.

A broad spectrum of metabolites is detected *in vivo* by an extensive repertoire of natural sensor devices (e.g. riboswitches, RNA-aptamers, transcription factors, enzymes, or periplasmic-binding proteins). Molecular constructs which make use of these biological molecules, linked to the optical readout of fluorescence proteins, represent versatile tools for small molecule detection. One approach to exploit the molecular recognition specificity of several protein classes is the development of FRET-based metabolite sensors. The modular architecture of FRET-sensors consists of interacting protein pairs or interacting protein domains linked to a pair of fluorescent proteins differing in their emission and excitation wavelengths. Binding of the substrate to the sensor domain of the protein pair triggers a conformational change which brings the chromophores of the fluorescence proteins into close proximity and enables FRET from the donor chromophore to the acceptor chromophore (Fig. 12.1C). Thus, effector binding can be detected and measured as change in the ratio of the emitted fluorescence intensities. Nowadays, a large diversity of FRET-based sensors is available for imaging of intrinsic and extrinsic conditions, such as pH, redox potential, ATP:ADP ratio, several sugars and metabolites, various ions, kinase activities and many more (Frommer *et al.*, 2009). However, a major drawback of FRET-based sensors is their inherent high sensitivity towards ionic strength and pH, which can significantly influence affinity parameters and fluorescence response of the sensor (Okumoto *et al.*, 2012).

Finally, genetic regulatory circuits have been extensively exploited for the construction of biosensors especially in the field of microengineering and bioremediation (van der Meer *et al.*, 2004; van der Meer and Belkin, 2010). Classically, the construction of these sensors is based on (i) a sensor protein, e.g. a transcription factor, (ii) a molecular switch which occurs upon ligand binding and results in transcriptional activation of the target gene, and (iii) production of a reporter protein (e.g. GFP) as sensor readout (Fig. 12.1D). This type of biosensors has a long tradition in microbial sensing systems for the detection of environmental pollutants (Belkin, 2003; Harms *et al.*, 2006; van der Meer and Belkin, 2010). For example, a biosensor for toluene-based environmental contaminants was developed by putting the *luc* gene for firefly luciferase under the control of the transcriptional activator XylR. In field tests, the sensor was shown to accurately detect the pollutants in deep aquifer water and soil samples with a sensitivity of about 40 μM (Willardson *et al.*, 1998). Despite the versatile applications in the field of chemical compounds and toxicity detection, only a few studies describe the use of transcription factor-based biosensors to report on production of industrially relevant metabolites (Schallmey *et al.*, 2014).

Two recent studies presented the construction of transcription factor-based biosensors for monitoring of amino acid production in *C. glutamicum* (Binder *et al.*, 2012; Mustafi *et al.*, 2012). These studies represent the first examples of the use of genetically encoded biosensors for this important platform organism and indicate the strong potential of this novel approach for strain development, HT screening and single-cell analysis. In the following chapter, we provide an overview of the recently developed *C. glutamicum* sensor systems and briefly sketch current efforts on single-cell analysis including live cell imaging and flow cytometry-based approaches.

Design of biosensors based on regulatory circuits

Biosensors based on regulatory circuits essentially consist of a target promoter responsive to a specific stimulus of interest fused to a promoterless reporter gene, which turns the particular response into a measurable output. The input signal can either be recognized by a single sensor protein or RNA or the promoter underlies control from a complex signal transduction network acting as an integral part of the particular cellular response. The latter principle has been successfully applied, e.g. for the construction of stress-responsive bioreporters (see the section 'Stress response' below) (van der Meer and Belkin, 2010).

Signal integration

In this section, we want to highlight particularly the use of transcription factors as versatile molecular devices to report on small molecule production in *C. glutamicum*. In fact, nature provides a variety of appropriate sensor devices in form of transcription factors. Typically, transcriptional regulators undergo a conformational change upon binding of their effector molecules, which in turn results in an altered binding affinity of the regulator for its target DNA sequence. Depending on the biosensors final application, the affinity of the sensor element towards its effector molecule is of major importance for assay design. In some cases, however, specificity and/or ligand affinity are not directly suitable for biosensor construction. Here, synthetic biology approaches can be applied to generate variants with altered specificity or transfer parameters (section 'Further screening applications of single-cell biosensors' below). This has been, for example, demonstrated by the work of Galvao *et al.* (2007), who were able to select mutated variants of the transcriptional regulator XylR, sensing toluene in its native form, that are responsive to 2,4-dinitrotoluene (DNT) for the detection of explosives. Even sensors with completely new effector specificities can be engineered as proven by the example of AraC, which was switched from an L-arabinose-induced regulator into a mevalonate-responsive protein (Tang and Cirino, 2011). Thus, the multitude of naturally available sensor devices in combination with current achievements in synthetic biology offers a broad, but almost unexploited, spectrum of opportunities for biosensor construction.

In order to decrease the rate of false-positives in FACS (fluorescence-activated cell sorting) screenings (see the section 'Application of biosensors in screening and single-cell analysis' below), a positive sensor readout is desirable. This can be easily achieved by setting the expression of the reporter protein under the control of a transcriptional activator. However, the response of a repressor protein can also be converted into a positive readout via the design of a coupled sensor. Here, the insertion of a strong, heterologous repressor and its tightly regulated promoter can be used to invert the signal polarity. A striking example was reported by Ohlendorf *et al.* (2012) who inverted the native light-repressed gene expression of the photoreceptor system YF1/FixK by the insertion of the λ phage repressor cI and promoter pR in front of the target gene (Fig. 12.2D).

Promoter fusion

A detailed knowledge of promoter architecture, transcription factor binding sites and translation start is an important prerequisite for successful sensor design. In general, the level of reporter gene expression and thus, reporter output can strongly be influenced by transcription and translation initiation rates. Therefore, choosing an adequate type of fusion between target promoter and reporter gene is a further crucial aspect. In general two types of reporter gene fusions can be distinguished. First, in transcriptional fusions, a promoter-less reporter gene is fused to the target promoter of the transcription factor. An effective ribosome binding (native or artificial) site has to be included in the construct (Fig. 12.2A). Thus, reporter output ideally reflects the transcriptional regulation of the target gene. Second, in a translational fusion, the promoter-less reporter gene is fused in frame to the target gene (first 5–10 codons or even whole open reading frame (orf)) and its upstream genomic region, (Fig. 12.2B). This design principle is especially of relevance when dealing with the high frequency of leaderless transcripts (approximately 30%) which are present in *C. glutamicum* (Pfeifer-Sancar *et al.*, 2013). In these cases, the mRNA completely lacks the binding region of the 16S-rRNA of the ribosomes, the so called Shine–Dalgarno sequence, which is normally located 5–13 base pairs upstream of the initiation codon in the untranslated mRNA sequence. In leaderless transcripts, as it is the case for example for the mRNA of *brnF* and *lysE* in *C. glutamicum*, the transcriptional start point is either identical with the first nucleotide of the start codon or is located in close proximity (1–3 bases). Regions downstream of the initiation codon were assumed to interact with the

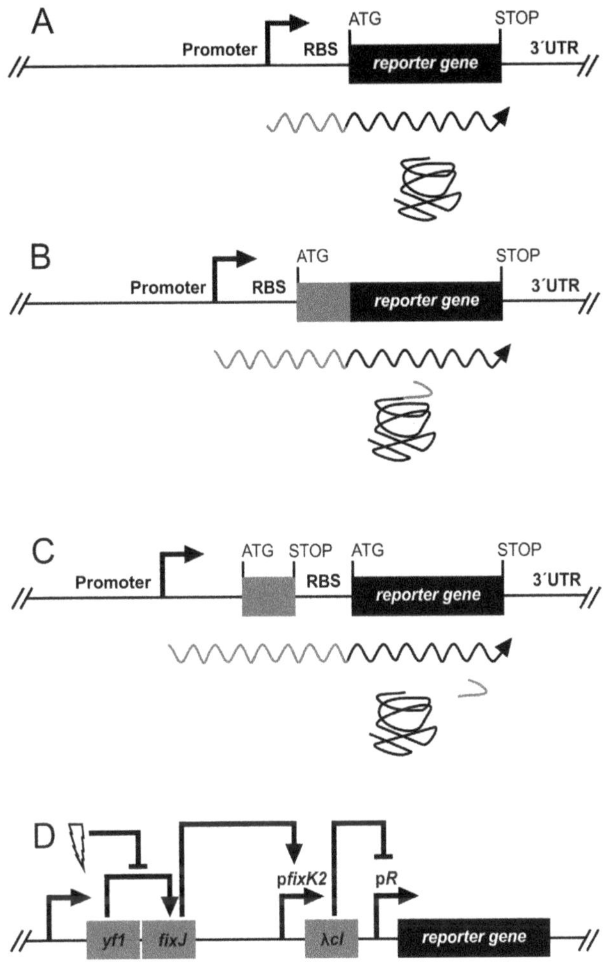

Figure 12.2 Architecture of reporter gene constructs. (A) Transcriptional fusion of reporter gene (black) and target promoter. (B) Translational fusion of reporter (black) and target gene (grey) resulting in the production of hybrid fusion proteins. (C) Transcriptional fusion design used for the construction of pSenLys and the Lrp-sensor. (D) Inversion of the signal polarity in the plasmid pDawn. In the absence of blue-light the two-component system YF1/FixJ represses gene expression from the p*fixK2* promoter. Signal polarity is inverted by setting expression of *cI* encoding the λ phage repressor under control of p*fixK2* and inserting of cI's target promoter pR in front of the respective target gene (Ohlendorf *et al.*, 2012).

16S-rRNA of the ribosomes (Moll *et al.*, 2002; Patek, 2005; Giliberti *et al.*, 2012). For this reason, the biosensor constructs of pSenLys and the Lrp-sensor, two recently described *C. glutamicum* sensors (see the sections 'Basic amino acids' and 'L-Methionine and branched-chain amino acids' below), were designed including the upstream promoter region, the first few codons of the corresponding orf (*lysE* and *brnF*, respectively) and a linker sequence introducing a ribosome binding site in front of the reporter gene *eyfp* (Fig. 12.2C) (Binder *et al.*, 2012; Mustafi *et al.*, 2012).

Choice of reporter protein

Finally, the construction of a biosensor demands an appropriate reporter protein, which provides an easily measurable signal output and is, ideally, compatible with single-cell analysis. In order to prevent a high background signal, a reporter protein should be chosen which is naturally not occurring in the host cell and non-toxic. Several reviews outline the variety of available reporter proteins, their applications and characteristics, advantages and disadvantages (Daunert *et al.*, 2000; Hakkila *et al.*, 2002; Shaner *et al.*, 2005;

Ghim *et al.*, 2010). Classical reporter systems which have been used in numerous promoter studies in *C. glutamicum* are, for example β-galactosidase (Cho *et al.*, 2012) and chloramphenicol acetyltransferase (Vasicova *et al.*, 1999). In recent years, autofluorescent proteins proved in practice as appropriate reporters for monitoring gene expression patterns in *C. glutamicum*. Among others, this has been successfully shown for GFP, CFP, and eYFP in studies of protein localization and interaction with DNA (Frunzke *et al.*, 2008; Donovan *et al.*, 2010; Burmann *et al.*, 2012), protein secretion (Teramoto *et al.*, 2011) and for eYFP and E2-Crimson in biosensors reporting on amino acid production (Binder *et al.*, 2012; Mustafi *et al.*, 2012). Recently, Hentschel *et al.* described the construction of unstable variants of eYFP and GFP_{uv} by using SsrA-mediated peptide tagging (Andersen *et al.*, 1998; Hentschel *et al.*, 2012). In comparison to the extremely long half-life of native fluorescent proteins (YFP>24 hours), the two presented eYFP-variants revealed protein half-lives of 22 and 8 minutes, respectively. First applications in transient gene expression analysis showed the general suitability of the destabilized eYFP variants for dynamic measurements in *C. glutamicum*.

A drawback of GFP-derived fluorescent reporter proteins is their dependency on molecular oxygen for chromophore maturation as well as time-consuming protein folding. Recently, flavin mononucleotide (FMN)-binding fluorescent proteins (FbFPs) were described as alternative option allowing quantitative real-time *in vivo* assays (Drepper *et al.*, 2007, 2010). This novel class of reporter proteins exhibits cyan-green fluorescence and does not depend on oxygen for chromophore formation. First proof of principle studies revealed their general suitability for *in vivo* studies in *C. glutamicum* (Nanda and Frunzke, unpublished).

Characterization of biosensor performance parameters

The final application of biosensors for process monitoring or screening approaches requires a detailed understanding and characterization of the sensor performance features. In simple terms, this is represented by an accurate description of the relationship between the effector molecule input and reporter output (e.g. fluorescence). This mathematical relationship, termed as transfer function, provides information on the biosensors sensitivity, the dynamic and linear range of detection as well as the detection threshold (Fig. 12.3). The sensitivity of a biosensor is described by the rate of increase in reporter output, which is depicted by the slope of the transfer curve within the linear range of detection (see below). The

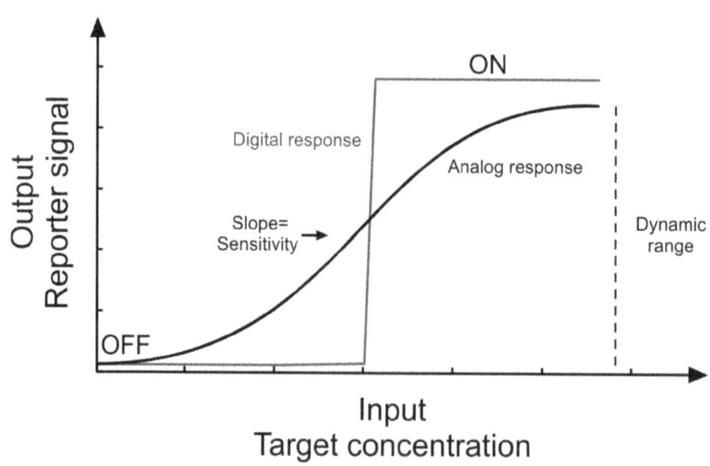

Figure 12.3 Biosensor transfer curve. The mathematical relationship of stimulus input and biosensor output provides information on product sensitivity, the dynamic range and the linear detection window (adapted from Dietrich *et al.*, 2010).

dynamic range of a sensor is defined by the maximum fold change of the reporter output resulting from a given sensory input. Furthermore, every sensor construct exhibits a detection window where signal input and sensor output display a linear correlation, the so-called linear detection range (Wall *et al.*, 2004; Bintu *et al.*, 2005a,b; Tabor *et al.*, 2009).

A detailed knowledge of the biosensor performance parameters is an important prerequisite for assay design since they illustrate the conditions under which the biosensor performs robust and reproducible measurements. In general, biosensor output can be classified as a digital on/off-type behaviour or an analogue-like response. The digital sensor type is characterized by a low detection threshold, a small-linear induction window, a large dynamic range and a high sensitivity. This tight control of response is typical for genetic circuits involved in antibiotic or quorum sensing (Dietrich *et al.*, 2010). One of the most widely used sensors, TetR, represents in its native form a good example for digital response behaviour. Lutz and Bujard (1997) engineered the already strong target promoter of the TetR repressor by adding further operator sides upstream of the native promoter version. This resulted in a tightly regulated promoter of TetR, showing no expression of target genes when the effector molecule anhydrotetracycline is absent and exhibits an up to ~5000-fold dynamic range. In contrast, a low affinity for the effector molecule is often characteristic for an analogue-like response showing a sigmoidal function as response to sensory input. Low levels of the effector molecule cause effectively no changes in reporter output signal. Increasing concentrations near the K_D of the sensor-element are responded by a linear increasing output. Through further increasing effector concentrations the sensor is saturated and a constant level of the output signal is reached (Tabor *et al.*, 2009). A linear correlation between input and output signal allows discrimination of cells or populations differing in their input concentrations. This performance feature makes analogue type biosensors more suitable for applications in screening studies for directed evolution of product yields for example.

Biosensors in *Corynebacterium glutamicum*

Biosensors for the detection of amino acids

Amino acids represent the most important product class produced at large-scale with *C. glutamicum;* the global market is constantly growing. In the last decades strain development and analysis followed traditional ways in terms of screening, selection and rational engineering. Amino acids are small, inconspicuous molecules, making them ideal targets for biosensor design in order to convert product formation into an easily accessible phenotype. Recently, two studies described the development and application of novel single-cell biosensors for the detection of several industrially important amino acids in *C. glutamicum* (Table 12.1) (Binder *et al.*, 2012; Mustafi *et al.*, 2012). These biosensors exploit the exquisite sensitivity of two *C. glutamicum* transcriptional regulators that act as natural sensing devices for elevated intracellular concentrations of the effector amino acids.

Basic amino acids

Amino acid export in bacteria is expected to prevent metabolic imbalances for the cell which can be caused by overflow metabolism or an accumulation of non-metabolizable amino acids (Marin and Krämer, 2007). Thus the export of amino acids is usually tightly regulated in order to avoid both wasting of metabolic energy and inhibitory effects of high intracellular amino acid concentrations. The recently described pSenLys biosensor uses the natural specificity and affinity of the *C. glutamicum* transcriptional regulator LysG to report on the production of the basic amino acids, L-lysine, L-histidine, L-arginine and L-citrulline. LysG was shown in previous studies to activate the expression of the *lysE* operon in the presence of increasing levels of the respective effector amino acids (Bellmann *et al.*, 2001). The transporter encoded by this operon, LysE, represents the first amino acid exporter described in literature and was shown to facilitate the specific export of L-lysine and L-arginine (Vrljic *et al.*, 1996). These key features of the LysG-LysE module make it highly suitable for biosensor

Table 12.1 Overview of *Corynebacterium glutamicum* biosensors

Sensor name	Sensor protein	Type of regulator	Targets	Promoter–reporter fusion	Sensitivity	Dynamic range (fold increase)	Linear range of detection (mM)	Reference
Lrp-sensor	Lrp	Lrp-type	L-Met[a], L-Ile L-Val, L-Leu	P_{brnF}-*eyfp*	2.6^{-2}	>78	>0.2–23.5	Mustafi *et al.* (2012), Schendzielorz *et al.* (2013)
pSenLys	LysG	LysR-type	L-Lys[a], L-His, L-Arg	P_{lysE}-*eyfp*	1.6^{-2}	2.6	7.2–12.0	Binder *et al.* (2012)
pSenSer	NCgl 0581	LysR-type	L-Ser	$P_{NCgl0580}$-*eyfp*	n.d.	n.d.	n.d.	Binder *et al.* (2012)
pSenOAS	CysR	ROK-type	*O*-acetyl-L-serine	P_{cysI}-*eyfp*	n.d.	7.5	n.d.	Hoffmann *et al.* (2013), Binder *et al.* (2012)
SOS sensor	RecA-LexA	LexA repressor	Single-stranded DNA	P_{recA}-*e2-crimson*	n.d.	n.d.	n.d.	Nanda *et al.* (2014)

[a]Performance features given in this table refer to the effector amino acid marked with this letter.
n.d., not determined.

design ensuring intrinsic molecular recognition specificity combined with appropriate substrate affinity in the low to medium mM range. In general, the sensor module of pSenLys consists of the orf *lysG*, the intergenic region of *lysG* and *lysE* including the regulator binding site, and the first 23 codons of *lysE* transcriptionally fused to *eyfp* (Fig. 12.2C). Between the truncated *lysE* and *eyfp* an artificial ribosomal binding site and a stop codon was introduced. By this means, eYFP production reflects the transcriptional control of *lysE* via LysG. First proof of principle experiments revealed the general capability of the system to monitor the intracellular accumulation of the basic amino acids L-lysine, L-arginine and L-histidine. For the characterization of pSenLys, *C. glutamicum* wild type, which does not excrete L-lysine in remarkable concentrations, as well as a set of defined mutant strains varying in their L-lysine productivity (DM1920, DM1919, DM1730, DM1800, and DM1728) were transformed with the plasmid-encoded sensor. In contrast to the wild-type cells, which showed no discernible fluorescence, L-lysine production strains exhibited a significant increase in eYFP fluorescence emission, which correlated with increased extracellular and intracellular L-lysine concentrations of the respective strains, with one exception. Strain DM1920, which produces comparable amounts of L-lysine as DM1919, displayed a much lower fluorescence signal than DM1919. In contrast to the other strains, DM1920 harbours two copies of the L-lysine exporter *lysE*, and thus accumulates lower cytoplasmic L-lysine concentrations due to an increased export. This observation demonstrates the adjustment of the reporter output due to varying effector input concentrations. The use of production strains differing in their cytoplasmic L-lysine concentration revealed a dynamic range of pSenLys of about 3-fold for this particular effector amino acid (Binder *et al.*, 2012).

L-Methionine and branched-chain amino acids

Lrp represents a further, well-characterized amino acid-sensing transcriptional regulator in *C. glutamicum*, which is therefore also well suitable as sensor device for the design of biosensors (Lange *et al.*, 2012). In response to increasing cytoplasmic concentrations of L-methionine or the branched-chain amino acids, L-leucine, L-isoleucine and L-valine, Lrp activates the expression of the *brnFE* operon, encoding the proton motive force-driven secondary transport system BrnFE involved in export of the respective amino acids (Fig. 12.4A) (Kennerknecht *et al.*, 2002; Trötschel *et al.*, 2005; Lange *et al.*, 2012). For the development of a sensor based on the native Lrp-BrnFE module of

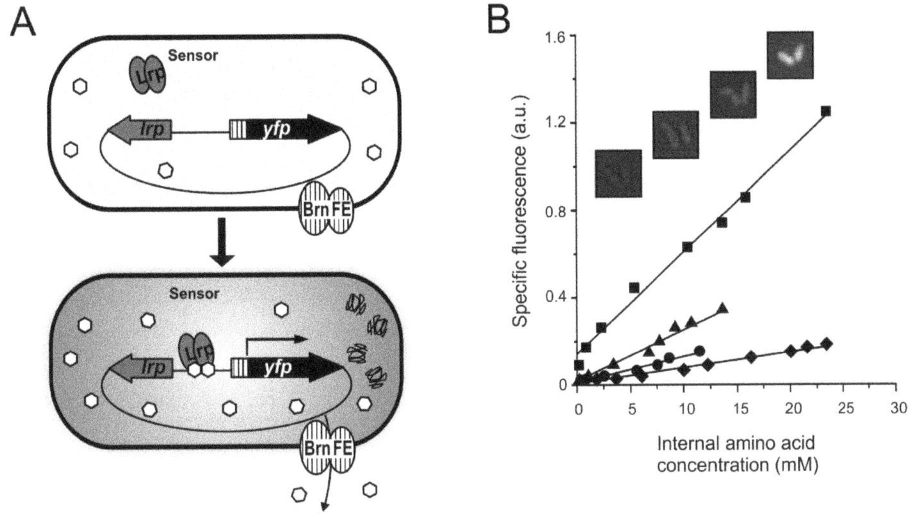

Figure 12.4 The Lrp-biosensor for intracellular amino acid detection. (A) A sensor cell with wild-type levels of the effector amino acids exhibits background levels of eYFP fluorescence (upper panel). Increased intracellular concentrations of the respective amino acids are sensed by the transcriptional regulator Lrp, which in turn induces *eyfp* expression from the promoter P_{brnF} resulting in fluorescence emission of the sensor cell (lower panel). (B) Characterization of the specificity and affinity of the Lrp-biosensor. The transfer curve displays the relationship between intracellular concentration of the effector amino acids L-methionine (■), L-leucine (▲), L-isoleucine (•), and L-valine (♦) and specific fluorescence output. The fluorescence microscopy images show *Corynebacterium glutamicum* sensor cells with increasing internal concentrations of L-methionine (1–20 mM) (adapted from Mustafi *et al.*, 2012).

C. glutamicum, a sensor cassette, including *lrp*, the intergenic region of *lrp* and *brnF*, and the first ten codons of *brnF* fused to *eyfp*, was constructed (see Fig. 12.2C).

As outlined in the section 'Characterization of biosensor performance parameters' above, biosensor assay design requires a detailed characterization of the relationship between effector input and reporter output. For the characterization of the Lrp-sensor transfer parameters defined intracellular amino acid concentrations were adjusted using the dipeptide feeding strategy (Vrljic *et al.*, 1996). The addition of dipeptides (e.g. L-alanyl-L-methionine) to the growth medium results in the uptake and subsequent hydrolysis of the dipeptides by cytoplasmic hydrolases and, thus, increases the intracellular concentration of the amino acids. To determine the transfer parameters, the intracellular amino acid concentrations were measured by silicon oil centrifugation and HPLC for all four effector amino acids (Met, Ile, Leu and Val) and set into relation with the maximal reporter output (eYFP fluorescence) (Mustafi *et al.*, 2014). The biosensor revealed highest sensitivity for L-methionine followed by L-leucine, L-isoleucine and L-valine (Fig. 12.4B). No signal was observed for any other amino acid tested. Owing to transport kinetic properties (rate of amino acid export is higher than peptide uptake rate), the system did not reach saturation. Thus, the presented values reveal only the biosensors minimal dynamic and linear range of detection (for L-methionine: 78-fold and 1–24 mM, respectively) (Mustafi *et al.*, 2014). However, the observed analogue-like response of the Lrp-biosensor makes it highly suitable for the biosensor-driven optimization of product yields.

To assess the suitability of the Lrp-sensor for online monitoring of amino acid formation in production strains, the sensor was tested in the L-valine producer *C. glutamicum ΔaceE*, lacking the E1 enzyme of the pyruvate dehydrogenase complex (Blombach *et al.*, 2007). For this purpose cultivations were performed in 48-well microtitre Flowerplates in the BioLector cultivation system (m2p-labs GmbH, Aachen, Germany), enabling the online measurement of growth as backscattered light and fluorescence. Cells transformed with the sensor plasmid displayed fluorescence emission as soon as L-valine production started after consumption of acetate (Mustafi *et al.*, 2012). Thus, the biosensor can be applied as convenient device to visualize the course of amino acid production using, for example, microcultivation systems, such as the BioLector or microfluidic chip devices (see below).

L-Serine and *O*-acetyl-L-serine

Besides pSenLys, Binder *et al.* reported on the construction and first proof of principle experiments of further metabolite sensors in *C. glutamicum* (pSenOAS and pSenSer) suitable for the detection of *O*-acetyl-L-serine and L-serine, respectively (Table 12.1) (Binder *et al.*, 2012; Hoffmann *et al.*, 2013). The pSenOAS sensor exploits the sensitivity of the dual regulator CysR, a member of the ROK family, towards *O*-acetyl-L-serine (OAS) and *O*-acetyl-L-homoserine (OAH). Among several other genes involved in the pathway for assimilatory sulfate reduction, CysR activates transcription of *cysI*, encoding sulfite reductase, in response to OAS and OAH (Rückert *et al.*, 2008). In recent studies the pSenOAS sensor was applied to visualize imbalances of sulfur assimilation and the synthesis of sulfur-containing amino acids in *C. glutamicum* (Hoffmann *et al.*, 2013).

The pSenSer sensor is based on the so far not characterized regulatory protein of the LysR superfamily NCgl0581 (cg0702) and its target gene, encoding NCgl0580 (cg0701) a carrier protein which shares about 40% sequence identity with RhtA (YbiF, b0813), which has been described as exporter of L-threonine and L-homoserine in *E. coli* (Livshits *et al.*, 2003; Marin and Krämer, 2007).

Stress responses

Cellular stress responses have long been exploited for the design of bacterial bioreporters used in assays exploring the toxicity or mutagenicity of specific chemicals or to monitor environmental contamination (van der Meer and Belkin, 2010). Many of these systems are directly or indirectly based on the bacterial SOS response; the cellular response to DNA damage conserved throughout all bacterial phyla (Friedberg *et al.*, 2005; Storz and Hengge, 2011). Here, the recombination and repair protein RecA functions as a direct sensor for single-stranded DNA, which occurs as a result of DNA damage, and activates autoproteolysis of the central repressor protein LexA. Proteolysis of LexA relieves repression of the SOS regulon, which typically comprises several genes and operons involved in DNA repair, error-prone polymerases, and cell division inhibitor proteins. For the construction of bacterial stress reporters, promoter fusions of prominent SOS genes, such as *recA*, *umuC* or *sulA*, and an appropriate reporter protein (e.g. *lacZ* or *gfp*) have been used in several studies.

Recent studies of our group aimed at the investigation of single-cell dynamics of the SOS response in *C. glutamicum* populations. For this purpose, we analysed cells containing a plasmid-based or genomically integrated reporter gene fusion of P_{recA} and further SOS target promoters to genes encoding autofluorescent proteins. Flow cytometry analyses and fluorescence microscopy revealed that a fraction of 0.1–0.3% of analysed single cells exhibited an up to 40-fold higher induction of the SOS response than the main population (Nanda *et al.*, 2014). The rate of this spontaneous SOS induction was shown to depend on the growth medium and the growth phase of the population. Induced cells exhibited a significantly decreased survival rate (40–60%) when sorted onto agar plates by FACS. Overall, this approach represents a convenient tool to assay for stress response induction and can be applied to optimize cultivation conditions in bioprocesses.

Application of biosensors in screening and single-cell analysis

In the section 'Biosensors in *Corynebacterium glutamicum*' above, several genetically encoded *C. glutamicum* biosensors, which have been developed in recent years, were summarized. Current efforts aim at the application and evaluation of this type of sensors in strain analysis and development of *C. glutamicum*. The central feature, namely the detection of a particular stimulus of interest (metabolite production, stress response, etc.) in single bacterial cells and the conversion thereof into an optical readout, opens the way for several new applications, such as:

- implementation in FACS HT screenings for the isolation of single metabolite-producing bacterial cells;
- biosensor-driven enzyme evolution/engineering;
- analysis of the complex phenotypic structure of microbial populations at single-cell resolution;
- live cell imaging of metabolite production – the time-resolved study of single-cell dynamics.

The following section summarizes the outcome of first biosensor applications in library screening and single-cell analysis of *C. glutamicum*.

Implementation of biosensors in FACS-based HT screenings

The last but crucial step in directed evolution is screening and selection. The success thereof clearly depends on the accessibility of the particular phenotype of interest. However, the majority of biotechnological-relevant molecules and products do not *per se* exhibit a conspicuous phenotype interfacing with HT screening methods. In traditional approaches, gas and liquid chromatography as well as NMR techniques have been ubiquitously applied for small molecule detection and quantification. These techniques are, however, tremendously limited in throughput. Genetically encoded biosensors can significantly improve screening throughput by converting an inconspicuous phenotype into a screenable output, e.g. fluorescence, enabling implementation in FACS HT screenings (Schallmey *et al.*, 2014).

FACS HT screenings using pSenLys and the Lrp-sensor

The *C. glutamicum* biosensors pSenLys and the Lrp-sensor have both already been applied in first screening approaches aiming at the isolation of amino acid producing mutants (Binder *et al.*, 2012; Mustafi *et al.*, 2012; Schendzielorz et *al.*, 2013). Owing to the activity of the respective biosensors, cells with enhanced cytosolic levels of the effector amino acid could be distinguished from cells with wild-type level. Experiments testing sensor-based sorting efficiency of the FACS system, revealed a >90% correct sorting efficiency and a survival rate of about 96% and 84% in case of the Lrp-sensor and the pSenLys sensor, respectively. These features allowed the implementation of these sensor systems in FACS HT screenings for amino acid producing cells. Fig. 12.5 provides a schematic overview of the different steps of the biosensor-based HT screen carried out with pSenLys and the Lrp-sensor.

For the generation of genetic diversity, *C. glutamicum* ATCC 13032 carrying a sensor plasmid (either pSenLys or the Lrp-sensor) was mutagenized by treatment with the chemical mutagen MNNG (*N*-methyl-*N'*-nitro-*N*-nitrosoguanidine). Subsequently, the obtained libraries were screened via FACS for cells showing a significantly increased fluorescent signal (Fig. 12.5). For pSenLys, the authors introduced an intermediate cultivation step and a second marker protein for screening (E2-Crimson) to increase the percentage of viable clones (Binder *et al.*, 2012). As both sensors report exclusively on cytosolic accumulation of the respective amino acids, information with respect to the excretion is not provided. Therefore, a second screen was introduced to measure amino acid accumulation in the supernatant. This step is crucial in order to discriminate 'false positive' clones, which only internally accumulate high concentrations of the

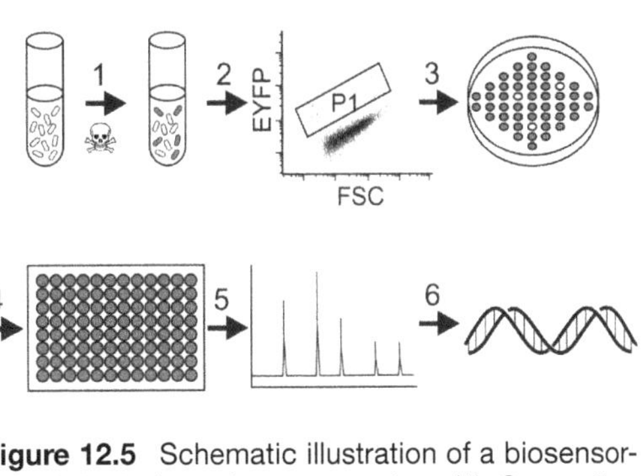

Figure 12.5 Schematic illustration of a biosensor-based high-throughput screening. (1) Generation of genetic diversity by chemical mutagenesis. (2) Screening of mutant library and isolation of single mutant cells via FACS. (3) Sorting of single cells on plates, (4) cultivation in microtitre plates, and (5) quantification of external amino acids via uHPLC. (6) Genome sequencing for target identification (adapted from Mustafi *et al.*, 2012).

amino acids (e.g. due to defects in export), from clones excreting significant amounts of the effector amino acids. Finally, either selected target genes or the whole genome of mutants were sequenced to identify novel and unexpected targets for strain development. In the following the results of these first screening attempts are summarized.

The screening approach using pSenLys started with chemically mutagenized *C. glutamicum* wild type and resulted in the isolation of 69% positive mutants (185 of 270 clones analysed via HPLC) accumulating L-lysine in the supernatant after 48-hour cultivation in CGXII minimal medium. Re-cultivation of 120 positively tested clones showed L-lysine concentrations ranging from 0.2 to 37 mM (Binder *et al.*, 2012). Sequence analysis of selected genes revealed several mutations in *lysC*, of which some were already known to prevent feedback inhibition of aspartate kinase activity. Besides, four new mutations in the regulatory subunit of the aspartate kinase, seven novel mutations in the *hom* gene, encoding the homoserine dehydrogenase and four new mutations in genes coding for L-threonine synthesis, *thrB* and *thrC*, were identified. Whole genome sequencing of selected mutants revealed a novel target gene of particular interest, namely *murE*, encoding UDP-*N*-acetylmuramyl-tripeptide synthetase. This enzyme utilizes the direct precursor of L-lysine D, L-diaminopimelate as substrate and channels it towards cell wall synthesis. Genomic replacement of the *murE* mutation in the wild type resulted in significant L-lysine production (up to 24 mM) and an increase of L-lysine production in defined producer strains, i.e. up to 27% in the best producer DM 1933 (Binder *et al.*, 2012).

The first screening approach using the Lrp-sensor was also started from randomly mutagenized *C. glutamicum* wild-type cells. This attempt resulted in the isolation of 21% (40 of 192 clones) positive clones that produced branched-chain amino acids (Mustafi *et al.*, 2012). Among these, the top five mutants produced up to 8 mM L-valine, up to 2 mM of L-isoleucine and up to 1 mM L-leucine after 48-hour cultivation in CGXII minimal medium. The conditions chosen for this initial screening obviously favoured the isolation of branched-chain amino acid producing mutants. None of the mutants accumulated significant amounts of L-methionine, which is surely a challenging target as it demands sulfur incorporation and precursors from C1-metabolism requiring high energy cost (synthesis of 1 mol L-methionine requires 7 mol ATP and 8 mol NADPH) (Neidhardt *et al.*, 1990). As L-methionine is currently of major interest for biotechnological strain development, following experiments aimed at the adaptation of the screen towards the isolation of L-methionine producing mutants. First studies revealed that the outcome of biosensor-based screenings depends on several aspects, including (i) cultivation conditions, (ii) the background strain used for mutagenesis, and (iii) the gating strategy of FACS (Mustafi and Frunzke, unpublished). For this reason mutant libraries based on *C. glutamicum* Δ*mcbR*, lacking the master regulator McbR of L-methionine and L-cysteine biosynthesis in *C. glutamicum* were screened in following studies (Rey *et al.*, 2003). Owing to the tight regulation of biosynthesis genes, mediated by McbR, removing transcriptional repression is suggested to be an important prerequisite for the isolation of L-methionine producing mutants. Overall, this screen resulted in 49% (47 of 96 clones) positive clones producing significant amounts of L-methionine (at least 2-fold more than the parental strain *C. glutamicum* Δ*mcbR*). Among these, the top three mutants produced up to 8 mM L-methionine after 48-hour cultivation in minimal medium (Mustafi and Frunzke, unpublished).

Further screening applications of single-cell biosensors

As mentioned above, the described sensor systems enable the intracellular measurement of amino acid concentrations in *C. glutamicum* cells. This feature can also be of benefit in a screening setup for the identification of novel transport systems. For instance in the case of L-methionine the existence of a second low-affinity export system besides BrnFE has been postulated by Krämer and co-workers in 2005, but it has not been identified to date (Trötschel *et al.*, 2005). Using the sensor system, mutant strains with defects in the export of the particular effector amino acid (e.g. mutation of LysE or BrnFE) will accumulate significantly increased internal amounts of the respective amino acids leading to a high fluorescence readout of the sensor. Indeed, detailed characterization of L-lysine producing mutants ($n=40$) using an expectation maximization algorithm revealed one cluster of mutants exhibiting extreme high fluorescence, but comparatively low L-lysine accumulation (Binder *et al.*, 2012). Targeted isolation of such mutants in screenings might be a promising approach for the identification of novel transport systems for the effector amino acids of the applied sensor.

For production of a specific metabolite of interest, researchers often rely on the performance of heterologously expressed pathways or enzymes. Directed evolution approaches have been used to optimize enzyme activity and specificity of a variety of different enzymes. Most assays are, however, carried out *in vitro* and cannot be adapted for enzyme evolution using whole cells (Arnold and Georgiou, 2003). As metabolite biosensors allow specific detection of small molecules, suchlike sensor systems can also be applied for *in vivo* enzyme evolution. The workflow for such an approach comprises the construction of an enzyme library using e.g. error-prone PCR, transformation of sensor cells with the resulting library, and finally FACS screening and validation of obtained variants. A striking example was recently reported by Michener *et al.* (2012) who developed an RNA aptamer-based sensor system to screen for improved enzyme activities of caffeine demethylase (Michener and Smolke, 2012). In contrast to conventional direct evolution techniques, this approach allows high-throughput analysis of mutant enzyme libraries, where enzyme activity is assayed *in vivo*.

Biosensor-based single-cell analysis

Nowadays, there is increasing evidence that clonal microbial populations display substantial variation in phenotypic traits. This common occurrence of phenotypic heterogeneity can originate, for example, from microenvironmental variations, differences in cell age or cell cycle, epigenetic variation or simply stochastic effects in gene expression (Elowitz, 2002; Lidstrom and Konopka, 2010). However, the aspect of population heterogeneity has hardly been studied in industrial productions strains, although inefficient subpopulations within a production culture might have a significant negative impact on the whole bioprocess (Müller *et al.*, 2010). For a detailed analysis of the phenotypic structure of microbial populations, novel techniques are required, which enable quantification and real-time monitoring of metabolites at the single cell level. In this context genetically encoded biosensors represent a powerful tool to analyse metabolite production in individual cells using flow cytometry, lab-on-a-chip devices, or time-lapse microscopy.

Flow cytometry

In recent years, more and more attention was drawn to the use of flow cytometry as a convenient HT-compatible method allowing the measurement of multiple parameters of a single microbial cell (Müller and Nebe-von-Caron, 2010; Tracy *et al.*, 2010). Parameters which can be assessed by flow cytometry include the cellular scattering properties (indicative for cell size and cellular complexity) as well as every phenotype which can be directly or indirectly linked to a fluorescent output (Müller and Nebe-von-Caron, 2010). Recently, Neumeyer *et al.* (2013) reported on the use of multiparameter flow cytometry to assess different phenotypic parameters of *C. glutamicum* populations. Application of the described procedures revealed, in fact, phenotypic heterogeneity in terms of viability, membrane potential and growth activity (DNA pattern) of *C. glutamicum* grown on standard minimal or complex media.

Current efforts aim at the establishment of downstream approaches for the analysis of subpopulations to not only monitor, but, more importantly, understand the molecular basis of the observed phenotypic variation. Müller and co-workers recently described a workflow for combining FACS cell sorting with mass spectrometry based proteomic analysis of subpopulations (Jehmlich *et al.*, 2010; Jahn *et al.*, 2012; Müller and Hiller, 2012). Current advances in mass spectrometry and sequencing technologies bring the analysis of small subpopulations or even single cells into reach. Furthermore, a variety of fluorescent dyes and staining protocols is available to assess multiple parameters at single-cell resolution. The implementation of metabolite-sensing biosensors into this framework will represent a further important step towards a systems level analysis of microbial populations in bioprocesses.

Microscopy and live cell imaging

In contrast to the snapshot analysis provided by flow cytometry, live cell imaging approaches offer the advantage of long-term cell observation and analysis during homogeneous and well-controllable cultivation conditions in microscale environments (Locke and Elowitz, 2009). Recently, Grünberger *et al.* (2012) reported on the fabrication and implementation of a microfluidic picoliter bioreactor allowing parallel studies of multiple microcolonies growing in a monolayer (Fig. 12.6A and B). First studies investigating growth of *C. glutamicum* in these microfluidic devices revealed a 50% higher growth rate of the wild-type strain than cultivation in shake flasks. Optimal growth conditions, such as continuous medium flow ensuring optimal supply with nutrients and oxygen as well as the removal of secreted by-products, offer an explanation for the observed increase in growth rate (Grünberger *et al.*, 2013).

This experimental setup was used in first live cell imaging studies to monitor L-valine production of pyruvate dehydrogenase complex-deficient *C. glutamicum* strains during growth in microfluidic chambers (Fig. 12.6C) (Mustafi *et al.*, 2014). Strains were transformed with the Lrp-sensor to study growth and amino acid production at single-cell resolution. Live cell imaging of microcolonies

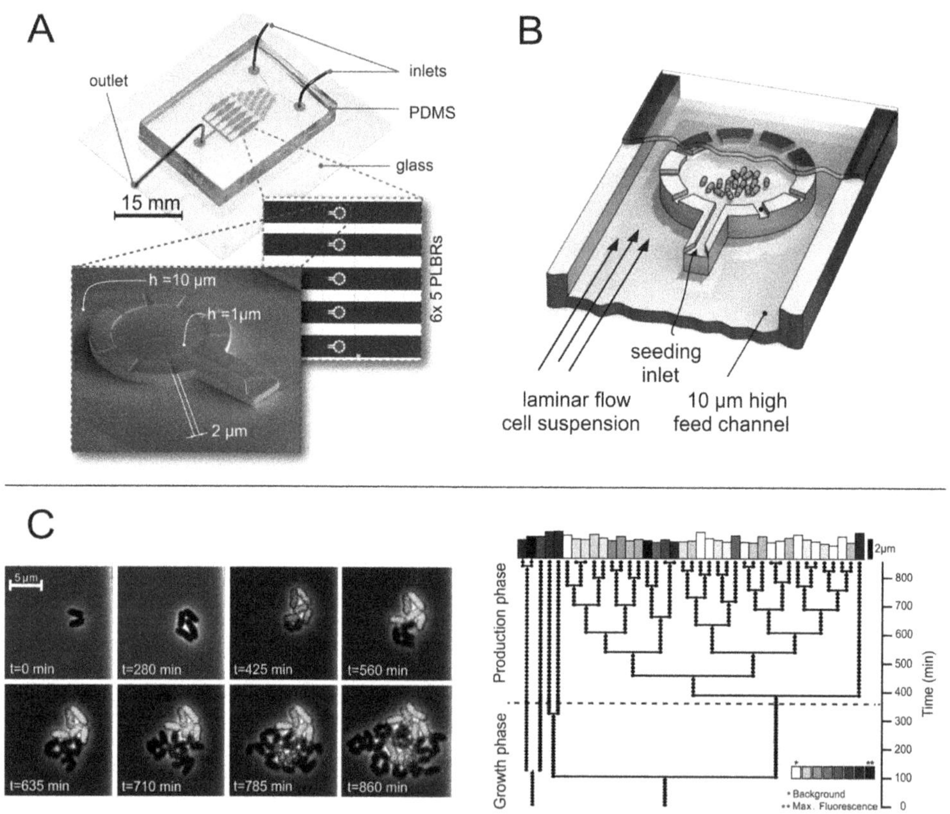

Figure 12.6 Biosensor-based analysis of a L-valine production strain in lab-on-a-chip devices. (A) Images showing the fabricated microfluidic chip devices (at the top), zooming into one growth array (in the middle). Scanning electron microscopy image (at the bottom) and (B) schematic illustration of a single picoliter bioreactor. (C) Microscopic image of an isogenic microcolony of *Corynebacterium glutamicum* Δ*aceE* containing the Lrp-sensor and the respective lineage tree, illustrating phenotypic heterogeneity upon initiation of the production phase (see, t = 425 minutes) (adapted from Grünberger *et al.*, 2012; Mustafi *et al.*, 2014). A colour version of this figure is available in the plate section at the back of the book.

(30–40 cells) grown from one single ancestor cell in the abovementioned picoliter bioreactor revealed variations in doubling time, cell size and, most interestingly, single-cell fluorescence, suggesting significant cell-to-cell variation with respect to L-valine production (Mustafi *et al.*, 2014). The observed phenotypic heterogeneity within isogenic microcolonies would have hardly been detected in typical shake flask experiments as here L-valine accumulates in the supernatant and thus influences internal L-valine levels. Overall this experimental setup in combination with a genetically encoded sensor system provided a first glimpse into phenotypic heterogeneity of a *C. glutamicum* strain in terms of growth and metabolite production.

Future perspectives and concluding remarks

In the last few years the development of biosensors and their implementation in HT screenings and single-cell analysis, is a rapidly emerging field of research in microbial biotechnology (Schallmey *et al.*, 2014). However, the great potential of biosensors based on regulatory circuits (transcription factors or RNA aptamers) for the detection and quantification of industrially relevant metabolites has not been exploited so far. First studies reporting on the design of *C. glutamicum* biosensors for the intracellular detection of amino acids already highlighted the impact this novel approach might have on future attempts regarding strain development and bioprocess monitoring. First biosensor-based FACS HT screenings resulted in the isolation of numerous mutant strains producing the respective amino acid of interest. Furthermore, current advances in the development of novel microfluidic devices for live cell imaging allow high resolution insights into phenotypic heterogeneity within clonal populations. However, also several critical issues need to be addressed in future studies: What are the limitations of the system? To which extent does the production of fluorescent reporter proteins impact the physiological state and, thus, the phenotype of the respective cell? Is this approach limited to a few selected success stories or can we believe the promises of synthetic biology that a specific and sensitive sensor system can be developed and adapted for almost every small molecule of interest? Time will tell.

References

Albrecht, M., Takaichi, S., Steiger, S., Wang, Z. Y., and Sandmann, G. (2000) Novel hydroxycarotenoids with improved antioxidative properties produced by gene combination in *Escherichia coli*. *Nat. Biotechnol.* **18**, 843–846.

An, G. H., Bielich, J., Auerbach, R., and Johnson, E. A. (1991) Isolation and characterization of carotenoid hyperproducing mutants of yeast by flow-cytometry and cell sorting. *Biotechnology* **9**, 70–73.

Andersen, J. B., Sternberg, C., Poulsen, L. K., Bjorn, S. P., Givskov, M., and Molin, S. (1998) New unstable variants of green fluorescent protein for studies of transient gene expression in bacteria. *Appl. Environ. Microbiol.* **64**, 2240–2246.

Arnold, F. H., and Georgiou, G. (2003) Directed Enzyme Evolution: Screening and Selection Methods. Humana Press, Totowa, NJ, USA.

Azuma, T., Harrison, G. I., and Demain, A. L. (1992) Isolation of a gramicidin-s hyperproducing strain of bacillus-brevis by use of a fluorescence activated cell sorting system. *Appl. Microbiol. Biotechnol.* **38**, 173–178.

Becker, J., and Wittmann, C. (2012) Bio-based production of chemicals, materials and fuels – *Corynebacterium glutamicum* as versatile cell factory. *Curr. Opin. Biotechnol.* **23**, 631–640.

Belkin, S. (2003) Microbial whole-cell sensing systems of environmental pollutants. *Curr. Opin. Microbiol.* **6**, 206–212.

Bellmann, A., Vrljic, M., Patek, M., Sahm, H., Krämer, R., and Eggeling, L. (2001) Expression control and specificity of the basic amino acid exporter LysE of *Corynebacterium glutamicum*. *Microbiology* **147**, 1765–1774.

Bernhardt, P., McCoy, E., and O'Connor, S. E. (2007) Rapid identification of enzyme variants for reengineered alkaloid biosynthesis in periwinkle. *Chem. Biol.* **14**, 888–897.

Bertels, F., Merker, H., and Kost, C. (2012) Design and characterization of auxotrophy-based amino acid biosensors. *PLoS One* **7**, e41349.

Binder, S., Schendzielorz, G., Stäbler, N., Krumbach, K., Hoffmann, K., Bott, M., and Eggeling, L. (2012) A high-throughput approach to identify genomic variants of bacterial metabolite producers at the single-cell level. *Genome Biol.* **13**, R40.

Bintu, L., Buchler, N. E., Garcia, H. G., Gerland, U., Hwa, T., Kondev, J., Kuhlman, T., and Phillips, R. (2005a) Transcriptional regulation by the numbers: applications. *Curr. Opin. Gen. Develop.* **15**, 125–135.

Bintu, L., Buchler, N. E., Garcia, H. G., Gerland, U., Hwa, T., Kondev, J., and Phillips, R. (2005b) Transcriptional regulation by the numbers: models. *Curr. Opin. Gen. Develop.* **15**, 116–124.

Blombach, B., and Eikmanns, B. J. (2011) Current knowledge on isobutanol production with *Escherichia coli*, *Bacillus subtilis* and *Corynebacterium glutamicum*. *Bioeng. Bugs* **2**, 346–350.

Blombach, B., Schreiner, M. E., Holatko, J., Bartek, T., Oldiges, M., and Eikmanns, B. J. (2007) L-Valine production with pyruvate dehydrogenase complex-deficient *Corynebacterium glutamicum*. *Appl. Environ. Microbiol.* **73**, 2079–2084.

Burkholder, P. R. (1951) Determination of vitamin-B12 with a mutant strain of *Escherichia coli*. *Science* **114**, 459–460.

Burkovski, A. (2008) *Corynebacteria*: Genomics and Molecular Biology. Caister Academic Press, Norfolk, UK.

Burmann, F., Sawant, P., and Bramkamp, M. (2012) Identification of interaction partners of the dynamin-like protein DynA from *Bacillus subtilis*. *Commun. Integr. Biol.* **5**, 362–369.

Chalova, V. I., Kim, W. K., Woodward, C. L., and Ricke, S. C. (2007) Quantification of total and bioavailable lysine in feed protein sources by a whole-cell green fluorescent protein growth-based *Escherichia coli* biosensor. *Appl. Microbiol. Biotechnol.* **76**, 91–99.

Cho, S., Yang, S., and Rhie, H. (2012) The gene encoding the alternative thymidylate synthase ThyX is regulated by sigma factor SigB in *Corynebacterium glutamicum* ATCC 13032. *FEMS Microbiol. Lett.* **328**, 157–165.

Daunert, S., Barrett, G., Feliciano, J. S., Shetty, R. S., Shrestha, S., and Smith-Spencer, W. (2000) Genetically engineered whole-cell sensing systems: coupling biological recognition with reporter genes. *Chem. Rev.* **100**, 2705–2738.

de la Pena Mattozzi, M., Tehara, S. K., Hong, T., and Keasling, J. D. (2006) Mineralization of paraoxon and its use as a sole C and P source by a rationally designed catabolic pathway in *Pseudomonas putida*. *Appl. Environ. Microbiol.* **72**, 6699–6706.

Dietrich, J. A., McKee, A. E., and Keasling, J. D. (2010) High-throughput metabolic engineering: advances in small-molecule screening and selection. *Annu. Rev. Biochem.* **79**, 563–590.

Donovan, C., Schwaiger, A., Krämer, R., and Bramkamp, M. (2010) Subcellular localization and characterization of the ParAB system from *Corynebacterium glutamicum*. *J. Bacteriol.* **192**, 3441–3451.

Drepper, T., Eggert, T., Circolone, F., Heck, A., Krauss, U., Guterl, J.-K., Wendorff, M., Losi, A., Gärtner, W., and Jäger, K.-E. (2007) Reporter proteins for *in vivo* fluorescence without oxygen. *Nat. Biotechnol.* **25**, 443–445.

Drepper, T., Huber, R., Heck, A., Circolone, F., Hillmer, A.-K., Büchs, J., and Jäger, K.-E. (2010) Flavin mononucleotide-based fluorescent reporter proteins outperform green fluorescent protein-like proteins as quantitative *in vivo* real-time reporters. *Appl. Environ. Microbiol.* **76**, 5990–5994.

Eggeling, L., and Bott, M. (2005) Handbook of *Corynebacterium glutamicum*. CRC Press, Boca Raton, FL, USA.

Elowitz, M. B. (2002) Stochastic gene expression in a single cell. *Science* **297**, 1183–1186.

Erickson, A. M., Diaz, I. B. Z., Kwon, Y. M., and Ricke, S. C. (2000) A bioluminescent *Escherichia coli* auxotroph for use in an *in vitro* lysine availability assay. *J. Microbiol. Methods* **40**, 207–212.

Friedberg, E. C., Walker, G. C., and Siede, W. (2005) DNA Repair and Mutagenesis. American Society for Microbiology, Washington, DC, USA.

Frommer, W. B., Davidson, M. W., and Campbell, R. E. (2009) Genetically encoded biosensors based on engineered fluorescent proteins. *Chem. Soc. Rev.* **38**, 2833.

Frunzke, J., Bramkamp, M., Schweitzer, J. E., and Bott, M. (2008) Population heterogeneity in *Corynebacterium glutamicum* ATCC 13032 caused by prophage CGP3. *J. Bacteriol.* **190**, 5111–5119.

Galvao, T. C., Mencia, M., and de Lorenzo, V. (2007) Emergence of novel functions in transcriptional regulators by regression to stem protein types. *Mol. Microbiol.* **65**, 907–919.

Ghim, C.-M., Lee, S. K., Takayama, S., and Mitchell, R. J. (2010) The art of reporter proteins in science: past, present and future applications. *BMB Rep.* **43**, 451–460.

Giliberti, J., O'Donnell, S., van Etten, W. J., and Janssen, G. R. (2012) A 5′-terminal phosphate is required for stable ternary complex formation and translation of leaderless mRNA in *Escherichia coli*. *RNA* **18**, 508–518.

Grünberger, A., Paczia, N., Probst, C., Schendzielorz, G., Eggeling, L., Noack, S., Wiechert, W., and Kohlheyer, D. (2012) A disposable picolitre bioreactor for cultivation and investigation of industrially relevant bacteria on the single cell level. *Lab Chip* **12**, 2060.

Grünberger, A., van Ooyen, J., Paczia, N., Rohe, P., Schiendzielorz, G., Eggeling, L., Wiechert, W., Kohlheyer, D., and Noack, S. (2013) Beyond growth rate 0.6: *Corynebacterium glutamicum* cultivated in highly diluted environments. *Biotechnol. Bioeng.* **110**, 220–228.

Hakkila, K., Maksimow, M., Karp, M., and Virta, M. (2002) Reporter genes *lucFF*, *luxCDABE*, *gfp*, and *dsred* have different characteristics in whole-cell bacterial sensors. *Anal. Biochem.* **301**, 235–242.

Harms, H., Wells, M. C., and van der Meer, J. R. (2006) Whole-cell living biosensors – are they ready for environmental application? *Appl. Microbiol. Biotechnol.* **70**, 273–280.

Heinemann, M., and Zenobi, R. (2011) Single cell metabolomics. *Curr. Opin. Biotechnol.* **22**, 26–31.

Hentschel, E., Will, C., Mustafi, N., Burkovski, A., Rehm, N., and Frunzke, J. (2012) Destabilized eYFP variants for dynamic gene expression studies in *Corynebacterium glutamicum*. *Microb. Biotechnol.* **6**, 196–201.

Hoffmann, K., Grünberger, A., Lausberg, F., Bott, M., and Eggeling, L. (2013) Visualization of imbalances in sulfur assimilation and synthesis of sulfur-containing amino acids at the single-cell level. *Appl. Environ. Microbiol.* **79**, 6730–6736.

Jahn, M., Seifert, J., von Bergen, M., Schmid, A., Bühler, B., and Müller, S. (2012) Subpopulation-proteomics in prokaryotic populations. *Curr. Opin. Biotechnol.* **24**, 79–87.

Jehmlich, N., Hübschmann, T., Salazar, M. G., Völker, U., Benndorf, D., Müller, S., von Bergen, M., and Schmidt, F. (2010) Advanced tool for characterization of microbial cultures by combining cytomics and proteomics. *Appl. Microbiol. Biotechnol.* **88**, 575–584.

Kennerknecht, N., Sahm, H., Yen, M. R., Patek, M., Saier, M. H., and Eggeling, L. (2002) Export of L-isoleucine from *Corynebacterium glutamicum*: a two-gene-encoded member of a new translocator family. *J. Bacteriol.* **184**, 3947–3956.

Kim, M. I., Yu, B. J., Woo, M.-A., Cho, D., Dordick, J. S., Cho, J. H., Choi, B.-O., and Park, H. G. (2010) Multiplexed amino acid array utilizing bioluminescent *Escherichia coli* auxotrophs. *Anal. Chem.* **82**, 4072–4077.

Lange, C., Mustafi, N., Frunzke, J., Kennerknecht, N., Wessel, M., Bott, M., and Wendisch, V. F. (2012) Lrp of *Corynebacterium glutamicum* controls expression of the *brnFE* operon encoding the export system for L-methionine and branched-chain amino acids. *J. Biotechnol.* **158**, 231–241.

Lidstrom, M. E., and Konopka, M. C. (2010) The role of physiological heterogeneity in microbial population behavior. *Nat. Chem. Biol.* **6**, 705–712.

Livshits, V. A., Zakataeva, N. P., Aleshin, V. V., and Vitushkina, M. V. (2003) Identification and characterization of the new gene *rhtA* involved in threonine and homoserine efflux in *Escherichia coli*. *Res. Microbiol.* **154**, 123–135.

Locke, J. C. W., and Elowitz, M. B. (2009) Using movies to analyse gene circuit dynamics in single cells. *Nat. Rev. Microbiol.* **7**, 383–392.

Lutz, R., and Bujard, H. (1997) Independent and tight regulation of transcriptional units in *Escherichia coli via* the LacR/O, the TetR/O and AraC/I-1-I-2 regulatory elements. *Nucleic Acids Res.* **25**, 1203–1210.

Marin, K., and Krämer, R. (2007) Amino acid transport systems in biotechnologically relevant bacteria. In: Wendisch V. F. (ed.) Amino Acid Biosynthesis – Pathways, Regulation and Metabolic Engineering. Microbiol Monographs, Springer Berlin Heidelberg, Germany, pp. 289–325.

Michener, J. K., and Smolke, C. D. (2012) High-throughput enzyme evolution in *Saccharomyces cerevisiae* using a synthetic RNA switch. *Metab. Eng.* **14**, 306–316.

Michener, J. K., Thodey, K., Liang, J. C., and Smolke, C. D. (2012) Applications of genetically-encoded biosensors for the construction and control of biosynthetic pathways. *Metab. Eng.* **14**, 212–222.

Moll, I., Grill, S., Gualerzi, C. O., and Blasi, U. (2002) Leaderless mRNAs in bacteria: surprises in ribosomal recruitment and translational control. *Mol. Microbiol.* **43**, 239–246.

Müller, S., Harms, H., and Bley, T. (2010) Origin and analysis of microbial population heterogeneity in bioprocesses. *Curr. Opin. Biotechnol.* **21**, 100–113.

Müller, S., and Hiller, K. (2012) From multi-omics to basic structures of biological systems. *Curr. Opin. Biotechnol.* **24**, 1–3.

Müller, S., and Nebe-von-Caron, G. (2010) Functional single-cell analyses: flow cytometry and cell sorting of microbial populations and communities. *FEMS Microbiol. Rev.* **34**, 554–587.

Mustafi, N., Grünberger, A., Kohlheyer, D., Bott, M., and Frunzke, J. (2012) The development and application of a single-cell biosensor for the detection of L-methionine and branched-chain amino acids. *Metab. Eng.* **14**, 449–457.

Mustafi, N., Grünberger, A., Mahr, R., Helfrich, S., Nöh, K., Blombach, B., Kohlheyer, D., and Frunzke, J. (2014) Application of a genetically encoded biosensor for live cell imaging of L-valine production in pyruvate dehydrogenase complex-deficient *Corynebacterium glutamicum* strains. *PLoS One* **9**, e85731.

Nanda, A. M., Heyer, A., Krämer, C., Grünberger, A., Kohlheyer, D., and Frunzke, J. (2014) Analysis of SOS-induced spontaneous prophage induction in *Corynebacterium glutamicum* at the single-cell level. *J. Bacteriol.* **196**, 180–188.

Neidhardt, F. C., Ingraham, J. L., and Schaechter, M., (1990) Physiology of the Bacterial Cell: A Molecular Approach. Sinauer Associates, Sunderland, MA, USA, p. 507.

Neumeyer, A., Hübschmann, T., Müller, S., and Frunzke, J. (2013) Monitoring of population dynamics of *Corynebacterium glutamicum* by multiparameter flow cytometry. *Microb. Biotechnol.* **6**, 157–167.

Ohlendorf, R., Vidavski, R. R., Eldar, A., Moffat, K., and Möglich, A. (2012) From dusk till dawn: one-plasmid systems for light-rRegulated gene expression. *J. Mol. Biol.* **416**, 534–542.

Okumoto, S., Jones, A., and Frommer, W. B., (2012) Quantitative imaging with fluorescent biosensors. *Ann. Rev. Plant Biol.* **63**, 663–706.

Patek, M., (2005) Regulation of gene expression. In: Eggeling, L., and Bott, M. (eds.) Handbook of *Corynebacterium glutamicum*. CRC Press, Boca Raton, FL, USA, pp. 81–98.

Payne, J. W., Bell, G., and Higgins, C. F. (1977) Use of an *Escherichia coli* lys-auxotroph to assay nutritionally available lysine in biological-materials. *J. Appl. Bacteriol.* **42**, 165–177.

Pfeifer-Sancar, K., Mentz, A., Rückert, C., and Kalinowski, J. (2013) Comprehensive analysis of the *Corynebacterium glutamicum* transcriptome using an improved RNAseq technique. *BMC Genomics* **14**, 888.

Rey, D. A., Pühler, A., and Kalinowski, J. (2003) The putative transcriptional repressor McbR, member of the TetR-family, is involved in the regulation of the metabolic network directing the synthesis of sulfur containing amino acids in *Corynebacterium glutamicum*. *J. Biotechnol.* **103**, 51–65.

Rückert, C., Milse, J., Albersmeier, A., Koch, D. J., Pühler, A., and Kalinowski, J. (2008) The dual transcriptional regulator CysR in *Corynebacterium glutamicum* ATCC 13032 controls a subset of genes of the McbR regulon in response to the availability of sulphide acceptor molecules. *BMC Genomics* **9**, 483.

Santos, C. N. S., and Stephanopoulos, G. (2007) Melanin-based high-throughput screen for L-tyrosine production in *Escherichia coli*. *Appl. Environ. Microbiol.* **74**, 1190–1197.

Schallmey, M., Frunzke, J., Eggeling, L., and Marienhagen, J. (2014) Looking for the pick of the bunch: high-throughput screening of producing microorganisms with biosensors. *Curr. Opin. Biotechnol.* **26C**, 148–154.

Schendzielorz, G., Dippong, M., Grünberger, A., Kohlheyer, D., Yoshida, A., Binder, S., Nishiyama, C., Nishiyama, M., Bott, M., and Eggeling, L. (2013) Taking control over control: use of product sensing in single cells to remove flux control at key enzymes in biosynthesis pathways. *ACS Synth. Biol.* **3**, 21–29.

Shaner, N. C., Steinbach, P. A., and Tsien, R. Y. (2005) A guide to choosing fluorescent proteins. *Nature Methods* **2**, 905–909.

Storz, G., and Hengge, R. (2011) Bacterial Stress Responses, 2nd edition. American Society for Microbiology, Washington, DC, USA.

Tabor, J. J., Groban, E. S., and Voigt, C. A. (2009) Performance characteristics for sensors and circuits used to program *E. coli*. In: Lee S. Y. (ed.) Systems Biology and Biotechnology of *Escherichia coli*. Springer, New York, NY, USA, pp. 402–433.

Tang, S.-Y., and Cirino, P. C. (2011) Design and application of a mevalonate-responsive regulatory protein. *Angewandte Chemie-International Edition* **50**, 1084–1086.

Teramoto, H., Watanabe, K., Suzuki, N., Inui, M., and Yukawa, H. (2011) High yield secretion of heterologous proteins in *Corynebacterium glutamicum* using its own Tat-type signal sequence. *Appl. Microbiol. Biotechnol.* **91**, 677–687.

Tracy, B. P., Gaida, S. M., and Papoutsakis, E. T. (2010) Flow cytometry for bacteria: enabling metabolic engineering, synthetic biology and the elucidation of complex phenotypes. *Curr. Opin. Biotechnol.* **21**, 85–99.

Trötschel, C., Deutenberg, D., Bathe, B., Burkovski, A., and Krämer, R. (2005) Characterization of methionine export in *Corynebacterium glutamicum*. *J. Bacteriol.* **187**, 3786–3794.

van der Meer, J. R. (2010) Bacterial Sensors: Synthetic Design and Application Principles. Morgan & Claypool, UK.

van der Meer, J. R., and Belkin, S. (2010) Where microbiology meets microengineering: design and applications of reporter bacteria. *Nat. Rev. Microbiol.* **8**, 511–522.

van der Meer, J. R., Tropel, D., and Jaspers, M. (2004) Illuminating the detection chain of bacterial bioreporters. *Environ. Microbiol.* **6**, 1005–1020.

Vasicova, P., Patek, M., Nesvera, J., Sahm, H., and Eikmanns, B. (1999) Analysis of the *Corynebacterium glutamicum dapA* promoter. *J. Bacteriol.* **181**, 6188–6191.

Vrljic, M., Sahm, H., and Eggeling, L. (1996) A new type of transporter with a new type of cellular function: L-lysine export from *Corynebacterium glutamicum*. *Mol. Microbiol.* **22**, 815–826.

Wall, M. E., Hlavacek, W. S., and Savageau, M. A. (2004) Design of gene circuits: lessons from bacteria. *Nature Rev. Genet.* **5**, 34–42.

Wendisch, V. F. (2007) Amino Acid Biosynthesis – Pathways, Regulation and Metabolic Engineering. Microbiol Monographs, Springer Berlin Heidelberg, Germany.

Wieschalka, S., Blombach, B., Bott, M., and Eikmanns, B. J. (2012) Bio-based production of organic acids with *Corynebacterium glutamicum*. *Microb. Biotechnol.* **6**, 87–102.

Willardson, B. M., Wilkins, J. F., Rand, T. A., Schupp, J. M., Hill, K. K., Keim, P., and Jackson, P. J. (1998) Development and testing of a bacterial biosensor for toluene-based environmental contaminants. *Appl. Environ. Microbiol.* **64**, 1006–1012.

Zhang, F., and Keasling, J. (2011) Biosensors and their applications in microbial metabolic engineering. *Trends Microbiol.* **19**, 323–329.

Index

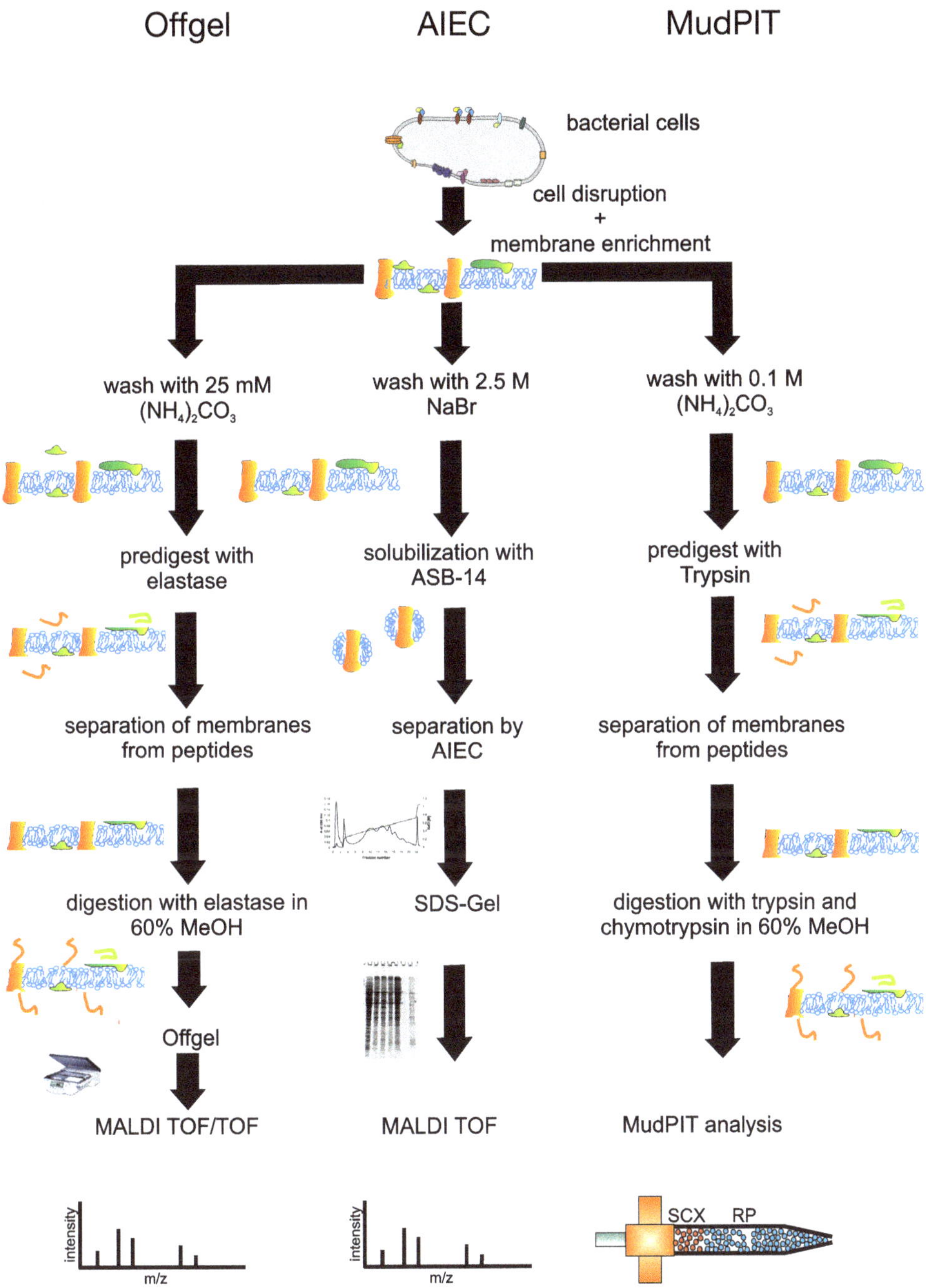

Figure 2.1 Schematic of different workflows for membrane enrichment and mass spectrometry-based analysis including peptide separation using Offgel, membrane protein separation with anion exchange chromatography (AIEC) and multidimensional protein identification technology (MudPIT).

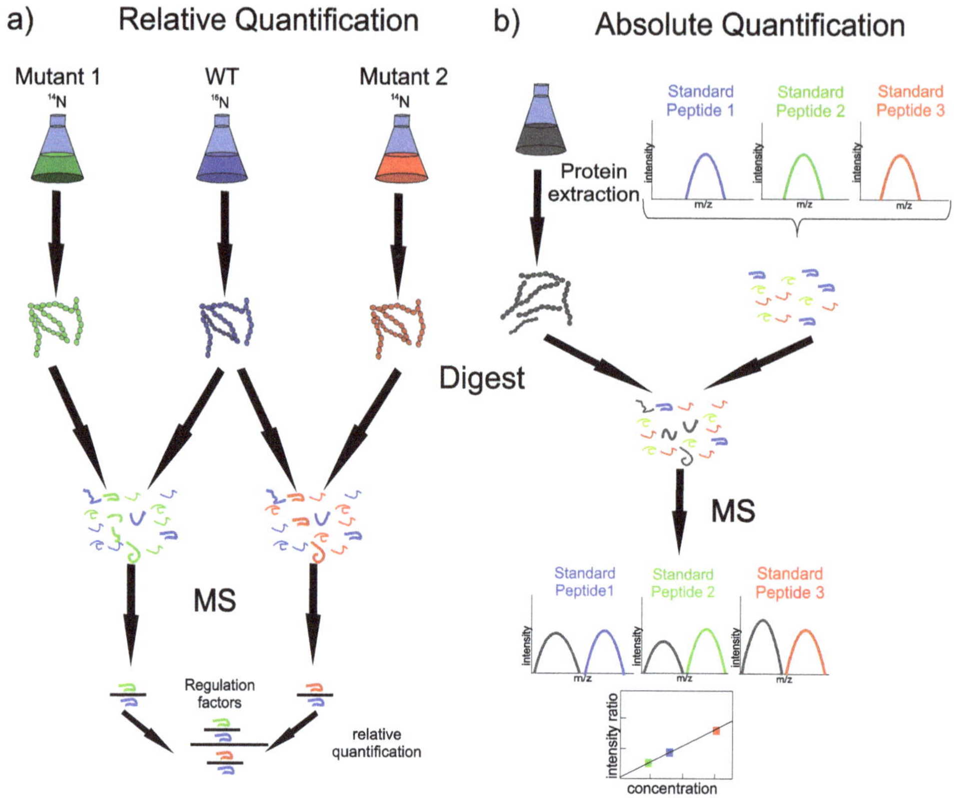

Figure 2.2 Schematic of workflows for quantification methods used for *Corynebacterium glutamicum*. (a) For relative quantification, mutant cultures are grown on ^{14}N medium and a wild-type (WT) culture grown on ^{15}N serves as internal standard. Cell extracts are mixed 1:1 and analysed using high-resolution mass spectrometry (MS). The ratios of unlabelled proteins to labelled proteins are calculated and called regulation factors. These factors can be used to elucidate if a protein is differentially regulated in two conditions. (b) Absolute quantification using standard peptides. Labelled peptides of the same sequence as found in the sample are synthesized and constant amounts are added to a digested sample of interest. Abundance of proteins is estimated using the peptides intensity ratios.

Figure 4.1 Graphical representation of the new *Corynebacterium glutamicum* genome-scale network *iEZ475*. The network comprises 475 reactions interconverting 408 metabolites. The reactions are classified in 25 pathways (see Table 4.1 for statistical data). Details of the reactions and metabolites are visually suppressed due to the high complexity but can be interactively explored using the Omix® network visualization tool. See the section 'Web resources' for availability.

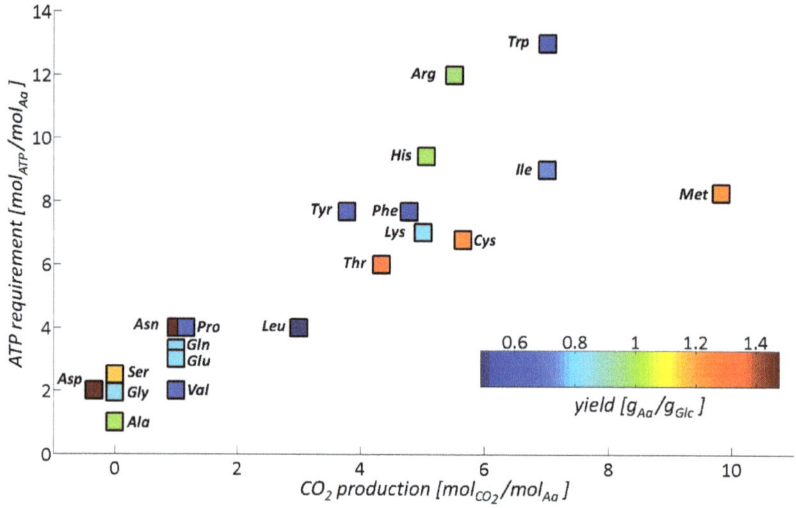

Figure 4.3 ATP requirement versus CO_2 production and yield (colour coded) of *in silico* amino acid production determined by FBA with the genome-scale *Corynebacterium glutamicum* network *iEZ475* using glucose as sole carbon source. Each square corresponds to a single FBA study that maximizes the associated amino acid export flux.

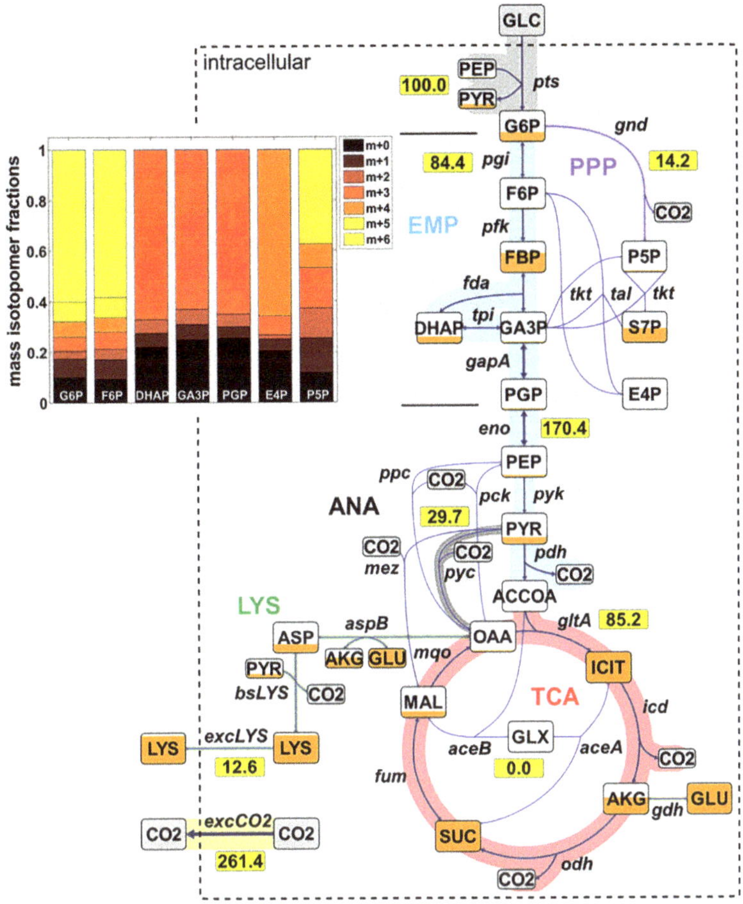

Figure 4.5 Flux map for the L-lysine producer strain *Corynebacterium glutamicum* DM 1730 under industrially relevant fed-batch conditions obtained with the INST-MFA method. The map shows a high activity of glycolysis and citric acid cycle. The flux in pentose phosphate pathway is known to be correlated to lysine production. Flux values are normalized to a glucose uptake rate of 100. Metabolite pool sizes (indicated by the filling heights of the nodes) are normalized to the concentration of glutamate. The inset shows labelling fractions for selected intracellular metabolites. For more details of the data set the reader is referred to Noack *et al.* (2011).

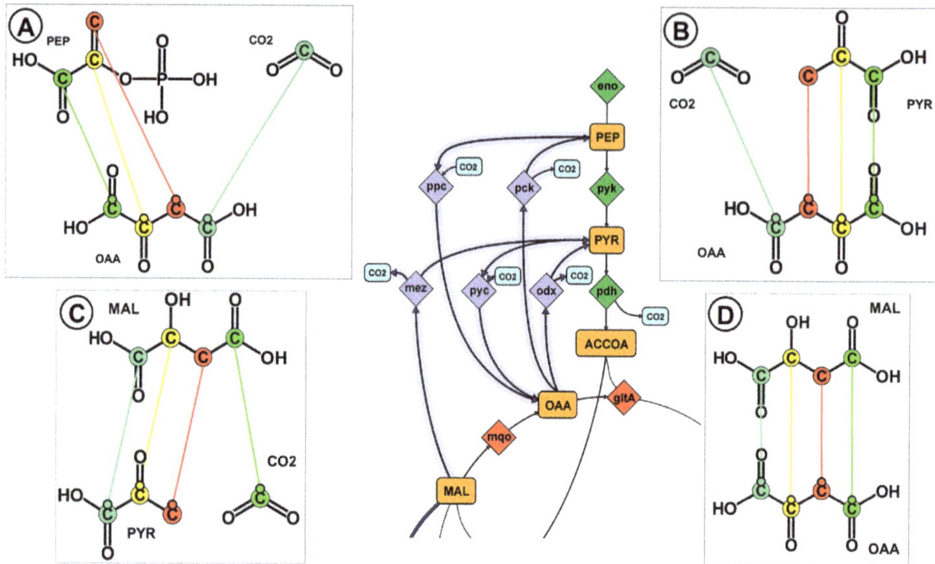

Figure 4.6 Carbon atom tansitions of the anaplerotic section in *C. glutamicum*. (A) PEP carboxylase (*pck*, EC:4.1.1.32)/PEP carboxykinase (*ppc*,EC:4.1.1.49), (B) pyruvate carboxylase (*pyc*, EC:6.4.1.1)/oxaloacetate decarboxylation (*odx*, EC 4.1.1.3), (C) malic enzyme (*mez*, EC:1.1.1.40), and (D) malate dehydrogenase (*mqo*, EC:1.1.5.4). Co-factors and energy equivalents are not shown for clarity. The same colors are used for substrate and target atom pairs.

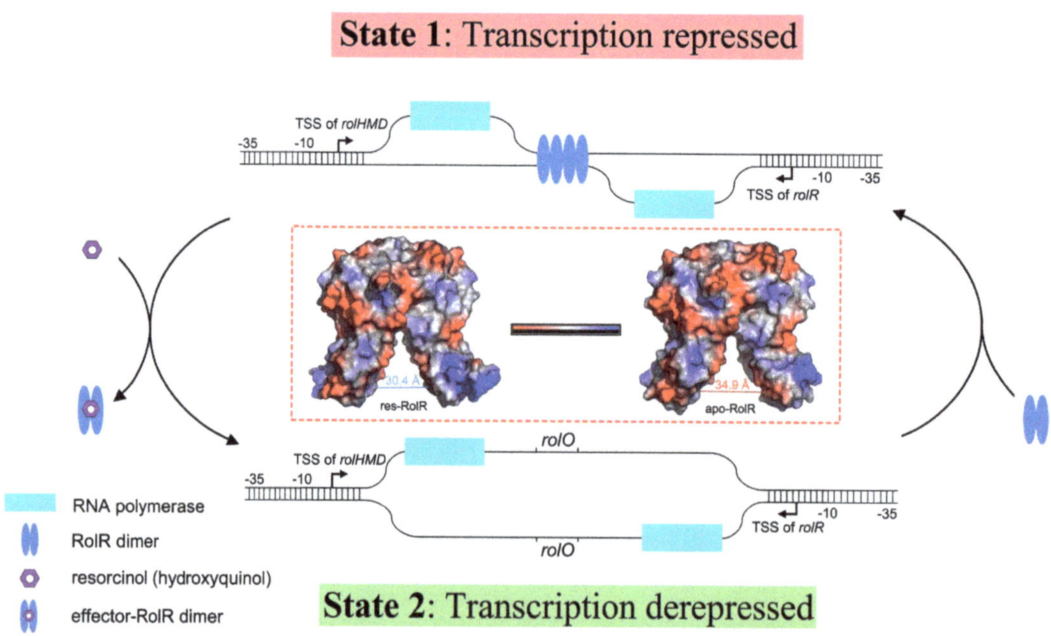

Figure 7.3 One stone two birds: a roadblock model to explain the regulation by RolR on resorcinol degradation. The crystal structures of RolR in its effector-bound (with resorcinol, res-RolR) and apo- (apo-RolR) forms are shown in the red box. With resorcinol-binding, the centre-to-centre separation of the recognition helices is decreased from 34.9Å to 30.4Å, thus not suitable for DNA binding (Li *et al.*, 2011, 2012).

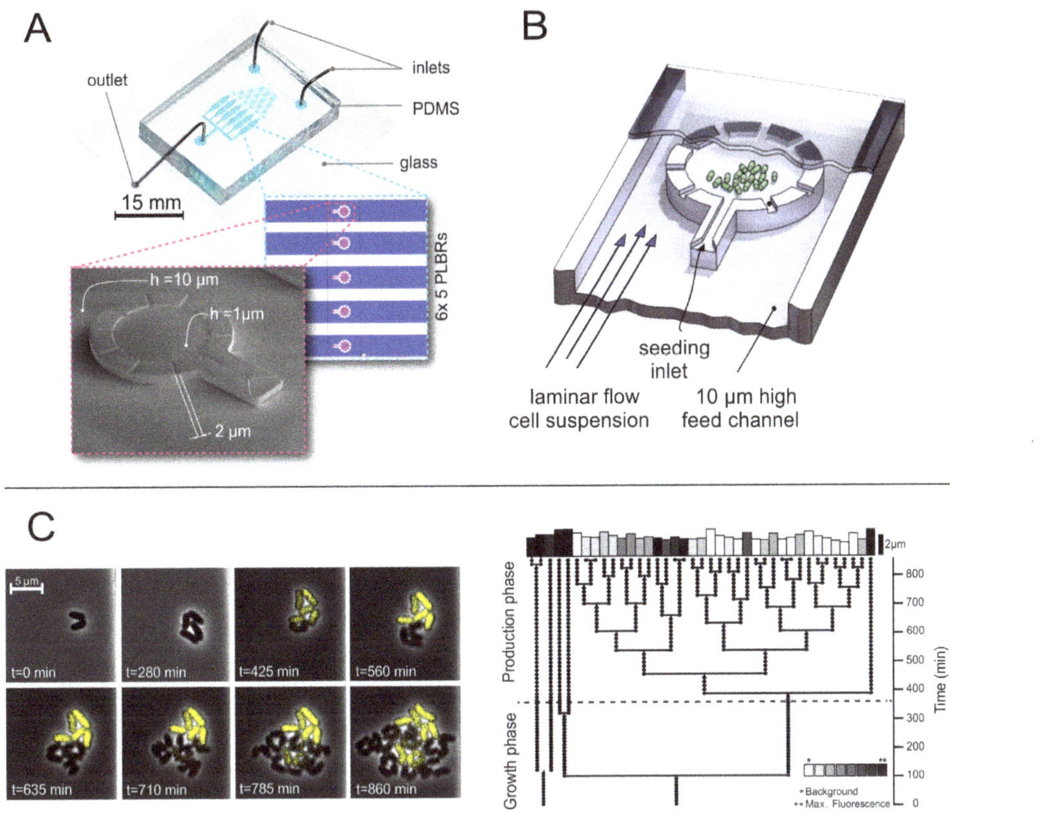

Figure 12.6 Biosensor-based analysis of a L-valine production strain in lab-on-a-chip devices. (A) Images showing the fabricated microfluidic chip devices (at the top), zooming into one growth array (in the middle). Scanning electron microscopy image (at the bottom) and (B) schematic illustration of a single picoliter bioreactor. (C) Microscopic image of an isogenic microcolony of *Corynebacterium glutamicum* Δ*aceE* containing the Lrp-sensor and the respective lineage tree, illustrating phenotypic heterogeneity upon initiation of the production phase (see, t = 425 minutes) (adapted from Grünberger *et al.*, 2012; Mustafi *et al.*, 2014).

www.ingramcontent.com/pod-product-compliance
Lightning Source LLC
LaVergne TN
LVHW071338210926
848486LV00003B/169
* 9 7 8 1 9 1 0 1 9 0 0 5 0 *